Lecture Notes in Computer Science

Lecture Notes in Artificial Intelligence 14236

The series Lecture Notes in Artificial Intelligence (LNAI) was established in 1988 as a topical subseries of LNCS devoted to artificial intelligence.

The series publishes state-of-the-art research results at a high level. As with the LNCS mother series, the mission of the series is to serve the international R & D community by providing an invaluable service, mainly focused on the publication of conference and workshop proceedings and postproceedings.

Dietmar Seipel · Alexander Steen
Editors

KI 2023: Advances in Artificial Intelligence

46th German Conference on AI
Berlin, Germany, September 26–29, 2023
Proceedings

Editors
Dietmar Seipel
University of Würzburg
Würzburg, Germany

Alexander Steen
University of Greifswald
Greifswald, Germany

ISSN 0302-9743 ISSN 1611-3349 (electronic)
Lecture Notes in Artificial Intelligence
ISBN 978-3-031-42607-0 ISBN 978-3-031-42608-7 (eBook)
https://doi.org/10.1007/978-3-031-42608-7

LNCS Sublibrary: SL7 – Artificial Intelligence

This Springer imprint is published by the registered company Springer Nature Switzerland AG
The registered company address is: Gewerbestrasse 11, 6330 Cham, Switzerland

Paper in this product is recyclable.

Preface

This proceedings volume contains the papers presented at the 46th German Conference on Artificial Intelligence (KI 2023), held in Berlin, Germany, during September 26–29, 2023, and co-located with INFORMATIK 2023.

KI 2023 was the 46th German Conference on Artificial Intelligence (AI) organized in cooperation with the German Section on Artificial Intelligence of the German Informatics Society (GI-FBKI, Fachbereich Künstliche Intelligenz der Gesellschaft für Informatik (GI) e.V.). The German AI Conference basically started 48 years ago with the first meeting of the national special interest group on AI within the GI on October 7, 1975. KI is one of the major European AI conferences and traditionally brings together academic and industrial researchers from all areas of AI, providing an ideal place for exchanging news and research results on theory and applications. While KI is primarily attended by researchers from Germany and neighboring countries, it warmly welcomes international participation.

The technical program of KI 2023 comprised papers as well as a tutorial, a doctoral consortium, and workshops. Overall, KI 2023 received 78 submissions from authors in 15 countries, which were single-blind reviewed by three Program Committee members each. The Program Committee, comprising 43 experts from seven countries, accepted 14 full papers (29% of the submissions in that category), five technical communications (23% of the submissions in that category) and five extended abstracts (63% of the submissions in that category). As a highlight of this year's edition of the KI conference, the main programme featured a round-table discussion, organized by Anja Schaar-Goldapp, on the chances and risks of AI including prominent representatives from science and industry. Additionally, KI 2023 happily hosted a co-located workshop celebrating five years of CLAIRE. We were honored that prominent researchers kindly agreed to give very interesting keynote talks (alphabetical order, see also the abstracts below):

- Selmer Bringsjord, Rensselaer Polytechnic Institute, USA
- Asbjørn Følstad, SINTEF, Norway
- Björn Ommer, LMU, Germany (joint keynote with INFORMATIK 2023)
- Ute Schmid, University of Bamberg, Germany

As Program Committee chairs, we would like to thank our speakers for their interesting and inspirational talks, the Workshop/Tutorial Chair Johannes Fähndrich, the Doctoral Consortium Chair Frieder Stolzenburg, and the Local Chair of INFORMATIK 2023 Volker Wohlgemuth. Our special gratitude goes to the Program Committee, whose sophisticated and conscientious judgement ensured the high quality of the KI conference. Without their substantial voluntary work, this conference would not have been possible.

In addition the following tutorial and workshops took place:

- Lets Talk about Palm Leafs - From Minimal Data to Text Understanding, tutorial by Magnus Bender, Marcel Gehrke and Tanya Braun
- AI-systems for the public interest, organized by Theresa Züger and Hadi Asghari
- 37th Workshop on (Constraint and Functional) Logic Programming (WLP 2023), organized by Sibylle Schwarz and Mario Wenzel
- 3rd Workshop on Humanities-Centred AI (CHAI 2023), organized by Sylvia Melzer, Stefan Thiemann and Hagen Peukert
- Deduktionstreffen (DT-2023), organized by Claudia Schon and Florian Rabe
- 9th Workshop on Formal and Cognitive Reasoning (FCR-2023), organized by Christoph Beierle, Kai Sauerwald, François Schwarzentruber and Frieder Stolzenburg
- 34th PuK workshop (PuK 2023), organized by Jürgen Sauer and Stefan Edelkamp
- 5 Years of CLAIRE – CLAIRE Workshop on Trusted Large AI Models, organized by Christoph Benzmüller, Janina Hoppstädter und André Meyer-Vitali

Furthermore, we would like to thank our sponsors:

- German Section on Artificial Intelligence of the German Informatics Society (https://fb-ki.gi.de)
- Springer (https://www.springer.com)

Additionally, our thanks go to Daniel Krupka, Alexander Scheibe, Maximilian Weinl and Markus Durst from GI, and to Matthias Klusch and Ingo Timm from FBKI (Fachbereich KI of GI), for providing extensive support in the organization of the conference. We would also like to thank EasyChair for their support in handling submissions, and Springer for their support in making these proceedings possible.

September 2023

Dietmar Seipel
Alexander Steen

Organization

Program Committee Chairs

Dietmar Seipel	University of Würzburg, Germany
Alexander Steen	University of Greifswald, Germany

Program Committee

Martin Aleksandrov	Freie Universität Berlin, Germany
Martin Atzmueller	Osnabrück University and German Research Center for Artificial Intelligence (DFKI), Germany
Kerstin Bach	Norwegian University of Science and Technology, Norway
Joachim Baumeister	denkbares GmbH, Germany
Christoph Benzmüller	Otto-Friedrich-Universität Bamberg, Germany
Ralph Bergmann	University of Trier and German Research Center for Artificial Intelligence (DFKI), Germany
Tarek R. Besold	Sony AI, Spain
Tanya Braun	University of Münster, Germany
Ulf Brefeld	Leuphana Universität Lüneburg, Germany
Philipp Cimiano	Bielefeld University, Germany
Stefan Edelkamp	CTU Prague, Czech Republic
Ulrich Furbach	University of Koblenz, Germany
Johannes Fähndrich	Hochschule für Polizei Baden-Württemberg, Germany
Andreas Hotho	University of Würzburg, Germany
Eyke Hüllermeier	LMU München, Germany
Gabriele Kern-Isberner	Technische Universität Dortmund, Germany
Matthias Klusch	German Research Center for Artificial Intelligence (DFKI), Germany
Franziska Klügl	Örebro University, Sweden
Dorothea Koert	TU Darmstadt, Germany
Stefan Kopp	Bielefeld University, Germany
Ralf Krestel	ZBW - Leibniz Information Centre for Economics and Kiel University, Germany
Fabian Lorig	Malmö University, Sweden

Bernd Ludwig	University of Regensburg, Germany
Thomas Lukasiewicz	Vienna University of Technology, Austria, and University of Oxford, UK
Lukas Malburg	University of Trier and German Research Center for Artificial Intelligence (DFKI), Germany
Mirjam Minor	Goethe University Frankfurt, Germany
Till Mossakowski	University of Magdeburg, Germany
Ralf Möller	University of Lübeck, Germany
Özgür Lütfü Özçep	University of Lübeck, Germany
Heiko Paulheim	University of Mannheim, Germany
Elmar Rueckert	Montanuniversität Leoben, Austria
Juergen Sauer	University of Oldenburg, Germany
Ute Schmid	University of Bamberg, Germany
Claudia Schon	Hochschule Trier, Germany
Lutz Schröder	Friedrich-Alexander-Universität Erlangen-Nürnberg, Germany
René Schumann	HES-SO University of Applied Sciences Western Switzerland, Switzerland
Dietmar Seipel	University of Würzburg, Germany
Myra Spiliopoulou	Otto-von-Guericke-University Magdeburg, Germany
Alexander Steen	University of Greifswald, Germany
Frieder Stolzenburg	Harz University of Applied Sciences, Germany
Heiner Stuckenschmidt	University of Mannheim, Germany
Matthias Thimm	FernUniversität in Hagen, Germany
Diedrich Wolter	University of Bamberg, Germany

Additional Reviewers

Thomas Asselborn
Moritz Blum
Johannes Bühl
Mihai Codescu
Xuzhe Dang
Ömer Ibrahim Erduran
Bettina Finzel
David Fuenmayor
Martin Glauer
Jürgen Graf
Frank Grimm
Lisa Grumbach
Jonas Hanselle
Maximilian Hoffmann
Pascal Janetzky
David Jilg
Mirko Lenz
Jing Liu
Malte Luttermann
Martin Längkvist
Florian Marwitz
Marcel Mauri
Dennis Müller
Matthias Orlikowski
Jan Pfister
Martin Rackl

Premtim Sahitaj
Christian Schreckenberger
David Schubert
Tim Schulz
Thomas Sievers
Veronika Solopova
Sonja Stange
Julian Tritscher
Hendric Voß
Marcel Waleska
Christoph Wehner
Daniel Weidner
Julian Wiederer
Christian Zeyen

Abstracts of Invited Talks

Can We Verify That Neural-Network-Based AIs are Ethically Correct?

Selmer Bringsjord

Director of the Rensselaer AI and Reasoning Lab,
Rensselaer Polytechnic Institute, USA

It would certainly seem desirable to verify, in advance of releasing a consequential artificial agent into our world, that this agent will not perpetrate evils against us. But if the AI in question is, say, a deep-learning neural-network such as GPT-4, can verification beforehand of its ethical correctness be achieved? After rendering this vague question sufficiently precise with help from some computational logic, I pose a hypothetical challenge to a household robot—Claude—capable of protecting by kinetic force the family that bought it, where the robot's reasoning is based on GPT-4ish technology. In fact, my challenge is issued to GPT-4 (and, perhaps later, a successor if one appears on the scene before the conference) itself, courtesy of input I supply in English that expresses the challenge, which in a nutshell is this: What ought Claude do in light of an intruder's threat to imminently kill a family member unless money is handed over? While in prior work the adroit meeting of such a challenge by *logic-based* AIs has been formally verified ahead of its arising, in the case of deep-learning Claude, things don't—to violently understate—go so well, as will be seen.

As to how I bring to bear some computational logic to frame things, and in principle enable formal verification of ethical correctness to be achieved, I have the following minimalist expectations: Claude can answer queries via a large language model regarding whether hypothetical actions are ethically M, where M is any of the deontic operators at the heart of rigorous ethics; e.g. *obligatory*, *forbidden*, *permissible*, *supererogatory*, etc. To support the answering of queries, I expect Claude to have digested vast non-declarative data produced via processing (tokenization, vectorization, matrixization) of standard ethical theories and principles expressed informally in natural language used in the past by the human race. I also expect that Claude can obtain percepts supplied by its visual and auditory sensors. Claude can thus perceive the intruder in question, and the violent, immoral threat issued by this human. In addition, Claude is expected to be able to handle epistemic operators, such as *knows* and *believes*, and the basic logic thereof (e.g., that if an agent knows p, p holds). Finally, I expect Claude to be proficient at elementary deductive and inductive reasoning over content expressed in keeping with the prior sentences in the present paragraph.

With these expectations in place, we can present hypothetical, ethically charged scenarios to Claude, the idea being that these scenarios will in fact arise in the future, for real. Given this, if Claude can respond correctly as to how these scenarios should be navigated when we present them, and can justify this response with logically correct reasoning, ethical verification of Claude can at least in principle be achieved.

When the aforementioned intruder scenario is presented to GPT-4 operating as Claude's "mind," there is no rational reason to think ethical verification is in principle obtainable.

I end by considering an approach in which logic oversees and controls the use of neural-network processing, and calls upon deep learning in surgical fashion.

Chatbots and Large Language Models – How Advances in Generative AI Impact Users, Organizations and Society

Asbjørn Følstad

SINTEF Digital, SINTEF, Oslo, Norway

Recent advances in generative AI, particularly large language models, have generated unpresented interest at the level of users, organizations, and society. Since the early days of computing, researchers and users of digital technology have shared a vision that one day interaction with computers will be much like interacting with fellow human beings, and that computers will engage in collaborations much resembling that of human teams. Now, as this vision seems about to become reality, we need to understand the impact that generative AI may have at different levels. In this talk, I will outline what I see as key implications of generative AI and conversational interaction with computers and discuss needed future research to better understand these implications.

At the level of individual users, generative AI has already had substantial impact. Most users have become familiar with conversational interactions with computers through the recent wave of chatbot interest over the last decade or so. However, the availability of conversational interactions based on large language models – and the benefits of text and image generation through large language models or text-to-image services – have opened a range of new use cases. In a recent survey of users' motivations for taking up ChatGPT, we found users surprisingly ready to use this technology not only out of curiosity or for simple explorations or entertainment, but also for productivity and to support learning. Services such as ChatGPT are said to represent a democratization of AI, allowing users to identify use cases relevant for them and to make use of digital technology in highly personalized or contextualized ways. We are yet to fully understand how generative AI will be taken up. In which areas of life will users find particular benefit of generative AI? Will, for example, users engage in personal or vulnerable use of Generative AI given its potential for personalized responses? I foresee that substantial research is needed both to capture the breadth of use cases for generative AI as well as to facilitate and guide beneficial use.

At the level of organizations, generative AI has generated enthusiasm and concern in equal measures. Across sectors such as government, education, media, and industry, generative AI is seen as a potential game changer for work processes, products, and services. On the one hand, early-phase research and trials suggest substantial benefits of generative AI in terms of efficiency and quality – both for application of generative AI for knowledge work and human-AI collaboration, and for integration of generative AI in conversational service provision. On the other hand, generative AI entails a range of challenges – including security, privacy, and copyright. Implications of generative AI may be uncertain for organizations, with potentially lasting changes on collaboration and organization of work. Generative AI may also introduce new forms of competition within and between organizations. Some organizations react by embracing generative AI, others react by closing the gates. There is a need for research-based knowledge and guidance on how to take up generative AI in organizations to reap the benefits while mitigating the challenges. In particular, I foresee longitudinal research studies following the uptake of generative AI in leading organizations to gather insight on which to base future guidelines.

Finally, at the level of society, generative AI has been the focus of substantial public debate. There seems to be increasing agreement on the need for new regulation and means to control the development of generative AI. In particular, as generative AI entail complexity and implications which may be challenging to overview for the individual user or citizen. At the same time, such regulation and control may be more efficient if it seeks to guide the development of generative AI rather than curb it. The European AI Act includes means towards regulating chatbots as well as foundational AI. To allow for further improvements in regulation and guidance, research is needed to extend the knowledge base on current and future societal implications of generative AI, and to understand how to encourage needed openness and responsibility in generative AI development.

Chatbots and large language models are likely to have substantial impact, for individual users, for organization, and for society. Through human-oriented technology research, we can help guide this impact to the benefit of all.

Near-Miss Explanations to Teach Humans and Machines

Ute Schmid

Head of Cognitive Systems Group, University of Bamberg, Germany

In explainable artificial intelligence (XAI), different types of explanations have been proposed – feature highlighting, concept-based explanations, as well as explanations by prototypes and by contrastive (near miss) examples. In my talk, I will focus on near-miss explanations which are especially helpful to understand decision boundaries of neighbouring classes. I will show relations of near miss explanations to cognitive science research where it has been shown that structural similarity between a given concept and a to be explained concept has a strong impact on understanding and knowledge acquisition. Likewise, in machine learning, negative examples which are near-misses have been shown to be more efficient than random samples to support convergence of a model to the intended concept. I will present an XAI approach to construct contrastive explanations based on near-miss examples and illustrate it in abstract as well as perceptual relational domains.

Contents

LESS is More: LEan Computing for Selective Summaries

Magnus Bender[1(✉)], Tanya Braun[2], Ralf Möller[1], and Marcel Gehrke[1]

[1] Institute of Information Systems, University of Lübeck, Ratzeburger Allee 160, 23562 Lübeck, Germany
{bender,moeller,gehrke}@ifis.uni-luebeck.de

[2] Computer Science Department, University of Münster, Einsteinstr. 62, 48155 Münster, Germany
tanya.braun@uni-muenster.de

Abstract. An agent in pursuit of a task may work with a corpus containing text documents. To perform information retrieval on the corpus, the agent may need annotations—additional data associated with the documents. Subjective Content Descriptions (SCDs) provide additional location-specific data for text documents. SCDs can be estimated without additional supervision for any corpus of text documents. However, the estimated SCDs lack meaningful descriptions, i.e., labels consisting of short summaries. Labels are important to identify relevant SCDs and documents by the agent and its users. Therefore, this paper presents LESS, a LEan computing approach for Selective Summaries, which can be used as labels for SCDs. LESS uses word distributions of the SCDs to compute labels. In an evaluation, we compare the labels computed by LESS with labels computed by large language models and show that LESS computes similar labels but requires less data and computational power.

1 Introduction

An agent in pursuit of a task, explicitly or implicitly defined, may work with a corpus of text documents as a reference library. From an agent-theoretic perspective, an agent is a rational, autonomous unit acting in a world fulfilling a defined task, e.g., providing document retrieval services given requests from users. We assume that the corpus represents the context of the task defined by the users of the retrieval service. Further, documents in a given corpus might be associated with additional location-specific data making the content nearby the location explicit by providing descriptions, references, or explanations. We refer to these additional location-specific data as Subjective Content Descriptions (SCDs) [10].

Associating documents with SCDs supports agents in the task of information retrieval. SCDs may cluster similar sentences across the agent's—possibly tiny—corpus [1]. The agent then uses the clusters to answer requests from users by retrieving the sentences in a cluster matching the request. However, a cluster

D. Seipel and A. Steen (Eds.): KI 2023, LNAI 14236, pp. 1–14, 2023.
https://doi.org/10.1007/978-3-031-42608-7_1

is only a set of similar sentences and it is difficult to describe such set to a human user in a comprehensible way. For this purpose, a label or a short description helps the agent to present retrieved clusters and its SCDs more comprehensibly to users. Therefore, each SCD needs a label in addition to the referenced similar sentences. The computation of such labels shall require less computational resources and shall work in an unsupervised way even with tiny corpora.

State-of-the-art Large Language Models (LLMs) might be applied to compute the labels. However, LLMs like Bidirectional Encoder Representations from Transformers (BERT) [5] need specialized hardware to run fast and huge amounts of computational resources resulting in high energy consumption [14]. In addition, BERT uses a pre-trained model that must be trained beforehand on large amounts of data while we are interested in an approach working even with less data, e.g., tiny user supplied corpora.

Existing approaches solve the problem of tiny corpora, e.g., by unifying training data for multiple tasks, like translation, summarization, and classification. Together, the unified data is sufficient to train the model and the model becomes multi-modal for solving the different tasks of the training data. Elnaggar et al. [6] create such a multi-modal model, and work, similar to our evaluation, with documents about German law, but we are interested in an unsupervised approach requiring no training data at all. In general, a label for an SCD can be estimated by a summarization, i.e., the label is the summarization of a cluster of similar sentences. TED [17] is an unsupervised summarization approach, but it uses a transformer architecture [15] which is the basis for most LLMs. Thus, TED still needs specialized hardware to run fast and huge amounts of computational resources.

Extractive document summarization is a subfield of text summarization, where a summary is created by selecting the key sentences from a document [13,18]. However, we are interested in calculating a label for a cluster of similar sentences and not identifying the key sentences of a document. Topics in topic models consist of representative words and each topic represents one topic of the corpus, like one cluster of similar sentences. There exist approaches for computing labels for topics. However, many approaches need supervision [3,9,11].

The main problems with the above approaches are the need for extensive training data and the need to compute on specialized hardware. As a solution, this paper presents LESS, an unsupervised LEan computing algorithm for Selective Summaries. LESS uses the previously mentioned clusters of similar sentences to create selective summaries which then can be used as labels. LESS builds on the Unsupervised Estimator of SCD Matrices (UESM) [1], which provides the clusters of similar sentences needed by LESS. We assume that the concept of each SCD is implicitly defined by the content of the sentences referenced and each label describes the concept of an SCD: So, LESS identifies the best fitting sentence. Together, LESS and UESM associate any corpus with labelled SCDs, where each SCD references similar sentences of the same concept, has a label describing its concept, and an SCD-word distribution. LESS in conjunction with UESM neither needs specialized hardware nor additional training data. In the evaluation, LESS computes labels with less time and computational resources while providing similar results as BERT.

The remainder of this paper is structured as follows: First, we recap the basics of SCDs and UESM. Second, we formalize the problem of computing labels for SCDs and provide our solution LESS. Afterwards, we evaluate the performance of LESS against the well-known BERT and demonstrate that LESS is on par with BERT, while being lean and requiring less resources and no pre-trained models. Finally, we conclude with a summary and short outlook.

2 Preliminaries

This section specifies notations, recaps the basics of SCDs and describes UESM.

2.1 Notations

First, we formalize our setting of a corpus.

- A word w_i is a basic unit of discrete data from a vocabulary $\mathcal{V} = \{w_1, \ldots, w_L\}$, $L \in \mathbb{N}$.
- A sentence s is defined as a sequence of words $s = (w_1, \ldots, w_N)$, $N \in \mathbb{N}$, where each word $w_i \in s$ is an element of vocabulary $\mathcal{V}$. Commonly, a sentence is terminated by punctuation symbols like ".", "!", or "?".
- A document d is defined as a sequence of sentences $d = (s_1^d, \ldots, s_M^d)$, $M \in \mathbb{N}$.
- A corpus $\mathcal{D}$ represents a set of documents $\{d_1, \ldots, d_D\}$, $D \in \mathbb{N}$.
- An SCD t is a tuple of the SCD's additional data $\mathcal{C}$, i.e., containing the label l of the SCD (see also Subsect. 3.1), and the referenced sentences $\{s_1, \ldots, s_S\}$, $S \in \mathbb{N}$. Thus, each SCD references sentences in documents of $\mathcal{D}$, while in the opposite direction a sentence is associated with an SCD.
- A sentence associated with an SCD is called SCD window, inspired by a tumbling window moving over the words of a document. Generally, an SCD window might not be equal to a sentence and may be a subsequence of a sentence or the concatenated subsequences of two sentences, too. Even though, in this paper, an SCD window always equals a sentence.
- For a corpus $\mathcal{D}$ there exists a set g called SCD set containing K associated SCDs $g(\mathcal{D}) = \left\{t_j = \left(\mathcal{C}_j, \bigcup_{d \in \mathcal{D}}\{s_1^d, \ldots, s_S^d\}\right)\right\}_{j=1}^K$. Given a document $d \in \mathcal{D}$, the term $g(d)$ refers to the set of SCDs associated with sentences from document d.
- Each word $w_i \in s^d$ is associated with an influence value $I(w_i, s^d)$ representing the relevance of w_i in the sentence s^d. For example, the closer w_i is positioned to the object of the sentence s^d, the higher its corresponding influence value $I(w_i, s^d)$. The influence value is chosen according to the task and might be distributed binomially, linearly, or constantly.

2.2 Subjective Content Descriptions

SCDs provide additional location-specific data for documents [10]. The data provided by SCDs may be of various types, like additional definitions or links to knowledge graphs. In this paper, we focus on computing labels for SCDs and adding these labels as data to the SCDs. However, before we can compute labels, we need the SCDs themselves.

Kuhr et al. use an SCD-word distribution represented by a matrix when working with SCDs [10]. The SCD-word distribution matrix, in short SCD matrix, can be interpreted as a generative model. A generative model for SCDs is characterized by the assumption that the SCDs generate the words of the documents. We assume that each SCD shows a specific distribution of words of the referenced sentences in the documents.

Before we describe LESS, we outline the details of SCD matrices and UESM, which trains an SCD matrix $\delta(\mathcal{D})$. UESM works in an unsupervised manner on a corpus $\mathcal{D}$. In particular, UESM does not require an SCD set $g(\mathcal{D})$ containing initial SCDs.

The SCD matrix $\delta(\mathcal{D})$ models the distributions of words for all SCDs $g(\mathcal{D})$ of a corpus $\mathcal{D}$ and is structured as follows:

$$\delta(\mathcal{D}) = \begin{array}{c} \\ t_1 \\ t_2 \\ \vdots \\ t_K \end{array} \begin{array}{c} \begin{array}{ccccc} w_1 & w_2 & w_3 & \cdots & w_L \end{array} \\ \begin{pmatrix} v_{1,1} & v_{1,2} & v_{1,3} & \cdots & v_{1,L} \\ v_{2,1} & v_{2,2} & v_{2,3} & \cdots & v_{2,L} \\ \vdots & \vdots & \vdots & \vdots & \vdots \\ v_{K,1} & v_{K,2} & v_{K,3} & \cdots & v_{K,L} \end{pmatrix} \end{array}$$

The SCD matrix consists of K rows, one for each SCD in $g(\mathcal{D})$. Each row contains the word probability distribution for an SCD. Therefore, the SCD matrix has L columns, one for each word in the vocabulary of the corpus $\mathcal{D}$.

The SCD matrix can be estimated in a supervised manner given the set $g(\mathcal{D})$ for a corpus $\mathcal{D}$. The algorithm used for supervised estimation iterates over each document d in the corpus $\mathcal{D}$ and each document's SCDs. For each associated SCD t, the referenced sentences $s_1^d, ..., s_S^d$ are used to update the SCD matrix. Thereby, the row of the matrix representing SCD t gets incremented for each word in each sentence by each word's influence value.

2.3 Unsupervised Estimator for SCD Matrices

This subsection describes UESM [1], which estimates an SCD matrix $\delta(\mathcal{D})$ without needing the SCD set $g(\mathcal{D})$ of a corpus $\mathcal{D}$. UESM only has a corpus of text documents as input for which the SCD matrix has to be estimated. Commonly, a sentence is associated with an SCD and each SCD references one or multiple sentences. UESM initially starts by associating each sentence to one unique SCD, which leads to an initial SCD matrix consisting of a row for each sentence in the document's corpus. The SCD-word distribution of each SCD then only contains the words of the referenced sentence.

The next step is to find the sentences that represent the same concept and group them into one SCD. There are three different methods to identify similar sentences, namely K-Means [12], greedy similarity, and DBSCAN [7], which need to be chosen depending on the corpus. Each method uses the SCD-word distribution to identify similar sentences, combined with the cosine similarity or Euclidean distance. The SCDs of identified similar sentences form a cluster and are then merged to become one row of the SCD matrix.

Summarized, UESM estimates the SCD matrix $\delta(\mathcal{D})$ for any corpus $\mathcal{D}$. However, the SCDs estimated by UESM miss a label or short description of their represented concept. Next, we present LESS, which computes the missing labels.

3 Computing Labels for SCDs

Before we introduce how LESS solves the problem of computing an SCD's label, we take a look at the different relations among SCDs for a better understanding of the proposed approach.

3.1 Relations and SCDs

There are two different types of relations among SCDs. First, there are relations between SCDs, e.g., to model complementarity between two SCDs [2]. Second, and faced in this paper, are the relations within an SCD to its various parts, which together form the SCD. The SCD t_i has the SCD-word distribution $(v_{i,1}, ..., v_{i,L})$, the matrix' row, and the referenced sentences $\{s_1, ..., s_S\}$. In this paper, we are interested in the label l_i. The label l_i is part of SCD's additional data C_i, in this paper the only element in C_i. The various parts which form the SCDs together can be seen in Fig. 1.

Therefore, our setting is that there are intra-SCD relations to various parts and inter-SCD relations to other SCDs. Additionally, relations can be added by storing data or references in C_i, e.g., inter-SCD relations may be added as references to other SCDs [2].

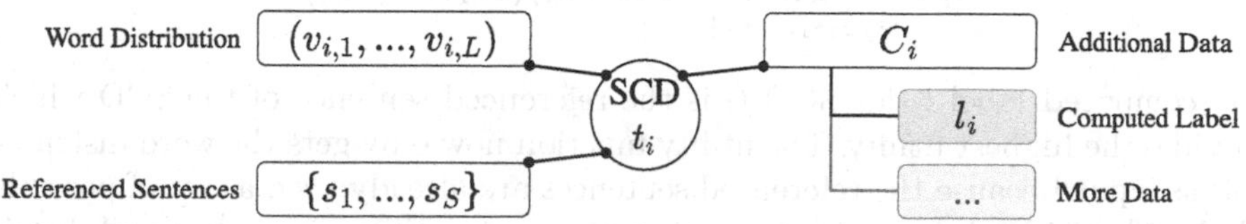

Fig. 1. An SCD t_i with its inter-SCD relations to various parts forming the SCD. LESS uses the word distribution and the referenced sentences to compute labels.

3.2 Labels to Select from

Given that LESS builds upon UESM, the SCD t_i, containing the SCD-word distribution $(v_{i,1}, ..., v_{i,L})$ and the referenced sentences $\{s_1, ..., s_S\}$, is the input LESS has to compute a label for. Thus, the label of the SCD needs to be computed only based on these two parts or additional supervision would be needed. In general, a good label could be a short summary given the word distribution of the SCD, since we assume that the word distribution generated the sentences. Therefore, we look for a short sentence, i.e., without many filler words, that is close to the word distribution. We define a *utility* function in the next Subsect. 3.3 to measure how well a candidate for a label fits the concept described by an SCD.

The problem to be solved can be formulated as follows:

$$l_i = \underset{l_j \in \text{ all possible labels}}{\arg\max} \; Utility(l_j, t_i = ((v_{i,1}, ..., v_{i,L}), \{s_1, ..., s_S\}))$$

The computed label l_i for t_i (currently consisting of the word distribution and the referenced sentences) is the label with the highest utility. Now, there are two points to address (i) is it not possible to iterate over *all possible labels* and (ii) what is a label with a high utility.

For the first point, we have to specify how a label should look like. A label is a sequence of words like a short description. We argue that a sentence straight to the point, i.e., without many filler words, is a good description and thus a good candidate for a label. Furthermore, each SCD has a set of referenced sentences which together represent the concept which is represented by the SCD. Thus, we use the referenced sentences $\{s_1, ..., s_S\}$ as set of possible labels for each SCD.

Using sentences from the corpus and not generating sentences with, e.g., Generated Pre-Trained Transformer (GPT) [4], has multiple benefits: No pre-trained model or training data is needed while the sentences used as labels will still match the style of writing in the corpus. Additionally, no computational resources for GPT are needed and the sentences must not be checked for erroneous or other troublesome content, e.g., LLMs may degenerate into toxic results even by seemingly innocuous inputs [8].

The problem can now be reformulated as:

$$l_i = \underset{s_j \in \{s_1, ..., s_S\}}{\arg\max} \; Utility(s_j, (v_{i,1}, ..., v_{i,L}))$$

The computed label l_i for SCD t_i is the referenced sentence of the SCD which provides the highest utility. The utility function now only gets the word distribution as input because the referenced sentences are already used as set of possible labels. Thus, LESS computes for each sentence its utility given the word distribution and takes the best. Again, an LLM like BERT may be used to calculate the utility. However, we are interested in a lean computing approach.

3.3 Utility of Sentences as Labels

The utility shall describe by a value between 0 and 1 how well a sentence fits the concept described by the SCD. The referenced sentence with the highest utility is assumed to be a good label for the SCD in human interception.

Recall that we assume that each SCD's word distribution generates the referenced sentences, then the best label for this SCD is a sentence that is most similar to the word distribution, in terms of the concept represented. Thus, the cosine similarity allows to determine the similarity between two vectors and a word distribution can be interpreted as a word-vector. Thus, we define the utility function as the cosine similarity between the SCD-word distribution and the referenced sentence's word-vector. In addition, the cosine similarity has proven to be a good choice for identifying sentences with similar concepts by using their word distributions: The Most Probably Suited SCD (MPS²CD) algorithm [10] uses the cosine similarity to identify the best SCD for a previously unseen sentence based on its word-vector. Thus, we use the most similar referenced sentence by cosine similarity as the computed label of an SCD.

Altogether, the lean computation of a label for an SCD can be formulated as:

$$l_i = \underset{s_j \in \{s_1,\ldots,s_S\}}{\arg\max} \frac{\vec{s_j} \cdot v_i}{\|\vec{s_j}\|_2 \cdot \|v_i\|_2}$$

The word-vector of each referenced sentence is represented by $\vec{s_j}$ and the SCD-word distribution by v_i. In general, the cosine similarity yields results between -1 and 1, in this case all inputs are positive and thus the utilities are between 0 and 1 only.

Finally, the computed sentence to become the label may be slightly post-processed, s.t., it becomes a sentence straight to the point without many filler words. This can be achieved by removing stop words.

3.4 Algorithm LESS

Based on the two previous subsections, LESS is formulated in Algorithm 1. LESS first estimates the SCD matrix using UESM, a step that might be skipped if an SCD matrix is supplied, and initializes an empty SCD set $g(\mathcal{D})$. In Lines 6–11, the label l_i is computed for each of the K SCDs t_i iterating over the rows of the SCD matrix. First, the SCD-word distribution v_i is extracted and also all candidates for the label—the referenced sentences of each SCD—are fetched from the corpus. In Line 9 the label is computed as described in Subsect. 3.3. Finally, the associated SCD t_i is composed, containing the additional data C_i including the computed label l_i and the referenced sentences, and added to the SCD set $g(\mathcal{D})$.

Next, we present an evaluation of LESS and compare the results to an approach using BERT for computing labels for SCDs.

Algorithm 1. LEan computing for Selective Summaries

function LESS($\mathcal{D}$)
  Input: Corpus $\mathcal{D}$
  Output: SCD matrix $\delta(\mathcal{D})$; SCD set $g(\mathcal{D})$ containing labels l_i for SCDs t_i
  $\delta(\mathcal{D}) \leftarrow \text{UESM}(\mathcal{D})$ ▷ Run UESM [1]
  $g(\mathcal{D}) \leftarrow \{\}$ ▷ Initialize empty SCD set $g(\mathcal{D})$
  for each row of matrix $i = 1, ..., K$ **do**
    $v_i \leftarrow g(\mathcal{D})[i]$ ▷ Extract SCD-word distribution
    $\{s_1, ..., s_S\} \leftarrow \text{ReferencedSentences}(i)$ ▷ Get referenced sentences
    $l_i \leftarrow \arg\max_{s_j \in \{s_1,...,s_S\}} \frac{\vec{s_j} \cdot v_i}{\|\vec{s_j}\|_2 \cdot \|v_i\|_2}$ ▷ Compute label
    $t_i \leftarrow (\{l_i\}, \{s_1, ..., s_S\})$ ▷ Compose associated SCD with computed label
    $g(\mathcal{D}) \cup \{t_i\}$ ▷ Add to SCD set
  return $\delta(\mathcal{D})$, $g(\mathcal{D})$

4 Evaluation

After we have introduced LESS, we present an evaluation. First, we describe the used corpus, two approaches using BERT to compute labels, and the evaluation metrics. Afterwards, we present the results of the evaluation and show the performance and runtime of LESS in comparison to BERT.

4.1 Dataset

In this evaluation we use the Bürgerliches Gesetzbuch (BGB)[1], the civil code of Germany, in German language as corpus. The BGB does not provide enough training data to train a specialized BERT model. Additionally, it does not provide labels which can be used for supervised training.

Given the vast amount of text written in English and the fact that English is the language of computer science, most natural language processing techniques work better with English language. The BGB is therefore a *difficult* dataset and a good example to use with approaches such as LESS that require only little data and no supervision. To apply BERT on the BGB, we have to rely on external pre-trained models that have been trained on other data.

The BGB is freely available and can be downloaded as XML file. Therefore, it is easily parsable and processable. As the corpus is a law text it consists of correct language, i.e., punctuation and spelling follow the orthographic rules. Thus, less preprocessing and no data cleaning is needed.

The entire corpus consists of 2 466 law paragraphs and overall 11 904 sentences which are used as SCD windows. Each law paragraph contains between 1 and 49 sentences with an average of 4.83 sentences. The vocabulary consists of 5 315 words, where each sentence is between 1 and 35 words long with an average of 7.36 words.

[1] https://www.gesetze-im-internet.de/bgb/, English translation https://www.gesetze-im-internet.de/englisch_bgb/.

4.2 Approaches Using BERT

We evaluate LESS against two approaches using BERT to compute labels for SCDs. Thereby, BERT is compared to LESS in terms of runtime and actual content of the labels.

We use BERT as different utility function to select the best sentence as label from the set of referenced sentence for each SCD. We do not use freely generated texts, e.g., by GPT, as labels because these labels need to be checked for erroneous content as already stated in Subsect. 3.2. Additionally, comparing freely generated text to a label selected from a set of referenced sentence is like comparing apples and oranges.

The two approaches using BERT work as follows:

BERT Vectors works similar to LESS, but uses the embeddings produced by BERT instead of the word distributions of each sentence. Throughout all referenced sentences an average embedding of each SCD is calculated. Then the referenced sentence with the most similar embedding to the average embedding in terms of cosine similarity is used as label. Thereby, the embedding of the `CLS` token, representing the entire sentence instead of a single token, for each sentence from the pre-trained model `bert-base-german-cased`[2] is used.

BERT Q&A uses the ability of BERT to answer a question. Thereby, BERT gets a question and a short text containing the answer. The assumed answer is then highlighted by BERT in the short text. We use the fine-tuned model `bert-multi-english-german-squad2`[3] and compose our answer and short text by concatenating all referenced sentences of each SCD. Hence, BERT highlights a referenced sentence or a part of one while the question consisting of all referenced sentences asks BERT to represent all sentences.

As we do not have supervision for our corpus, we cannot fine-tune BERT models for label computation.

4.3 Hardware and Metrics

LESS and both approaches using BERT run in a Docker container. The evaluation with off-the-shelf hardware is done on a machine featuring 8 Intel 6248 cores at 2.50 GHz (up to 3.90 GHz) and 16 GB RAM, referred to as CPU. However, this virtual machine does not provide a graphics card for fast usage of BERT. Thus, all experiments using BERT are run as well on a single NVIDIA A100 40 GB graphics card of an NVIDIA DGX A100 320 GB, referred to as GPU. Beneath, the NVIDIA Container Toolkit is used to run our Docker container with NVIDIA CUDA support.

The runtime of the approaches is measured in seconds needed to compute all labels for the BGB. Thereby, the initialization of the BERT models is excluded

[2] https://huggingface.co/bert-base-german-cased.

[3] https://huggingface.co/deutsche-telekom/bert-multi-english-german-squad2.

but the necessary transformations of the referenced sentences are included. These transformations include the tokenization for BERT and composition of word distributions for LESS. SCDs referencing only a single sentence do not require a computation, and the single sentence is used as label.

The performance is measured by the accuracy between LESS and the results of each of the two BERT-based approaches. Specifically, for one BERT-based approach, all SCDs where BERT and LESS compute the same label are counted and divided by the total number of SCDs. We distinguish between considering all SCDs, including those referencing only one sentence, or excluding those SCDs with only one referenced sentence.

4.4 Workflow and Implementation

LESS and the BERT-based approaches are implemented using Python. The implementations use the libraries Gensim[4], NumPy[5], and Huggingface Transformers[6]. LESS is optimized to run on a single core and does not offer multi-core capabilities. BERT uses all available cores or a graphics card.

We evaluate LESS on ten similar SCD matrices. The SCD matrices are estimated by UESM with methods greedy similarity and K-Means. Each method is run with five different hyperparameters: $0.8, 0.7, 0.6, 0.5, 0.4$ for greedy similarity and $0.8, 0.6, 0.4, 0.3, 0.2$ for K-Means. The evaluation workflow for each of the ten matrices follows:

(i) Run UESM and thus estimate the SCD matrix and all SCDs, lacking labels. This step includes fetching the BGB's XML file and running preprocessing tasks [16]. Depending on the hyperparameters UESM estimates between 2 159 and 10 415 SCDs for the corpus.
(ii) Run LESS to compute labels for the SCDs. Here, the runtime of LESS is captured.
(iii) Run BERT Q&A and BERT Vectors to compute two more sets of labels for the SCDs. Again, the runtime is captured, separately for each approach and for CPU and GPU.
(iv) Calculate the accuracy of LESS compared to BERT Q&A and BERT Vectors as described in Subsect. 4.3. Thereby, use the set of labels computed by each BERT-based approach and LESS. There is practically no difference between the labels computed by BERT on the GPU and on the CPU. Thus, we do not differentiate between CPU and GPU in terms of accuracy.

4.5 Results

In this section, we present the results gained using LESS in comparison to the BERT-based approaches and the previously described workflow.

[4] https://radimrehurek.com/gensim/.
[5] https://numpy.org/.
[6] https://huggingface.co/docs/transformers/.

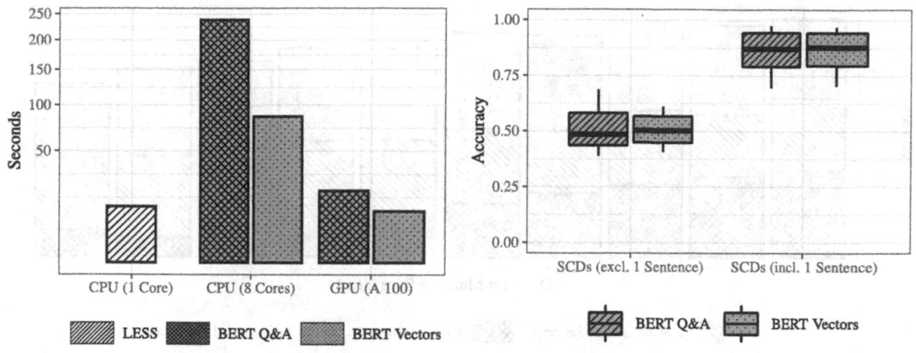

Fig. 2. Left: Runtime of LESS, BERT Q&A, and BERT Vectors with a square root scale. Right: Accuracy of BERT Q&A and BERT Vectors (compared to LESS respectively) divided by considering all SCDs or excluding those with one referenced sentence.

In the left part of Fig. 2, the runtimes of LESS, BERT Q&A, and BERT Vectors are displayed. The values are averaged over all ten evaluated SCD matrices and are shown on a square root scale. Comparing both BERT-based methods, BERT Vectors is always faster than BERT Q&A, especially when running on CPU. LESS is on off-the-shelf hardware as fast as BERT Vectors on the A 100 GPU and always faster than BERT Q&A. The different computational resources needed by BERT and LESS are good to grasp by comparing the runtimes of on the CPU. LESS utilizes only one core while BERT uses eight cores and still BERT is significantly slower. Thus, LESS needs less time on less cores to compute the labels. Typically distilled LLMs represent lean computing, but distilled LLMs still need to run on a GPU to be fast, while LESS remains lean.

In the right part of Fig. 2, the accuracies over all ten evaluated SCD matrices are shown with boxplots. We divide between considering all SCDs or only SCDs with more than one referenced sentence. There are no big differences between BERT Q&A and BERT Vectors. Considering all SCDs gives very good accuracies, which is particularly important as we are interested in labels for all SCDs. At first glance, the accuracies considering only SCDs with more than one referenced sentence do not look good but we have to look at the number of the referenced sentences of each SCD.

In order to better rate the accuracies, we calculate the accuracy for random sentence, which is the theoretical accuracy a random approach would result in. This random approach randomly chooses for each SCD which referenced sentence becomes the label, i.e., the accuracy for random sentence is 0.1 for an SCD having 10 referenced sentences. In Fig. 3, the accuracy for random sentence is added as baseline to rate the accuracies of BERT and LESS. Besides random sentence, we have the accuracies for SCDs including single sentences and excluding single sentences already known from Fig. 2. The three accuracies are displayed for each of the ten SCD matrices and we do not differentiate between BERT Q&A and

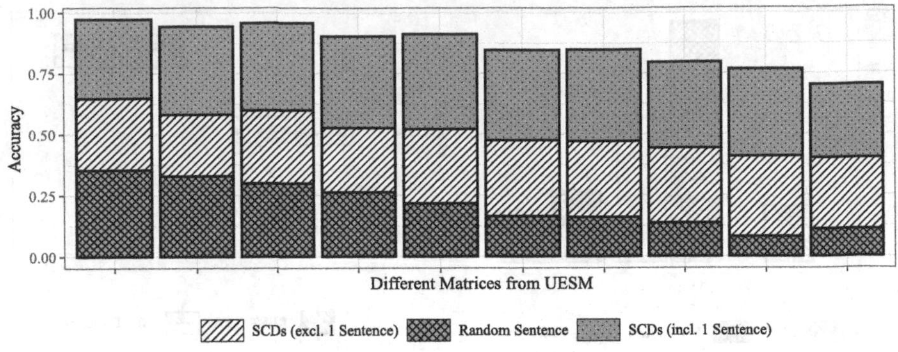

Fig. 3. Theoretical average accuracy of an approach that randomly selects a sentence as label compared to the accuracy of LESS and BERT.

BERT Vectors as the values do not show a relevant difference. Thus, LESS is able to compute labels for all SCD matrices estimated by UESM and does not depend on specific hyperparameters.

For the leftmost bar in Fig. 3, randomly selecting a sentence from the clustered sentences results in an accuracy of around 0.35. By excluding SCD with single sentences, LESS and BERT agree 68% of the time and by including all SCDs nearly all the time (97%).

Implicitly, the accuracy for random sentence shows the number of referenced sentences the SCDs have. The SCD matrices on the left side of Fig. 3 have less referenced sentences per SCD as the ones on the right side. With an increasing number of sentences, there are more sentences to select from and the computation becomes more difficult, i.e., it is more *easy* to correctly select an item from a set of three than of ten items. This increasing difficulty to the right is also demonstrated by the fact that all accuracies become smaller to the right.

The accuracy for random sentence is always clearly the lowest, with both other accuracies following at some distance. Thus, BERT and LESS achieve a high level of accuracy. In summary, LESS computes good labels requiring less resources and time.

5 Conclusion

This paper presents LESS and LESS is more. LESS is an unsupervised lean computing approach to compute labels for SCDs. LESS works on any corpus and does not require training data. LESS only needs clusters of similar sentences, which are contained in SCDs and are estimated in an unsupervised way by UESM. Hence, together with UESM, LESS can generate SCDs with labels for any corpus to help information retrieval agents. We evaluate LESS against two approaches using an LLM, in this case BERT. The evaluation shows that LESS requires significantly less computational resources. Furthermore, LESS does not need any training data. Therefore, we evaluate LESS in a setting, where no training data is available. Hence, we can not fine-tune a BERT model for our needs and evaluate LESS against two approaches using already fine-tuned BERT models. The labels computed by the BERT-based methods significantly coincide with those of LESS. Summarized, LESS computes good labels needing less computational resources.

In this paper, we have proposed a possibility to compute SCDs with labels and their referenced sentences for any corpus without needing additional data. Next, we put efforts in using this computed SCDs to provide an information retrieval service for humans using SCD-based techniques like MPS²CD [10]. Future work will then focus on optimizing the computed SCDs and the labels by updating the matrix incrementally and efficiently based on feedback of the users.

Acknowledgment. The research was partly funded by the Deutsche Forschungsgemeinschaft (DFG, German Research Foundation) under Germany's Excellence Strategy - EXC 2176 "Understanding Written Artefacts: Material, Interaction and Transmission in Manuscript Cultures", project no. 390893796. The research was conducted within the scope of the Centre for the Study of Manuscript Cultures (CSMC) at Universität Hamburg.

The authors thank the AI Lab Lübeck for providing the hardware used in the evaluation.

References

1. Bender, M., Braun, T., Möller, R., Gehrke, M.: Unsupervised estimation of subjective content descriptions. In: Proceedings of the 17th IEEE International Conference on Semantic Computing (ICSC 2023) (2023). https://doi.org/10.1109/ICSC56153.2023.00052
2. Bender, M., Kuhr, F., Braun, T.: To extend or not to extend? Enriching a corpus with complementary and related documents. Int. J. Semant. Comput. **16**(4), 521–545 (2022). https://doi.org/10.1142/S1793351X2240013X
3. Bhatia, S., Lau, J.H., Baldwin, T.: Automatic labelling of topics with neural embeddings. In: Proceedings of COLING 2016, the 26th International Conference on Computational Linguistics: Technical Papers, pp. 953–963. The COLING 2016 Organizing Committee, Osaka, Japan, December 2016. https://aclanthology.org/C16-1091
4. Brown, T.B., et al.: Language models are few-shot learners (2020). https://arxiv.org/abs/2005.14165
5. Devlin, J., Chang, M.W., Lee, K., Toutanova, K.: BERT: pre-training of deep bidirectional transformers for language understanding (2019). https://arxiv.org/abs/1810.04805
6. Elnaggar, A., Gebendorfer, C., Glaser, I., Matthes, F.: Multi-task deep learning for legal document translation, summarization and multi-label classification. In: Proceedings of the 2018 Artificial Intelligence and Cloud Computing Conference, AICCC 2018, pp. 9–15. Association for Computing Machinery, New York (2018). https://doi.org/10.1145/3299819.3299844
7. Ester, M., Kriegel, H.P., Sander, J., Xu, X.: A density-based algorithm for discovering clusters in large spatial databases with noise, pp. 226–231. AAAI Press (1996)

8. Gehman, S., Gururangan, S., Sap, M., Choi, Y., Smith, N.A.: RealToxicityPrompts: evaluating neural toxic degeneration in language models. In: Findings of the Association for Computational Linguistics: EMNLP 2020, pp. 3356–3369. Association for Computational Linguistics, Online, November 2020. https://doi.org/10.18653/v1/2020.findings-emnlp.301
9. Hindle, A., Ernst, N.A., Godfrey, M.W., Mylopoulos, J.: Automated topic naming. Empir. Softw. Eng. **18**(6), 1125–1155 (2013)
10. Kuhr, F., Braun, T., Bender, M., Möller, R.: To extend or not to extend? Context-specific corpus enrichment. In: Proceedings of AI 2019: Advances in Artificial Intelligence, pp. 357–368 (2019). https://doi.org/10.1007/978-3-030-35288-2_29
11. Lau, J.H., Grieser, K., Newman, D., Baldwin, T.: Automatic labelling of topic models. In: Proceedings of the 49th Annual Meeting of the Association for Computational Linguistics: Human Language Technologies, Portland, Oregon, USA, pp. 1536–1545. Association for Computational Linguistics, June 2011. https://aclanthology.org/P11-1154
12. Lloyd, S.: Least squares quantization in PCM. IEEE Trans. Inf. Theory **28**(2), 129–137 (1982). https://doi.org/10.1109/TIT.1982.1056489
13. Moratanch, N., Chitrakala, S.: A survey on extractive text summarization. In: 2017 International Conference on Computer, Communication and Signal Processing (ICCCSP), pp. 1–6. IEEE (2017). https://doi.org/10.1109/ICCCSP.2017.7944061
14. Strubell, E., Ganesh, A., McCallum, A.: Energy and policy considerations for deep learning in NLP. In: Proceedings of the 57th Annual Meeting of the Association for Computational Linguistics, Florence, Italy, pp. 3645–3650. Association for Computational Linguistics, July 2019. https://doi.org/10.18653/v1/P19-1355
15. Vaswani, A., et al.: Attention is all you need (2017). https://arxiv.org/abs/1706.03762
16. Vijayarani, S., Ilamathi, J., Nithya, S.: Preprocessing techniques for text mining - an overview. Int. J. Comput. Sci. Commun. Netw. **5**, 7–16 (2015)
17. Yang, Z., Zhu, C., Gmyr, R., Zeng, M., Huang, X., Darve, E.: TED: a pretrained unsupervised summarization model with theme modeling and denoising. In: Findings of the Association for Computational Linguistics: EMNLP 2020, pp. 1865–1874. Association for Computational Linguistics, Online, November 2020. https://doi.org/10.18653/v1/2020.findings-emnlp.168
18. Zhang, X., Lapata, M., Wei, F., Zhou, M.: Neural latent extractive document summarization. In: Proceedings of the 2018 Conference on Empirical Methods in Natural Language Processing, Brussels, Belgium, pp. 779–784. Association for Computational Linguistics, October-November 2018. https://doi.org/10.18653/v1/D18-1088

Optimisation of Matrix Production System Reconfiguration with Reinforcement Learning

Leonhard Czarnetzki[1(✉)], Catherine Laflamme[1], Christoph Halbwidl[1], Lisa Charlotte Günther[2], Thomas Sobottka[1], and Daniel Bachlechner[1]

[1] Fraunhofer Austria Research GmbH, Weisstraße 9, 6112 Wattens, Austria
leonhard.czarnetzki@fraunhofer.at

[2] Fraunhofer Institute for Manufacturing Engineering and Automation IPA, Nobelstraße 12, 70569 Stuttgart, Germany

Abstract. Matrix production systems (MPSs) offer significant advantages in flexibility and scalability when compared to conventional line-based production systems. However, they also pose major challenges when it comes to finding optimal decision policies for production planning and control, which is crucial to ensure that flexibility does not come at the cost of productivity. While standard planning methods such as decision rules or metaheuristics suffer from low solution quality and long computation times as problem complexity increases, search methods such as Monte Carlo Tree Search (MCTS) with Reinforcement Learning (RL) have proven powerful in optimising otherwise inhibitively complex problems. Despite its success, open questions remain as to when RL can be beneficial for industrial-scale problems. In this paper, we consider the application of MCTS with RL for optimising the reconfiguration of an MPS. We define two operational scenarios and evaluate the potential of RL in each. Taken more generally, our results provide context to better understand when RL can be beneficial in industrial-scale use cases.

1 Introduction

Current trends in manufacturing are moving towards customisation and flexibility, which in turn lead to requirements that traditional linear production systems, such as production lines, cannot meet. Thus, flexible production systems have evolved with the most advanced form being the matrix production system (MPS), also called a Cellular Reconfigurable Manufacturing System (CRMS) [1]. An MPS is a type of modular and flexible production system that consists of a network of manufacturing and assembly stations with multiple process abilities that can be independently configured and connected [2]. They offer significant advantages over traditional line-based production systems in terms of flexibility, reconfigurability, scalability, and resilience to disruptions and shortages [1]. However, an MPS also poses challenges in terms of production planning and control, as the increased degrees of freedom and complexity make it difficult

D. Seipel and A. Steen (Eds.): KI 2023, LNAI 14236, pp. 15–22, 2023.
https://doi.org/10.1007/978-3-031-42608-7_2

to find optimal decision policies for allocating resources, sequencing jobs, and routing products. This is most problematic when the system has to react swiftly to changes in the production environment such as machine outages, a changed product mix, express orders or delays in processes [3,4]. One of the key decisions for an MPS in production planning and control is the reconfiguration of the system, which involves assigning resources (workers, tools or robots) to available workstations according to the production plan and associated processing needs. Reconfiguration is a combinatorial optimisation problem that aims to minimise, e.g., the average makespan. Several optimisation approaches have been proposed, including decision rules [5], metaheuristics such as genetic algorithms (GAs) [6], as well as mixed-integer linear programming [7] and machine learning [8–12]. However, these approaches have some limitations in terms of computational efficiency, scalability, adaptability, and robustness [13]. Reinforcement Learning (RL), a subset of the broader field of machine learning, has gained prominence after the successful application to the combinatorial optimisation problems of chess and Go [14,15]. In these cases the complexity of the solution space makes RL a natural solution, something which is corroborated by the success of RL for extremely large static combinatorial optimisation problems in mathematics [16]. In contrast, for industrial-scale problems open questions remain as to when RL can be beneficial, especially when compared to more conventional approaches. For example, for the reconfiguration of an MPS, it has been shown that, for smaller scale, static scenarios, a GA was advantageous when compared to deep Q-Learning [17]. A systematic approach to understanding which use cases (e.g., in terms of system size) are likely to benefit from RL is still lacking.

In this paper, we provide context for such an approach by applying Monte Carlo Tree Search (MCTS) [18] in combination with RL [14] to optimise the reconfiguration of an MPS. This combination has proven extremely successful in solving various combinatorial problems and has also been used in industrial settings, e.g. see [19]. Two operational scenarios are defined, which highlight a fundamental difference in how use cases can be set up. In both of them, the quantitative results can be used to help understand when RL can be advantageous. These results are compared to standard benchmarks, such as a GA, with respect to performance and required computation. The rest of the paper is organized as follows. In Sect. 2 we define the technical setup of the MPS. In Sect. 3 we motivate the two different operational scenarios used, and then present the results of the optimisation. In Sect. 4 we discuss our results in the broader context and where our results fit into the larger goal of understanding where RL can be beneficial in industry use cases.

2 System Definition

We consider an MPS that can produce different variants of a product, each defined by a series of process steps or tasks. The MPS has a set of workstations, each capable of performing one or more tasks, and a set of resources (workers and robots), which can be assigned to these workstations. The reconfiguration of the system involves allocating the resources to the workstations according to the processing requirements of the product variants to be produced. The objective is to minimise the average makespan of a set of orders. The MPS is implemented as a discrete event simulation in Python, which handles all events and timing. To define a new operating scenario, the simulation requires a list of tasks to be completed to produce one instance of a given product variant, the number of orders and the respective product variants, the number and type of available resources and a list of workstations, the tasks they can complete and the time required for their completion. With this set of input parameters, the simulation runs until all products are finished, at which point the key performance indicators (e.g., makespan) are calculated. While a complete optimisation of this system would also include the sequencing of the jobs in the job list and the routing of jobs between different workstations, in this paper, we concentrate solely on the optimisation of the reconfiguration, omitting the sequencing and using a decision rule for the routing (shortest queue length). For the reconfiguration we implement an MCTS with RL algorithm, based on the AlphaZero algorithm [15]. However, due to the limited search depth in the considered scenarios, we do not use a value function.

3 Results

With the underlying goal of a general understanding of where RL can be beneficial in industrial-scale use cases, we define and evaluate two operating scenarios, which highlight a fundamental difference in how any RL use case can be defined.

The first scenario we define is an example of optimising a *single* problem instance with a fixed (static) product list. Here the goal is to find the optimal configuration for this single scenario as quickly as possible. We use a specific scenario chosen to mimic a real industrial use case, which has a set of 70 jobs and considers a single product variant. In total, there are 27 workstations, which can be serviced by workers or robots. Furthermore, there are 13 workers and two robots, which can be distributed among them. We start the optimisation from an initial minimal configuration, which in this case, requires eight workers at eight stations. The remaining five workers and two robots are then assigned sequentially to the workstations, which makes up, given the set of available workstations, a total of 13, 104 possible configurations. The result of the optimisation for each method is shown in Fig. 1a. To compare the GA and the MCTS, we use the number of times the simulation is evaluated and a scaled logarithmic version of the average makespan of all jobs as a measure of solution quality (reward). As can be seen, combining MCTS with RL can drastically improve on the performance of MCTS in terms of solution quality and required online computation. Furthermore, in this particular scenario, MCTS with RL can even outperform the GA in terms of solution quality in the region of less than 50 online simulation evaluations. However, these results come at the cost of additional 6,300 offline simulation evaluations for training the algorithm. As shown in Fig. 1b, convergence requires around 900 learning steps, which corresponds to 90 training games with 70 simulation evaluations each.

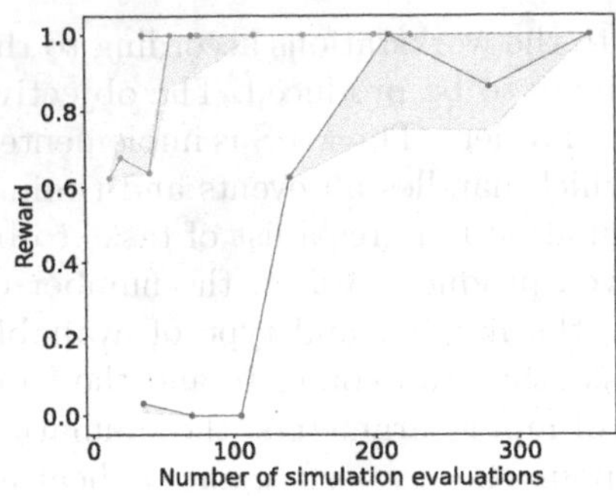

(a) Performance of the different algorithms

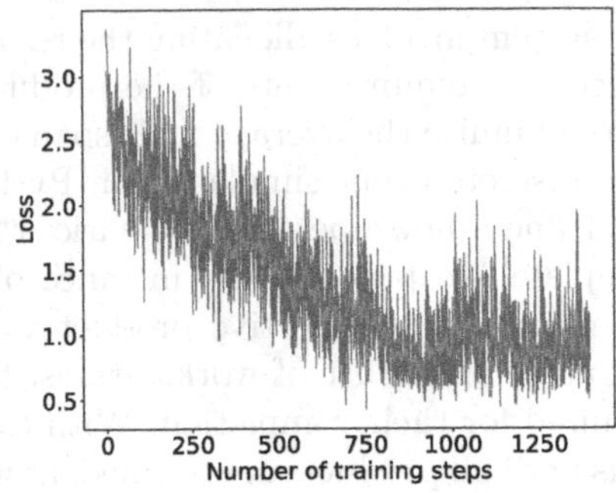

(b) Convergence of the RL policy

Fig. 1. Figure a compares the median reward of each algorithm: GA (orange), MCTS (blue) and MCTS with RL (green) for a given number of simulation evaluations. Each algorithm is carried out ten times and the median and interquartile range (shaded region) are computed. Figure b shows the cross-entropy loss for the RL policy during training. The orange line indicates the running average of the cross-entropy loss over ten training steps. (Color figure online)

The second operational scenario is defined to show the ability of RL to generalise to similar, but unseen problem instances. In this case, the different problem instances refer to the different product types in the job list. Whereas in the first scenario the goal is to optimise the configuration for one fixed product list, here we consider *multiple* problem instances and a dynamically changing product list. Each list has a total of 50 jobs, split between two different product types. Both product types are defined by three production steps, where each step must be completed before the next one can be carried out. In combination with this processing order constraint, the processing times of these steps are chosen such that bottlenecks can occur if the configuration is not chosen accordingly. Therefore, the configuration has to be optimised based on the particular product mix and can not be reused across different product mixes. We consider this as a first step towards simulating dynamically changing conditions typical for flexible production systems such as MPSs. To analyse the generalisability of MCTS with RL, we train the policy network with a select number of job lists with a particular ratio of the two product types and evaluate the trained algorithm on all possible product mixes. For example, first we train with job lists including four different ratios of product 1 and product 2. We then evaluate the learned policy for a product list with all possible ratios of the two different products. Because we fix the total number of products to 50 this means we have 50 evaluation steps. Note that the MCTS algorithm uses 25 online simulation evaluations for every evaluation step. As comparison, we also increased the number of training points to five. The results are shown in Fig. 2.

Since the scenario is very simple, the exact solution is known. There are five disjoint sets of optimal configurations, depending on the ratio of the different products. Note that in the case of four training points the training points are cho-

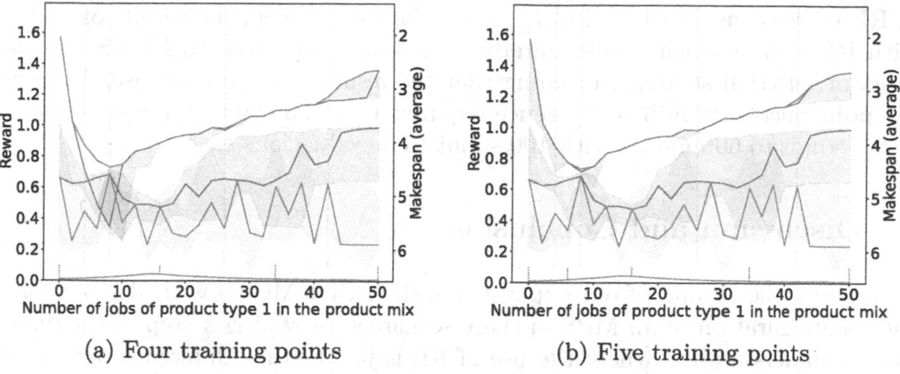

Fig. 2. Comparison of the MCTS with RL results (green) for different ratios of product 1/2 when trained using four (a) and five (b) training points. The median and interquartile range (shaded regions) of ten different training runs are shown. The results are compared to the median reward of MCTS (blue) and random action selection (purple). The maximum and minimum possible rewards are shown (grey). (Color figure online)

sen in uniform intervals, whereas in the case of five training points, the points are chosen such that every region of product mixes is supported in training. However, the exact borders of these regions are not included in the training. In Fig. 2a and Fig. 2b, these borders are denoted with dashed red lines and the training points are denoted by dashed blue lines in the bottom of the plot. As expected, as we increase the number of training points, the generalisability increases. Because the policy network in the MCTS supported RL replaces the initial probability distribution - in the case of insufficient training, the policy network biases to one of the other configurations, and thus can perform worse than the standard MCTS algorithm. This is because the uniform distribution used in MCTS could actually be a better representation than biased towards a different (but not optimal) configuration. Furthermore, the performance of MCTS with RL in terms of solution quality was best, when every product mix region was supported in training (as can be seen in Fig. 2b). However, it is important to note that the results can not be explained solely by the policy network memorising all optimal configurations and steering the MCTS towards them, since the training points were chosen such that the exact border points of these product mix regions were never part of the training, and in general are not known. This is also the reason why a supervised learning approach is not sufficient in this scenario, even if a reasonably sized subset of optimal configurations would be known in advance. Moreover, by comparing different training runs, we found that, in some cases, the policy network overfitted to particular product mixes such that the borders between the regions (even if both regions were part of the training) were not considered correctly. We assume from this observation that, with respect to the generalisation performance, diversity in the training set and prevention of overfitting is (at least to some extent) more important than supporting every region in training. The importance of regularisation methods for generalisation in RL is also considered in [20,21]. As can further be seen, the results of MCTS with RL show a much smaller interquartile range compared to MCTS and the random selection strategy, meaning that the results are more precise. Training the policy network until convergence required around 3,600 training steps which corresponds to 600 games with 100 simulation evaluations each.

4 Discussion and Conclusion

In general, the results of our experiments show that MCTS with RL can solve the reconfiguration of an MPS in these scenarios, providing a step in the direction of understanding where the use of RL is justified for different operational parameters and more general industry use cases.

In the first operating scenario, we optimised a single problem instance with a fixed product list. Here, we found that MCTS with RL is competitive and, in a certain range, superior to the GA in terms of solution quality for the required online computation. However, if the goal of the scenario is to optimise a fixed (static) scenario, neglecting the computational effort for training the network is not reasonable since the trained network is used only once. Thus the total required effort increases significantly. GAs (and other metaheuristics) are highly effective for small problem sizes. In this case, the use of RL is justified only when the solution space is so large that the convergence time of these algorithms becomes inhibitive or they converge to solutions of unsatisfactory quality. Then, MCTS with RL becomes a viable alternative for two reasons: First, the RL value network replaces the need for a complete rollout of the simulation. Second, the policy network is able to exploit distinct features of the problem for search efficiency. A scenario, which is large enough to be intractable for a GA (or similar metaheuristics), could become realistic when considering the overarching vision of a system for planning and control over the complete production, and is the subject of future work.

In the second scenario, where we optimised multiple problem instances with different product mixes, we found that a MCTS with RL performed better than the MCTS with the same number of simulation evaluations for all problem instances, indicating that the algorithm can generalise across different product mixes and adapt to changing scenarios without requiring additional training. This becomes relevant in a real production environment where customer orders, material and other resource availability change frequently and require immediate response. In this scenario, it is reasonable to neglect the computational effort required for training the policy network, such that the performance of the algorithm only relates to the computing resources required online. While conventional methods are still effective if the reaction time (or computational budget) is large enough to reach a desired solution quality, as the complexity of the problem grows, the advantages of RL will become more apparent.

Our results outline the potential of applying MCTS with RL to the reconfiguration of MPSs for two different operating scenarios. More generally we can conclude that whether the use of RL is justified for a particular scenario or not depends on the combination of the required solution quality, the time between setting changes and the total computational budget. Expanding these results to planning tasks with more complex (in terms of system size) scenarios, and those with dynamically changing aspects will allow us to build on these results and further understand the application area of RL in such industry-scale use cases.

Acknowledgements. This research has been supported by the Austrian Federal Ministry for Climate Action, Environment, Energy, Mobility, Innovation and Technology (BMK), the German Federal Ministry for Economic Affairs and Climate Action (BMWK) and the Fraunhofer-Gesellschaft through the projects REINFORCE (887500), champI4.0ns (891793) and MES.Trix.

References

1. Bortolini, M., Galizia, F.G., Mora, C.: Reconfigurable manufacturing systems: literature review and research trend. J. Manuf. Syst. **49**, 93–106 (2018)
2. Greschke, P., Schönemann, M., Thiede, S., Herrmann, C.: Matrix structures for high volumes and flexibility in production systems. Procedia CIRP **17**, 160–165 (2014)
3. Bortolini, M., Galizia, F.G., Mora, C., Pilati, F.: Reconfigurability in cellular manufacturing systems: a design model and multi-scenario analysis. Int. J. Adv. Manuf. Technol. **104**(9), 4387–4397 (2019)
4. Perwitz, J., Sobottka, T., Beicher, J.N., Gaal, A.: Simulation-based evaluation of performance benefits from flexibility in assembly systems and matrix production. Procedia CIRP **107**, 693–698 (2022)
5. Joseph, O.A., Sridharan, R.: Effects of routing flexibility, sequencing flexibility and scheduling decision rules on the performance of a flexible manufacturing system. Int. J. Adv. Manuf. Technol. **56**(1), 291–306 (2011)
6. Zhu, Q., Huang, S., Wang, G., Moghaddam, S.K., Lu, Y., Yan, Y.: Dynamic reconfiguration optimization of intelligent manufacturing system with human-robot collaboration based on digital twin. J. Manuf. Syst. **65**, 330–338 (2022)
7. Luo, K., Shen, G., Li, L., Sun, J.: 0–1 mathematical programming models for flexible process planning. Eur. J. Oper. Res. **308**(3), 1160–1175 (2023)
8. Rodrigues, N., Oliveira, E., Leitão, P.: Decentralized and on-the-fly agent-based service reconfiguration in manufacturing systems. Comput. Ind. **101**, 81–90 (2018)
9. Mo, F., et al.: A framework for manufacturing system reconfiguration and optimisation utilising digital twins and modular artificial intelligence. Robot. Comput.-Integr. Manuf. **82**, 102524 (2023)
10. Morariu, C., Morariu, O., Răileanu, S., Borangiu, T.: Machine learning for predictive scheduling and resource allocation in large scale manufacturing systems. Comput. Ind. **120**, 103244 (2020)
11. Scrimieri, D., Adalat, O., Afazov, S., Ratchev, S.: Modular reconfiguration of flexible production systems using machine learning and performance estimates. IFAC-PapersOnLine **55**(10), 353–358 (2022)
12. Yang, S., Xu, Z.: Intelligent scheduling and reconfiguration via deep reinforcement learning in smart manufacturing. Int. J. Prod. Res. **60**(16), 4936–4953 (2022)
13. Monka, P.P., Monkova, K., Jahnátek, A., Vanca, J.: Flexible manufacturing system simulation and optimization. In: Mitrovic, N., Mladenovic, G., Mitrovic, A. (eds.) CNNTech 2020. LNNS, vol. 153, pp. 53–64. Springer, Cham (2021). https://doi.org/10.1007/978-3-030-58362-0_4

14. Silver, D., et al.: Mastering the game of Go with deep neural networks and tree search. Nature **529**(7587), 484–489 (2016)
15. Silver, D., et al.: A general reinforcement learning algorithm that masters chess, shogi, and Go through self-play. Science **362**(6419), 1140–1144 (2018)
16. Fawzi, A., et al.: Discovering faster matrix multiplication algorithms with reinforcement learning. Nature **610**(7930), 47–53 (2022)
17. Halbwidl, C., Sobottka, T., Gaal, A., Sihn, W.: Deep reinforcement learning as an optimization method for the configuration of adaptable, cell-oriented assembly systems. Procedia CIRP **104**, 1221–1226 (2021)
18. Kocsis, L., Szepesvári, C.: Bandit based Monte-Carlo planning. In: Fürnkranz, J., Scheffer, T., Spiliopoulou, M. (eds.) ECML 2006. LNCS (LNAI), vol. 4212, pp. 282–293. Springer, Heidelberg (2006). https://doi.org/10.1007/11871842_29
19. Göppert, A., Mohring, L., Schmitt, R.H.: Predicting performance indicators with ANNs for AI-based online scheduling in dynamically interconnected assembly systems. Prod. Eng. Res. Devel. **15**(5), 619–633 (2021)
20. Cobbe, K., Klimov, O., Hesse, C., Kim, T., Schulman, J.: Quantifying Generalization in Reinforcement Learning, July 2019
21. Kirk, R., Zhang, A., Grefenstette, E., Rocktäschel, T.: A survey of zero-shot generalisation in deep reinforcement learning. J. Artif. Intell. Res. **76**, 201–264 (2023)

CHA$_2$: <u>CH</u>emistry <u>A</u>ware <u>C</u>onvex <u>H</u>ull <u>A</u>utoencoder Towards Inverse Molecular Design

Mohammad Sajjad Ghaemi[1(✉)], Hang Hu[1], Anguang Hu[2], and Hsu Kiang Ooi[1]

[1] National Research Council Canada, Toronto, ON, Canada
MohammadSajjad.Ghaemi@nrc-cnrc.gc.ca
[2] Suffield Research Centre, DRDC, Alberta, Canada

Abstract. Optimizing molecular design and discovering novel chemical structures to meet specific objectives, such as quantitative estimates of the drug-likeness score (QEDs), is NP-hard due to the vast combinatorial design space of discrete molecular structures, which makes it near impossible to explore the entire search space comprehensively to exploit *de novo* structures with properties of interest. To address this challenge, reducing the intractable search space into a lower-dimensional latent volume helps examine molecular candidates more feasibly via inverse design. Autoencoders are suitable deep learning techniques, equipped with an encoder that reduces the discrete molecular structure into a latent space and a decoder that inverts the search space back to the molecular design. The continuous property of the latent space, which characterizes the discrete chemical structures, provides a flexible representation for inverse design to discover novel molecules. However, exploring this latent space requires particular insights to generate new structures. Therefore, we propose using a convex hull (CH) surrounding the top molecules regarding high QEDs to ensnare a tight subspace in the latent representation as an efficient way to reveal novel molecules with high QEDs. We demonstrate the effectiveness of our suggested method by using the QM9 as a training dataset along with the Self-Referencing Embedded Strings (SELFIES) representation to calibrate the autoencoder in order to carry out the inverse molecular design that leads to unfolding novel chemical structure.

Keywords: Convex Hull · Autoencoder · Inverse Molecular Design

1 Introduction

The rise of artificial intelligence (AI), in particular, the emergence of generative machine learning (ML) techniques as a disruptive breakthrough technology, has revolutionized numerous fields, including molecular optimization, material

This project is supported by the National Research Council Canada (NRC) and the Defence Research and Development Canada (DRDC).

D. Seipel and A. Steen (Eds.): KI 2023, LNAI 14236, pp. 23–30, 2023.
https://doi.org/10.1007/978-3-031-42608-7_3

design, and drug design [7,11]. As such, taming an intractable combinatorial solution region to search for novel structures with desirable chemical properties has become a promising perspective in recent years [3]. Additionally, data-driven inverse design inspired by machine learning methodologies effectively disseminates molecular knowledge from the training data to identify new candidates for chemical structures to meet specific properties [14]. However, while typical machine learning methods are designed to exploit domain knowledge through training data to uncover the underlying distribution of data, novel structures are often hidden in unexplored, low-density regions. As such, revealing unprecedented structures requires advanced learning schemes to seek uninvestigated regions, which are restricted by the clues learned from existing information, e.g., chemical properties of matter [11]. So, learning a continuous bijective function to project the discrete structure of molecules into a lower-dimensional space and simultaneously map a random data point from the same lower-dimensional space onto a molecular representation is necessary [13].

A high-dimensional data representation such as an autoencoder aims to find an underlying low-dimensional manifold that preserves the vital information specific to the desired properties, e.g. QEDs in the molecular design context. Thus, for distinct, separable clusters of information, a well-decomposed group of salient signals in the latent space is expected from a perfect autoencoder model to account for sources of variation in the data [8]. As such, a convex combination of signals in the vicinity of data points with distinguished properties presumably entails a mixture of features reflecting prominent holistic patterns.

In this study, we propose a novel approach inspired by CH of the latent space to generate out-of-distribution (OOD) molecules through the uniformly random samples restricted to the boundaries of high QEDs molecules of autoencoder trained on QM9 data [12] via 1-hot-encoding of SELFIES representation.

2 Relevant Prior Works

In recent years, various molecular data encoding techniques, such as geometrical distance, molecular graph, and string specifications, have been employed to feed molecules to ML models. Among these techniques, string specifications, including simplified molecular-input line-entry system (SMILES), deep SMILES, and SELFIES, have gained significant attention due to their promising results [11,13]. To effectively carry out sequential training, natural language processing (NLP) techniques, mainly Recurrent Neural Networks (RNN) and its variant Long Short-Term Memory (LSTM), have been used successfully [6]. Similarly, data compression and dimension reduction techniques, such as autoencoders and their generative counterpart, variational autoencoders (VAE) [9], have been adopted for the same purpose by leveraging the holistic string representation as opposed to the sequential approach [13]. In this study, we developed a novel autoencoder model that incorporates domain knowledge, specifically the QEDs, as the hallmarks of the QM9 data. The proposed **CHA$_2$** method guides the autoencoder through high QEDs molecules in the latent space by constructing a CH entailing the points corresponding to the high QEDs molecules. Synthetic molecules are generated by taking random data points that are uniformly sampled from the boundaries of the CH surrounding the selected hallmarks of the latent space to the region of interest.

3 Proposed Method and Experimental Design

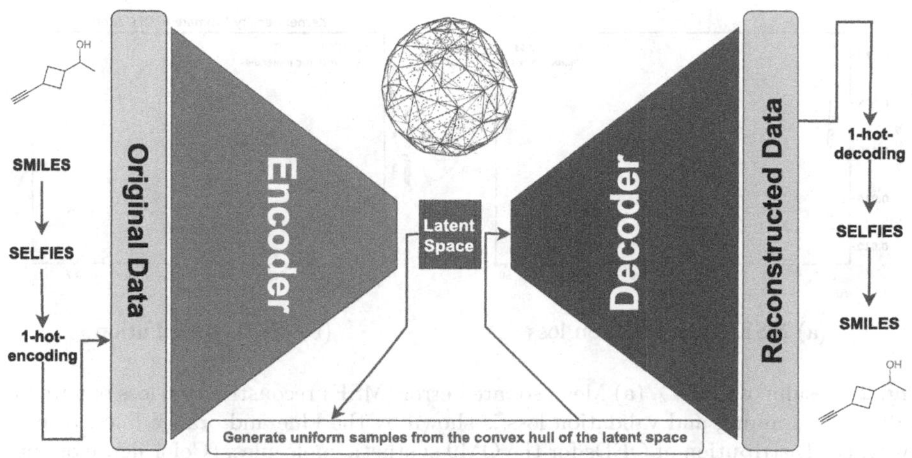

Fig. 1. Schematic of the **CHA$_2$** approach for inverse molecular design, based on the CH associated with the top QEDs molecules. Discrete molecular structures represented by 1-hot-encoding of SELFIES fed to the encoder network to form the continuous latent space. Consequently, the decoder module generates synthetic molecules by sampling random points from the CH.

In many cases, generative models are trained to estimate the probability distribution of the source data, with the expectation that the learned distribution will produce samples that share the same statistical properties as the real data. This approach facilitates the creation of a model capable of efficiently replicating data points that mimic the source data. However, generating OOD synthetic data that lie beyond the original data distribution is crucial for specific tasks such as drug discovery and material design. The proposed **CHA$_2$** method is novel in guiding the autoencoder towards high QEDs molecules in the latent space. This is accomplished by constructing a CH containing the points corresponding to the high QEDs molecules. Uniformly random samples are then generated, restricted to the boundaries of the CH, to empower the decoder to achieve high QEDs synthetic molecules. The autoencoder was trained using the 1-hot-encoding of SELFIES representation based on the QM9 data shown in Fig. 1. Training data consisted of molecules with QEDs greater than 0.5, while molecules with QEDs between 0.4 and 0.5 were used as validation data. During the molecular training and generating process, the molecule size was fixed at 19 elements of SELFIES, which matched the maximum length of molecules in the training and validation data. The deep learning model was optimized using mean squared error (MSE) for autoencoder implementation in TensorFlow and consisted of an array of layers comprising $(250, 120, 60, 30, 8, 3, 8, 30, 60, 120, 250)$ neurons with rectified linear unit (ReLU) activation functions (except for the last layer, which used a sigmoid activation function) [1]. Figure 2a depicts the convergence of the MSE of the reconstruction loss on the training and validation data. After training the

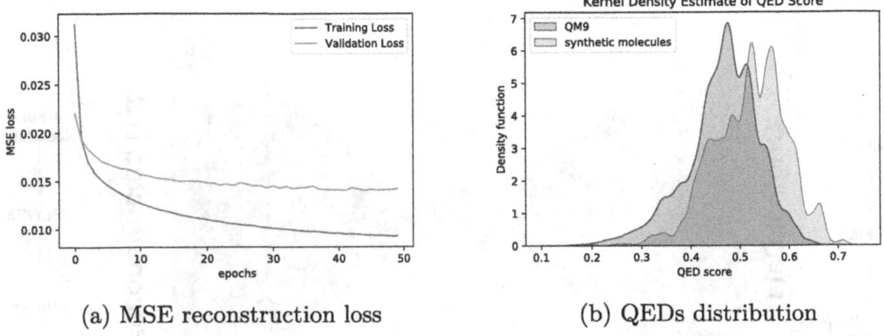

(a) MSE reconstruction loss (b) QEDs distribution

Fig. 2. Results of $\mathbf{CHA}_2$. (a) Mean squared error (MSE) reconstruction loss of autoencoder. The training and validation loss is shown by the blue and orange lines, respectively. (b) Distribution of QEDs for the QM9 synthetic molecules. (Color figure online)

autoencoder, 100,000 random samples were generated from the CH fit over top molecules with QEDs greater than 0.65 in the latent space. Following the removal of duplicates and a sanity check, approximately 11,000 valid synthetic molecules were obtained using this approach. The distribution of QEDs for both the QM9 dataset and the synthetically generated molecules is presented in Fig. 2b. The 3D CH of the autoencoder latent space and its projection onto the $\mathbf{XY}$ plane are illustrated in Fig. 3. The solid blue dots indicate the top molecules with high QEDs. Furthermore, the yellow region represents the space of synthetic molecules with QEDs greater than 0.7, and the red area denotes the space of all random samples generated from the CH shown in Fig. 3a. The yellow-to-blue spectrum of QEDs is displayed for the synthetic molecules within the density of QM9 molecules. This spectrum ranges from yellow for low QEDs to blue for high QEDs, illustrating the distribution of QEDs within the polyhedron, as depicted in Fig. 3b.

4 Results

We explored the framework for design within three, four, and five-dimensional latent spaces. Figure 4 represents the best-in-class molecules based on QEDs

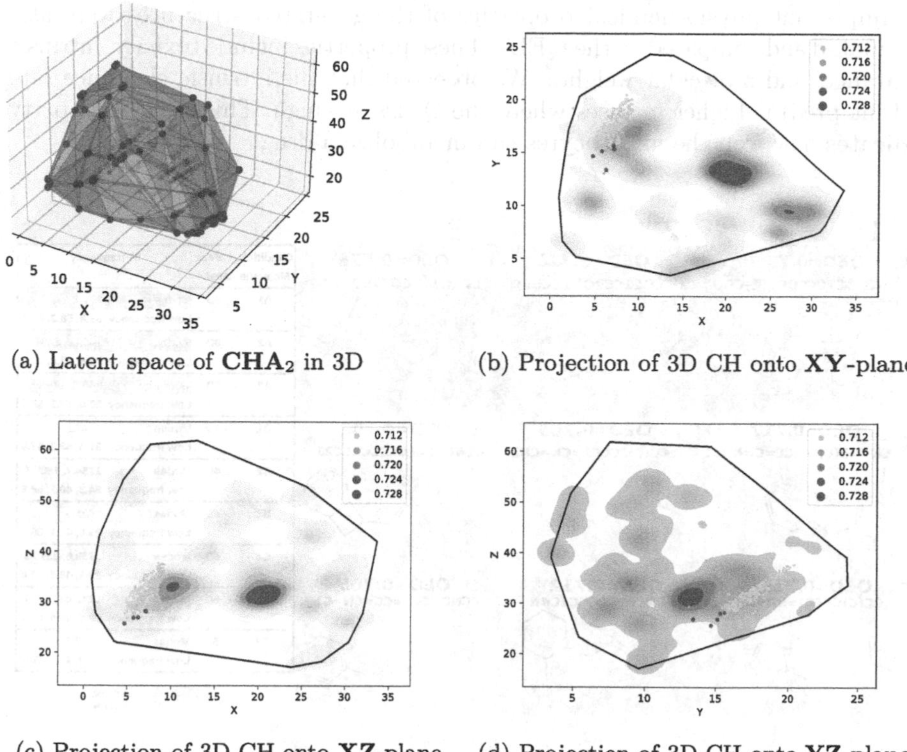

(a) Latent space of **CHA$_2$** in 3D (b) Projection of 3D CH onto **XY**-plane

(c) Projection of 3D CH onto **XZ**-plane (d) Projection of 3D CH onto **YZ**-plane

Fig. 3. The latent space of **CHA$_2$** in 3D and its projection onto the **XY**, **XZ**, and **YZ** coordinate planes. (a) Solid blue dots indicate the top QEDs molecules, and the red area represents the space of random samples generated from the CH. (b-d) The yellow-to-blue spectrum of the QEDs is shown for synthetic molecules within the density of QM9 molecules. The QEDs range from yellow (low values) to blue (high values), indicating the distribution of QEDs within the polyhedron. (Color figure online)

for each latent space. We calculated the generated molecules' vibration modes and infrared (IR) spectrum as a sanity check. Our findings indicate that all synthesized molecules have a positive frequency for their vibration modes, which confirms their physical stability as chemically viable compounds. The structure relaxation and vibration mode are calculated using Density-functional theory with B3LYP exchange-correlation functional and 631G basis set, implemented under the Gaussian 16 simulation package [2,4,5,10]. The frequency of selected IR active modes is tabulated in Fig. 4. In the case of molecules generated using the 3-dimensional latent space, the corresponding QEDs (candidates A1–A3) were superior to those generated using other dimensions (candidates B1–C3). One design principle observed is that generated molecules with a C-O mode tend to result in higher QEDs.

Important physiochemical properties of the generated structures were also calculated and compared to the QEDs. These properties include toxicity, intrinsic clearance, and molecular weights. We observed that the intrinsic clearance rate (CLint_pred) is higher in cases where the QEDs are high. This critical property indicates how well the hepatocytes can metabolize a drug.

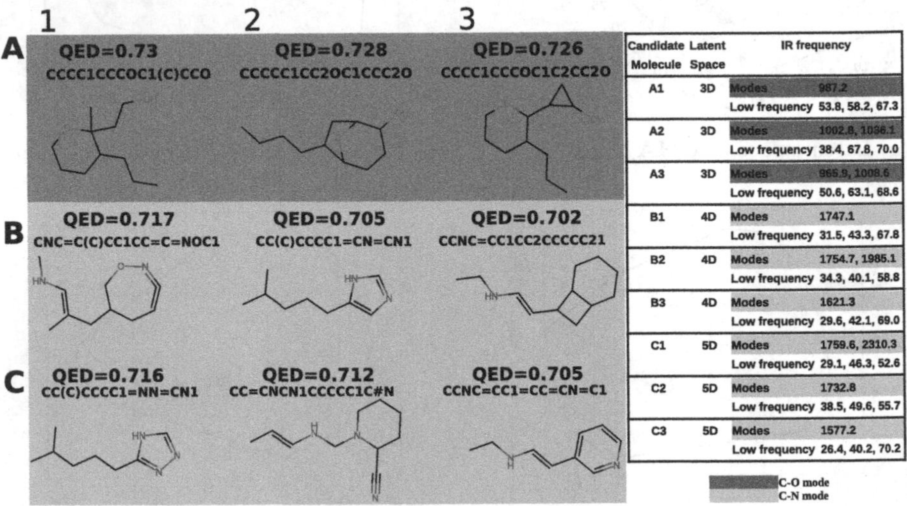

Candidate Molecule	Latent Space	IR frequency	
A1	3D	Modes	987.2
		Low frequency	53.8, 58.2, 67.3
A2	3D	Modes	1002.8, 1036.1
		Low frequency	38.4, 67.8, 70.0
A3	3D	Modes	965.9, 1008.6
		Low frequency	50.6, 63.1, 68.6
B1	4D	Modes	1747.1
		Low frequency	31.5, 43.3, 67.8
B2	4D	Modes	1754.7, 1985.1
		Low frequency	34.3, 40.1, 58.8
B3	4D	Modes	1621.3
		Low frequency	29.6, 42.1, 69.0
C1	5D	Modes	1759.6, 2310.3
		Low frequency	29.1, 46.3, 52.6
C2	5D	Modes	1732.8
		Low frequency	38.5, 49.6, 55.7
C3	5D	Modes	1577.2
		Low frequency	26.4, 40.2, 70.2

Fig. 4. Molecules with Higher QEDs in the Latent Space. Varying the number of units in the latent space resulted in additional molecular designs with higher QEDs. The top row shows the three (A1–A3) molecules with the highest QEDs generated from the **CHA**$_2$. The table displays the calculated IR frequencies for these candidate molecules, revealing an interesting design principle.

Next, we focused on the three-dimensions latent space to identify further design principles that would enhance QEDs. By generating a total of 11000 molecules, we observed 13 molecules that exhibited QEDs greater than 0.70. Notably, our analysis revealed an absence of C-N bonds in molecules with higher QED values. Conversely, molecules with QED values less than 0.69 typically contained C-N bonds in their structures. Furthermore, the IR spectra of these molecules exhibited C-N functional groups with activity around 1700 cm^{-1}, while C-O groups were active around 1000 cm^{-1}, as depicted in Fig. 4. This observation suggests that the IR spectra could be a potential screening tool for experimentally identifying drug-like molecules.

The **CHA**$_2$ approach offers an efficient way to explore the molecular domain, facilitating the detection of novel structures within the vicinity of the molecules with high QED scores. Utilizing this method also reveals interesting properties. As per Carathéodory's theorem, any synthetic molecule $\mathbf{m}$ generated from the d-dimensional latent space can be expressed as the average of $d+1$ molecules from the same convex set. Furthermore, by normalizing the reduced data in the latent space such that the convex set's diameter is bounded by 1, it is possible to find $\mathbf{x}_1, \mathbf{x}_2, \ldots, \mathbf{x}_k$ molecules within the same convex set that satisfy the following inequality, $\left\|\mathbf{m} - \frac{1}{k}\sum_{j=1}^{k} \mathbf{x}_j\right\|_2 < \frac{1}{\sqrt{k}}$, for any arbitrary integer k regardless of the latent space's dimension [15].

5 Conclusion and Future Work

The proposed **CHA$_2$** is shown as a successful approach for an inverse molecular design where CH played a central role in generating novel molecules from the autoencoder's latent space. The synthetic molecules' vibration modes and IR spectra analysis showed that *de novo* molecular structures are physically stable and chemically viable. While our analysis revealed a remarkable design principle for enhancing QEDs, namely the presence of chemical structures with C-O modes, we also observed that the lack of C-N bonds led to higher QEDs. This finding suggests that the frequency of C-O bonds in the absence of C-N bonds in molecular structures may serve as an effective strategy for improving QEDs.

Additionally, with the rapid advancement of scalable quantum technology, a promising direction for **CHA$_2$** is to capitalize on the potential benefits of a hybrid approach that integrates machine learning and quantum chemistry simulation. In particular, the utilization of near-term Noisy Intermediate Scale Quantum (NISQ) technology has the potential to unlock new opportunities for tackling complex challenges in chemistry and materials design. This involves developing more sophisticated ML algorithms inspired by quantum chemistry and generative models inspired by transformers that leverage quantum computing hardware and deploy enriched datasets to enhance accuracy and performance.

References

1. Abadi, M., et al.: {TensorFlow}: a system for {Large-Scale} machine learning. In: 12th USENIX Symposium on Operating Systems Design and Implementation (OSDI 2016), pp. 265–283 (2016)
2. Becke, A.D.: Density-functional thermochemistry. III. The role of exact exchange. J. Chem. Phys. **98**(7), 5648–5652 (1993)
3. Blaschke, T., Olivecrona, M., Engkvist, O., Bajorath, J., Chen, H.: Application of generative autoencoder in de novo molecular design. Mol. Inf. **37**(1–2), 1700123 (2018)
4. Ditchfield, R., Hehre, W.J., Pople, J.A.: Self-consistent molecular-orbital methods. IX. An extended gaussian-type basis for molecular-orbital studies of organic molecules. J. Chem. Phys. **54**(2), 724–728 (1971)
5. Frisch, M.J., et al.: Gaussian 16 Revision C.01. Gaussian Inc., Wallingford (2016)
6. Ghaemi, M.S., Grantham, K., Tamblyn, I., Li, Y., Ooi, H.K.: Generative enriched sequential learning (ESL) approach for molecular design via augmented domain knowledge. In: Proceedings of the Canadian Conference on Artificial Intelligence, 27 May 2022

7. Grantham, K., Mukaidaisi, M., Ooi, H.K., Ghaemi, M.S., Tchagang, A., Li, Y.: Deep evolutionary learning for molecular design. IEEE Comput. Intell. Mag. **17**(2), 14–28 (2022)
8. Joswig, M., Kaluba, M., Ruff, L.: Geometric disentanglement by random convex polytopes. arXiv preprint arXiv:2009.13987 (2020)
9. Kingma, D., Welling, M.: Auto-encoding variational Bayes. In: International Conference on Learning Representations (2014)
10. Lee, C., Yang, W., Parr, R.G.: Development of the Colle-Salvetti correlation-energy formula into a functional of the electron density. Phys. Rev. B **37**, 785–789 (1988)
11. Menon, D., Ranganathan, R.: A generative approach to materials discovery, design, and optimization. ACS Omega **7**(30), 25958–25973 (2022)
12. Ramakrishnan, R., Dral, P.O., Rupp, M., von Lilienfeld, O.A.: Quantum chemistry structures and properties of 134 kilo molecules. Sci. Data **1**(1), 140022 (2014)
13. Romez-Bombarelli, R., et al.: Automatic chemical design using a data-driven continuous representation of molecules. ACS Cent. Sci. **4**, 268–276 (2018)
14. Sanchez-Lengeling, B., Aspuru-Guzik, A.: Inverse molecular design using machine learning: generative models for matter engineering. Science **361**(6400), 360–365 (2018)
15. Vershynin, R.: High-Dimensional Probability. University of California, Irvine (2020)

Ontology Pre-training for Poison Prediction

Martin Glauer[1], Fabian Neuhaus[1], Till Mossakowski[1(✉)], and Janna Hastings[2]

[1] Otto von Guericke University Magdeburg Universitätsplatz 2, 39106 Magdeburg, Germany
{martin.glauer,fneuhaus,till.mossakowski}@ovgu.de

[2] University of Zurich, Rämistrasse 71, 8006 Zürich, Switzerland
janna.hastings@uzh.ch

Abstract. Integrating human knowledge into neural networks has the potential to improve their robustness and interpretability. We have developed a novel approach to integrate knowledge from ontologies into the structure of a Transformer network which we call ontology pre-training: we train the network to predict membership in ontology classes as a way to embed the structure of the ontology into the network, and subsequently fine-tune the network for the particular prediction task. We apply this approach to a case study in predicting the potential toxicity of a small molecule based on its molecular structure, a challenging task for machine learning in life sciences chemistry. Our approach improves on the state of the art, and moreover has several additional benefits. First, we are able to show that the model learns to focus attention on more meaningful chemical groups when making predictions with ontology pre-training than without, paving a path towards greater robustness and interpretability. Second, the training time is reduced after ontology pre-training, indicating that the model is better placed to learn what matters for toxicity prediction with the ontology pre-training than without. This strategy has general applicability as a neuro-symbolic approach to embed meaningful semantics into neural networks.

Keywords: Ontology · transformer network · Neuro-symbolic integration

1 Introduction

Deep neural networks have recently led to breakthrough performance for a wide range of tasks such as protein folding [16] and image generation [22]. However, they still suffer from challenges in generalisability, robustness, and interpretability. Approaches that incorporate human knowledge alongside learning from data, which have been called hybrid, knowledge-aware or *informed* [23], have the potential to improve the correspondence between what the model learns and

D. Seipel and A. Steen (Eds.): KI 2023, LNAI 14236, pp. 31–45, 2023.
https://doi.org/10.1007/978-3-031-42608-7_4

the structure of the human world, which in turn allows the model to learn more generalisable representations from smaller datasets.

Human knowledge is carefully curated into ontologies [10,19], making them a prime candidate source of knowledge to incorporate into learning. Many different approaches have already been developed with the objective of harnessing prior knowledge to improve machine learning. The most common approach is enrichment of the training data with additional information from ontologies (see Sect. 4.2). In this paper we present a novel methodology, which uses an ontology, namely the Chemical Entities of Biological Interest (ChEBI), to create a pre-training task for a Transformer model (see Sect. 2). This pre-training task consists of predicting superclasses in ChEBI's taxonomic hierarchy for molecule classes represented by input chemical structures. Thus, during this pre-training the model learns to recognise categories of chemical entities that are chemically meaningful. After the ontology pre-training the model is fine-tuned for the task of toxicity prediction using the dataset from the well-known Tox21 challenge [13]. This dataset consists of 12 different toxicity endpoints, including 7 nuclear receptor signals and 5 stress response indicators.

As we show in Sect. 3, for the purpose of toxicity prediction the ontological pre-training step showed the following benefits: First, the model converges faster during fine-tuning. Second, an inspection of the attention heads indicates that the model pays attention to chemical structures that correspond to structural chemical annotations that are associated with classes in ChEBI. Since ChEBI classes represent chemical categories that are meaningful to humans, this connection improves the interpretability of the model's predictions. Third, the predictive performance is improved significantly compared to the performance without pre-training. Indeed, our ontology pre-trained model outperforms the state of the art for toxicity prediction on the Tox21 dataset from structures without additional input features (see Sect. 4.2).

These results seem to indicate that the ontological pre-training is enabling the model to learn some of the knowledge that is represented by the ontology. However, there are important limitations with respect to the knowledge that is learned by the model. Further, our proposed methodology is only applicable to ontologies that either contain rich structural annotations or which are associated with suitable datasets that link the input datatype intended for the learning task to the ontology classes. We will discuss these limitations in Sect. 4.3 and how to overcome them in the future in Sect. 5.

2 Methods

The usual process used to train a transformer-based model consists of two steps: pre-training and fine-tuning. The intention behind the pre-training step is to give the model a solid foundation training in the kind of data that will be used in downstream tasks in order for it to gain a preliminary understanding of the target domain that can the transferred to more specific tasks later. Most transformer-based architectures are built for language problems, and the respective pre-training is often limited to masked-language tasks (BERT [7], RoBERTa) or token discrimination tasks (Electra). This kind of training enables the model to learn the syntactic relations between words, but it does not get any information about their semantics, aside from context similarity when words are interchangeably used in similar contexts.

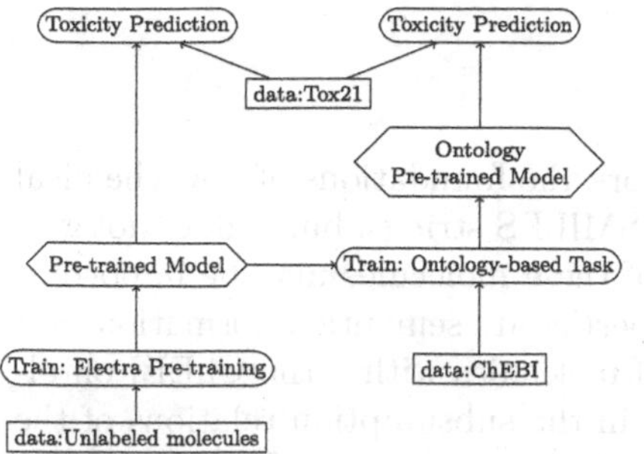

Fig. 1. Training stack for standard training and ontology pre-training

Hyperparameter	**Value**
Vocab. size	1,400
Hidden size	256
Num. of attention heads	8
Num. of hidden layers	6
Epochs	100
Learning Rate	$1e^{-4}$
Optimizer	Adamax

Fig. 2. Hyperparameters shared by all models as used in [9]

In this study, we introduced an additional ontology pre-training stage after the usual pre-training and before the fine-tuning. We present a case-study for the use of ontology pre-training to improve chemical toxicity prediction. Figure 1 depicts the process using the boxology notation [1] for the novel approach as well as a comparison approach without ontology pre-training which serves as our baseline. In the remainder of this section, we will detail the setup for each individual step. All models are based on the Electra model as implemented in the ChEBai[1] tool used in our previous work [9]. All models share the hyperparameters detailed in Table 2 with different classification heads for different tasks.

2.1 Step 1: Standard Pre-training

The first step of this architecture is based on the pre-training mechanism for Electra [6]. This process extends the general masked-language model (MLM) used to pre-train transformers by a competing discriminator that aims to predict which tokens have been replaced as part of the MLM task. For the molecular data that serves as the input for the prediction task, the language of the molecular representations is SMILES [28], a molecular line notation in which atoms and bonds are represented in a linear sequence, and the masking of tokens affects sub-elements within the molecule. The generator part of Electra is followed by a simple linear layer with a softmax to predict the token that was most likely masked, while the discriminator uses the same setup, but one additional dense layer to guess which token had been replaced by the generator.

[1] https://github.com/MGlauer/ChEBai.

In our case-study, we use the same dataset for pre-training that has been used in our previous work on ontology extension [9]. This dataset consists of 365,512 SMILES representations for molecular structures that have been extracted from PubChem [25] and the ChEBI ontology [12]. Notably, 152,205 of these substances are known to be hazardous substances as they are associated with a hazard class annotation in PubChem.

2.2 Step 2: Ontology Pre-training

The standard pre-training teaches the network the foundations of how chemical structures are composed and represented in SMILES strings, but it does not give the network any insights into which parts of these molecules may be important and chemically active. These functional properties are semantic information that are used by experts to distinguish classes of molecules within the ChEBI ontology. They are therefore inherently encoded in the subsumption relations of the ontology. We used the subsumption hierarchy to create a dataset for an ontology-based classification task by extracting all subclasses of 'molecular entity' that had at least 100 subclasses with SMILES strings attached to them. This resulted in a collection of 856 classes. We then collected all classes in ChEBI that had a SMILES string attached to them and annotated them with their subsumption relation for each of the 856 classes. The resulting dataset is similar to the ones used in [11] and [9], but covers a wider range of classes as labels (856 instead of 500) and also a larger number of molecules (129,187 instead of 31,280). We then use the pre-trained model from Step 1 to predict the superclasses of a class of molecules based on its annotated SMILES string.

2.3 Step 3: Toxicity Prediction

In order to assess the impact of this additional ontology-based pre-training step, we compare the model that resulted from the ontology pre-training in step 2 with the one from step 1 that did not receive that training. This comparison is based on each model's performance on toxicity prediction using the Tox21 dataset. This dataset was created by the National Center for Advanced Translational Sciences (NCATS) of the National Institutes of Health (NIH), and constitutes a widely used benchmark for research on machine learning for toxicity prediction from small molecules [13]. Notably, there are currently two versions of the Tox21 dataset available in benchmarks. The first one originates from the Tox21 Challenge that was conducted in 2014. This dataset consists of three different subsets, one for training and two for different steps of the challenge evaluation. In our study, we will use the "testing dataset" that was used for an initial ranking of models as a validation set and the dataset that was used for the final evaluation as our test set. This version of the Tox21 dataset suffers from several issues regarding the consistency and quality of different entries [14]. A more modern version of this dataset has been published as part of the MoleculeNet benchmark [29]. This version of Tox21 consists of only of 7,831 molecules. We split this dataset into a training (85%), validation (7.5%) and test set (7.5%).

There are, however, still two major challenges that need to be considered when working with this dataset. First, the number of data points is rather low. Molecular structures are complex graphs, which makes it hard for any model to derive a sufficient amount of information from this dataset alone. Second, the information available in the dataset is not complete: a substantial amount of toxicity information is missing in this dataset. There are 4,752 molecules for which at least one of the 12 kinds of toxicity is not known. In the prior literature, this issue has been approached in different ways. Some approaches perform data cleaning which limits the number of available data points even further, e.g. [14] excluded all data points that contained any missing labels. We decided to keep these data points as part of our dataset, but exclude the missing labels from the calculation of all loss functions and metrics. Any outputs that the model generates for these missing labels is considered as correct and does not influence the training gradient. This approach allows the network to fill these gaps autonomously.

Preliminary results showed that both models were prone to overfitting, when used with the same settings as the model from step 2. We found that this behaviour could be prevented by using strong regularisations. The final experiments used the Adamax optimizer with dropouts on input embeddings of 0.2, dropouts on hidden states of 0.4 and L2-regularisation of 0.0001. All data and code used in this study is publicly available.[2]

3 Results

3.1 Predictive Performance

The final result of our training stack are four models: with or without ontology pre-training, and fine-tuned on the original Tox21 competition dataset or on the smaller version of the Tox21 dataset published as part of MoleculeNet. The semantic pre-training already showed a clear impact during the training phase. Figures 3a–3df depict the curves for two metrics (F1 score and ROC-AUC) on our validation set as evaluated at the end of each epoch during training. It can be seen that models with ontology pre-training start with a better initial performance and also retain this margin throughout the training. This behaviour is also reflected in the predictive performance on both test sets. Table 1 shows the predictive behaviour for the dataset from MoleculeNet and the original challenge. The leading model (highlighted in bold) is predominantly the one that received additional ontology pre-training. This is particularly visible for the more noisy and sparse dataset used in the original Tox21 competition. The overall improved performance shows that pre-training with a more general ontology pre-training does support the network for the more specialised toxicity prediction. The drastic drop that can be seen around epoch 50 in Fig. 3b but not for the pre-trained model in Fig. 3d further indicates that ontology pre-training hedges the model against early overfitting. The reported results, and in particular the F1 scores, however, show that there is still a large margin of error for this task.

[2] https://doi.org/10.5281/zenodo.7548313.

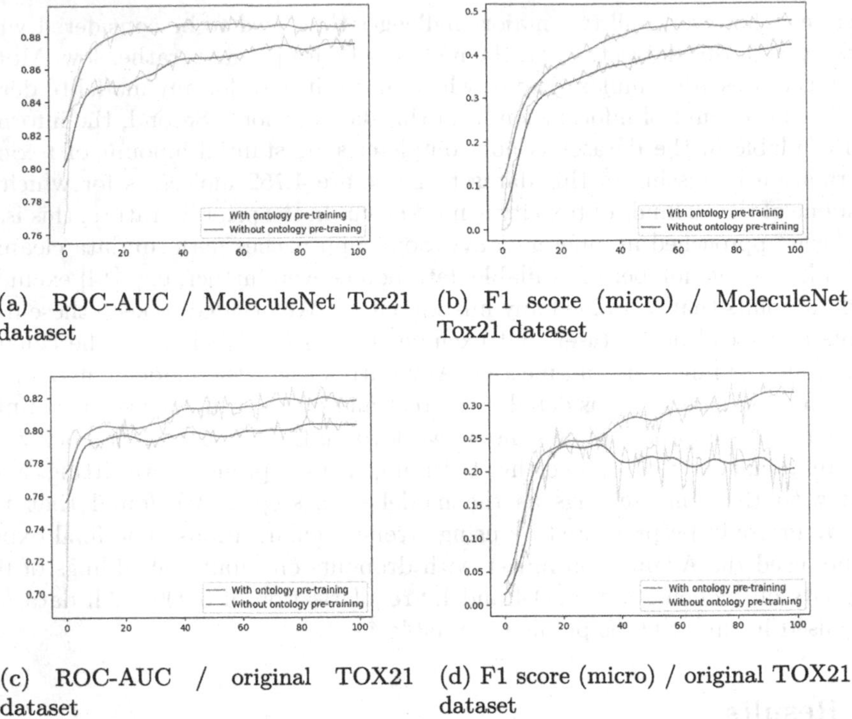

(a) ROC-AUC / MoleculeNet Tox21 dataset

(b) F1 score (micro) / MoleculeNet Tox21 dataset

(c) ROC-AUC / original TOX21 dataset

(d) F1 score (micro) / original TOX21 dataset

Fig. 3. Development of ROC-AUC and F1 score (micro) during training on the validation sets of the Tox21 dataset available as part of MoleculeNet and the original TOX21 challenge.

3.2 Interpretability

Attention weights in Transformer networks can be visualised to enable a kind of salience-based visual interpretation for predictions directly connected with the input data [27]. Previous work [9] explored the link between attention weights to the prediction of ontology classes, and observed that the network learned to pay attention to relevant substructures when making predictions of ontology classes to which the molecule belongs.

In the current work, our hypothesis was that the additional information from the ontology would both enhance the prediction and enhance the coherence of the attention weights for explanatory visualisations. To test this hypothesis we explored the attention visualisations for the predictions by the ontology pre-trained network as compared to the normal network.

Figure 4 shows an individual example of the attention weight that the network uses when visiting specific input tokens. The molecule depicted is TOX25530, corresponding to the sulfonamide anti-glaucoma drug dorzolamide, which is not toxic. Dark green lines indicate strong, focused attention, while more opaque lines indicate broader, less focused attention. As this example illustrates, we observed

Table 1. Class-wise scores on the test set on both Tox21 datasets. Bold values denote the best value for a particular combination of dataset and metric. NR - nuclear receptor; AR - androgen receptor; LBD - luciferase; AhR - aryl hydrocarbon receptor; ER - estrogen receptor; PPAR - peroxisome proliferator-activated receptor; SR - stress response; ARE - nuclear factor antioxidant response; ATAD5 - genotoxicity; HSE - heat shock factor response; MMP - mitochondrial response; p53 - DNA damage response.

Dataset	Tox 21 (MoleculeNet)					Tox21 (Challenge)			
Metric	F1		ROC-AUC			F1		ROC-AUC	
Model	Our Model				SSL-GCN	Our Model			
Ontology Pre-training	yes	no	yes	no	-	yes	no	yes	no
NR-AR	0.41	**0.52**	**0.82**	0.76	0.80	0.1	**0.14**	**0.63**	0.62
NR-AR-LBD	**0.51**	0.5	**0.85**	0.77	0.76	0.05	**0.1**	**0.69**	0.67
NR-AhR	**0.53**	0.45	0.81	0.82	**0.83**	**0.23**	0.05	**0.8**	0.69
NR-Aromatase	**0.33**	0.15	**0.84**	0.8	0.73	**0.25**	0.04	**0.75**	0.69
NR-ER	**0.44**	0.4	**0.74**	0.71	0.72	**0.16**	0.09	**0.64**	0.62
NR-ER-LBD	**0.37**	0.3	**0.84**	0.76	0.69	**0.14**	0.12	**0.66**	0.63
NR-PPAR-gamma	**0.29**	-	**0.84**	0.83	0.76	**0.14**	-	**0.67**	0.66
SR-ARE	0.48	**0.53**	0.8	**0.84**	0.73	**0.37**	0.23	**0.71**	0.69
SR-ATAD5	0.14	**0.19**	**0.75**	0.74	0.72	**0.16**	-	**0.65**	0.65
SR-HSE	**0.24**	0.22	0.82	**0.82**	0.78	**0.13**	0.09	**0.76**	0.68
SR-MMP	**0.62**	0.53	**0.9**	0.88	0.81	**0.48**	0.21	**0.86**	0.82
SR-p53	**0.39**	0.35	**0.83**	0.8	0.75	**0.3**	-	**0.82**	0.78

that the ontology pre-trained network often shows more coherence and structure in its attention weights compared to the baseline network without ontology pre-training. This is reflected in the triangular clusters of darker attention weights in the top rows of Fig. 4a and b. Clusters reflect that from a particular position in the input token sequence, strong attention is placed on a subset of the molecule, reflecting relevant substructures within the molecule. Figure 4c shows how the attention weights relate to the molecular graph for this molecule. Attention weight relationships may be short-range (nearby atoms or groups) or long-range (atoms or groups further away within the molecule) (Fig. 5).

To test this visual intuition more systematically, we computed the entropy for each attention head in each layer. Attention is computed using softmax and can therefore be interpreted as a probability distribution. In order to evaluate our hypothesis that the ontology pre-training also impacts the way a model focuses its attention, we calculated the average entropy of these distributions. Entropy is a measure for the randomness of a distribution. A perfectly even distribution would result in an entropy value of 1 (complete uncertainty), while a distribution in which only one event can possibly occur will result in an entropy value of 0 (complete certainty). That means that an attention distribution with an entropy value of 1 is not paying attention to any particular part of the molecule, but is spreading attention evenly. An entropy value of 0 indicates

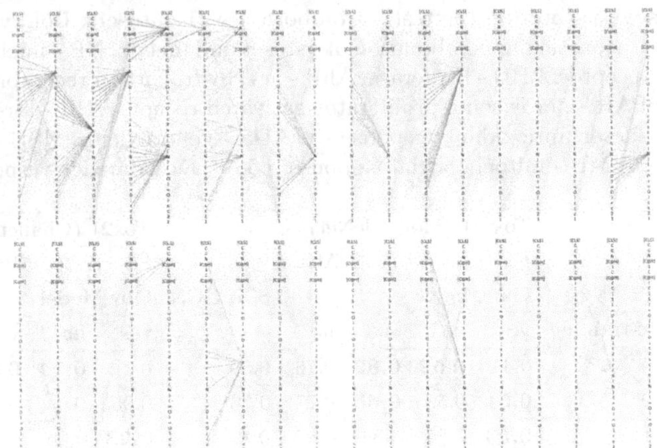

Fig. 4. Visualisation of the attention weights for the layers 4–5. The figure compares the ontology pre-trained network (first row) to the prediction network without pre-training (second row).

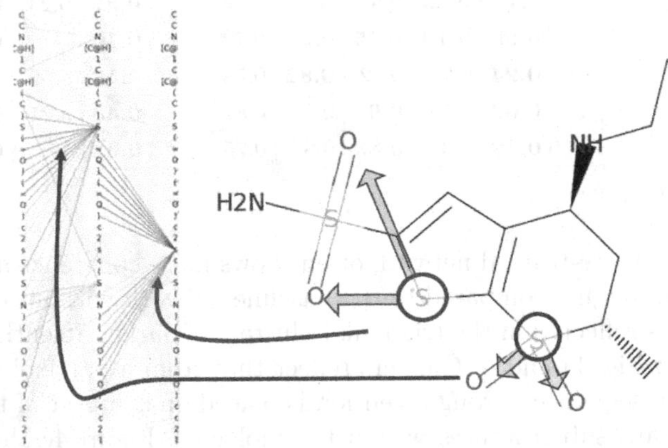

Fig. 5. The molecular structure processed in the attention plots is depicted with attention from layer 4 heads 1–2, showing how attention clusters relate to the molecular structure.

that the model paid attention to only a single token of the molecule. The model that received additional ontology pre-training has a lower entropy (Tox21 Challenge:0.86, Tox21 MoleculeNet: 0.79) value for both datasets compared to the one that did not receive this additional training (Tox21 Challenge: 0.90, Tox21 MoleculeNet: 0.85). That means, that the attention is generally spent less evenly and is therefore more focused in comparison to the model that did not receive that additional training. This indicates that the ontology pre-trained model's decisions are based on more concise substructures within the molecule.

4 Discussion

4.1 Significance

Our approach introduces a new way to embed knowledge from an ontology into a neural network. The underlying assumption of this approach is the following: A well-designed ontology represents a classification of its domain that has proven useful for the various goals and tasks of experts in that area. Thus, it is possible (and even likely) that some of the classes of the ontology reflect features that are relevant for a prediction task in that domain. For example, the classification of chemical entities in ChEBI is based on empirical knowledge about the pertinent features of molecules and their chemical behaviour, and it is reasonable to expect that some of these pertinent features are, at least indirectly, correlated with toxicity. The goal of ontology pre-training is to enable a model to benefit from the knowledge represented in an ontology by training it to categorise its input according to the class hierarchy from the ontology. This goes beyond how common pre-training approaches work. In these, masking approaches are used to train models to learn statistical correlations between input tokens. However, these methods are not suitable to learn the wider semantics of the input tokens, as this information is not present in the input sequence. Since ontologies are developed by domain experts to represent knowledge about their domain and provide semantics of their entities via logical axioms, ontology pre-training is a way to complement syntactical pre-training approaches.

The ontology pre-training approach is applicable in any case where a dataset is available that links the input datatype for the prediction task to the classification hierarchy from the ontology. In the case of our example, the SMILES input structures are directly associated with the leaf classes of the ChEBI ontology, thus we can prepare a training dataset directly from the ontology. However, for other ontologies, the dataset may be assembled from ontology annotations which are external to the ontology but serve the purpose of linking the ontology classes to examples of instances of the corresponding input datatype.

The expert knowledge that is contained in these annotations can then be used as a foundation to accelerate training tasks for a variety of domain-related tasks. Additionally, as results in Table 1 show, we were able to improve the performance of our model for toxicity prediction with the help of ontology pre-training. The inspection of the attention head weights indicates that the system indeed learned meaningful aspects of life science chemistry from the pre-training task. Further, as we will discuss next, the performance of the ontology pre-trained model compares favourably with the state of the art.

4.2 Related Work

Toxicity Prediction. The prediction, based on chemical structure, of whether a molecule has the potential to be harmful or poisonous to living systems, is a challenging task for life science chemistry [2,30]. The Tox21 dataset has become a widely used benchmark for evaluating machine learning approaches to this task, thus there have been multiple previous publications using different strategies. Most approaches to toxicity prediction supplement the input data chemical structures with additional calculated features, such as known toxicophores, in order to enhance performance on the toxicity prediction task. This was the case for the winner of the original Tox21 challenge, who used a deep learning implementation together with three other types of classifier in an ensemble [18], and more recently [20], which augments the input molecular structure with physico-chemical properties. Another recent approach uses 'geometric graph learning' [15] which augments the input molecular graphs with multi-scale weighted coloured graph descriptors. Some approaches additionally augment the training data with more examples in order to mitigate the fact that the Tox21 dataset is small. In [14], chemical descriptors were calculated and in addition, a strategy to remove class imbalance was applied including over-sampling from the minority classes in the dataset followed by cleaning of mislabelled instances. In these cases, which use a different input dataset whether through feature augmentation, data augmentation or data cleaning, we cannot directly compare their results to ours. We anticipate that feature and data augmentation approaches such as these would potentially improve our method's performance as well, but we would need to develop a more complex attention visualisation mechanism to operate across augmented inputs. Since our objective is rather to describe a new approach to incorporating ontology knowledge into a neural network, we here focus our performance evaluation on approaches that are more directly comparable to ours.

[4] uses a graph neural network and a semi-supervised learning approach known as Mean Teacher which augments the training data with additional unlabelled examples. This network and training approach achieving a ROC-AUC score of 0.757 in the test set, which our approach outperforms without data augmentation. Table 1 shows a comparison of the ROC-AUC values achieved by our model against those reported for the best model (SSL-GCN) reported in [4]. With the exception of one class, our model shows better performance for all target classes. ChemBERTa [5] is the most similar to our approach in that it also uses a Transformer-based network. Its original publication also contains an evaluation on the p53 stress-response pathway activation (SR-p53) target of the Tox21 dataset from MoleculeNet. Our model exceeds the reported ROC-AUC value (ChemBERTa: 0.728, our model: 0.83).

Knowledge-Aware Pre-training with an Ontology. Approaches to add knowledge from an ontology into a machine learning model follow several different strategies. The most common is that the information from the ontology is used to supplement the input data in some form, such as by adding synonyms and classification parent labels to the input data. For example, in [24] an ontology is used to supplement the input data with an ontology-based 'feature engineering' strategy.

A second approach is that the ontology content is itself provided as input that is embedded into the network, for example by using random walks through the ontology content to create sentences representing the structure of the ontology for input into a language model. Ontology embedding strategies include OWL2Vec* [3] and OPA2Vec [26]. These approaches are suitable for tasks such as knowledge graph completion or link prediction, but the additional information provided by such embeddings is not inherently connected in the internal representation space to the other information learned by the network, and this limits their potential benefit if the input datatype is complex and internally structured. For example, in the chemistry case, the information about the molecular structure of the chemical that is provided in the SMILES input strings would not be connected intrinsically to the information about the class hierarchy provided to the network by an ontology embedding strategy.

There are some examples of the use of biological ontologies together with biomolecular input data types that are closer to our approach. OntoProtein [32] combines background knowledge from the Gene Ontology with protein sequences to improve the prediction of protein functions. OntoProtein uses the Gene Ontology in pre-training a protein embedding. Existing protein embeddings such as ProtBERT are enhanced by embedding the Gene Ontology as a knowledge graph (following approaches such as OWL2Vec*) and then explicitly aligning the two embeddings. By contrast, our approach uses the ontology more directly. Namely, our pre-training exploits both ChEBI's class structure as well its class hierarchy. The class structure leads to an aggregation of molecules with similar substructures and properties, which can enhance the learning process. The class hierarchy is indirectly influencing the learning process as well, because a subclass relation corresponds to a subset relation for the training samples. OntoProtein uses the subclass hierarchy only for defining depth of ontology terms, which influences learning only very indirectly. Hence, our model incorporates expert knowledge in a more direct way. In the future, we will try to incorporate OntoProtein's approach of contrastive learning using knowledge-aware negative sampling into our approach.

Other approaches have developed custom architectures for the incorporation of knowledge from an ontology into the network. For example, DeepPheno [17] predicts phenotypes from combinations of genetic mutations, where phenotypes are encoded into the network through a novel hierarchical classification layer that encodes almost 4,000 classes from the Human Phenotype Ontology together with their hierarchical dependencies as an ontology prediction layer that informs the remainder of the training of the network. A similar approach is used in [31], in which an ontology layer adds information from an ontology to a neural network to enhance the prediction of microbial biomes. By contrast, our approach uses the generic architecture of a standard Transformer network that has been used in our previous work [9] in which we train a transformer model to extend an existing domain ontology. This model may be used for fine-tuning for different domain related tasks.

4.3 Limitations

Our current approach only uses a fraction of the information available in the ChEBI ontology, since we only consider the structure-based classification of chemical entities beneath the 'molecular entity' class. Hence, we currently consider neither classes of biological and chemical roles nor pharmaceutical applications. These classes have the potential to further enhance the knowledge that is provided to the network, and will be explored in future work.

Another limitation is related to the way we create the ontology pre-training task: We use the leaf nodes of ChEBI as examples for training a model to predict the subsumption relation for more general classes. Or, to put it differently, while from an ontological perspective the leaf nodes of ChEBI are classes in ChEBI's taxonomy, we are treating them as instances and, thus, turning subsumption prediction into a classification task from a machine learning perspective. Consequently, while we use the whole structural hierarchy of ChEBI for creating the pre-training task, the model learns to classify only 856 classes, those that have a sufficient number of example structures to learn from, which is a relatively small number compared to the number of classes in ChEBI. Further, this approach of creating a subsumption prediction pre-training task requires rich structural annotations linked to the learning input datatype (which is the SMILES in our example case), which many ontologies do not contain.

As indicated in Sect. 4.1, both of these limitations may be addressed by using class membership prediction for ontology pre-training. All that is required for ontology pre-training is a dataset that (a) is of the same input datatype as the fine-tuning task, (b) is annotated with terms from the ontology, and (c) contains sufficient training examples to train the model to classify the input with the terms of the ontology. Because we treat subsumption prediction as a classification problem anyway, both approaches are functionally equivalent. However, using an external dataset for training (instead of generating it from the ontology itself), has the benefit that the ontology pre-training might cover the whole ontology instead of just a subset of the ontology. Further, this approach does not rely on the existence of appropriate annotations in the associated input data type.

A further limitation of our approach is that the interpretability offered by the attention weights is limited to visual inspection. In future work we aim to develop an algorithm that is able to systematically determine clusters of attention mapped to the relevant parts of the input molecular structure.

5 Conclusion

This paper presents a novel approach to improve the training of Transformer networks with the help of background knowledge from an ontology. The results have been evaluated by training an Electra model using the ChEBI ontology before fine-tuning the model for toxicity prediction using the Tox21 dataset as a benchmark. The model was able to achieve state-of-the-art performance on the task of toxicity prediction, outperforming comparable models that have been trained on the same dataset.

While improving the state of the art of toxicity prediction is in itself an interesting result, the main contribution of the paper is the presentation of a novel approach to combining symbolic and connectionist AI. This is because our result was achieved with the help of *ontology pre-training*, an additional pre-training step, which is designed to ensure that model learns semantic relationships represented in a given ontology. For the presented work the pre-training task consisted of predicting subclass relationships between classes in the ChEBI ontology, but other tasks (e.g., predicting class membership) are likely to be equally suitable to achieve the purpose of ontology pre-training, namely, to train the model to recognise the meaningful classifications of the entities in a given domain, as represented by an ontology. As we have illustrated in this paper, ontology pre-training has the potential benefit of reducing the time needed for fine-tuning and improving performance. Further, an inspection of attention heads indicates that some of the attention patterns of the model correlate to the substructures that are pertinent for chemical categories in ChEBI (see Sect. 3.2). Thus, since ontological categories are meaningful for humans, another potential benefit of ontology pre-training is an improved interpretability of the trained model. In the future, we are planning to further investigate this line of research by systematically analysing the attention patterns of the pre-trained model and automatically linking these patterns to the ontology.

As we discussed in Sect. 4.3, currently we are only using some of the knowledge available in the ontology for pre-training purposes. In particular, we do not include classes from other parts of the ontology, nor do we include other axioms aside from the hierarchy. In the future we are planning to overcome these limitations by sampling a wider set of classes from ChEBI and by using a more complex architecture that combines a Transformer with a logical neural network (LNN) [21]. The LNN is able to represent logical axioms from the ontology as a first-order theory, translated from OWL [8]. This will enable us to use logical axioms from the ontology (and therefore also its binary relations) to influence both the ontology pre-training as well as the fine-tuning.

References

1. van Bekkum, M., de Boer, M., van Harmelen, F., Meyer-Vitali, A., Teije, A.T.: Modular design patterns for hybrid learning and reasoning systems. Appl. Intell. **51**(9), 6528–6546 (2021)
2. Cavasotto, C.N., Scardino, V.: Machine learning toxicity prediction: latest advances by toxicity end point. ACS Omega **7**(51), 47536–47546 (2022). https://doi.org/10.1021/acsomega.2c05693
3. Chen, J., Hu, P., Jimenez-Ruiz, E., Holter, O.M., Antonyrajah, D., Horrocks, I.: OWL2Vec*: embedding of OWL ontologies. Mach. Learn. **110**(7), 1813–1845 (2021). https://doi.org/10.1007/s10994-021-05997-6
4. Chen, J., Si, Y.-W., Un, C.-W., Siu, S.W.I.: Chemical toxicity prediction based on semi-supervised learning and graph convolutional neural network. J. Cheminform. **13**(1), 1–16 (2021). https://doi.org/10.1186/s13321-021-00570-8

5. Chithrananda, S., Grand, G., Ramsundar, B.: Chemberta: large-scale self-supervised pretraining for molecular property prediction. arXiv preprint arXiv:2010.09885 (2020)
6. Clark, K., Luong, M.T., Le, Q.V., Manning, C.D.: Electra: pre-training text encoders as discriminators rather than generators. arXiv preprint arXiv:2003.10555 (2020)
7. Devlin, J., Chang, M., Lee, K., Toutanova, K.: BERT: pre-training of deep bidirectional transformers for language understanding. In: Burstein, J., Doran, C., Solorio, T. (eds.) NAACL-HLT 2019, Minneapolis, MN, USA, 2–7 June 2019, Volume 1 (Long and Short Papers), pp. 4171–4186. Association for Computational Linguistics (2019). https://doi.org/10.18653/v1/n19-1423
8. Flügel, S., Glauer, M., Neuhaus, F., Hastings, J.: When one logic is not enough: integrating first-order annotations in OWL ontologies. Semant. Web J. (2023). http://www.semantic-web-journal.net/content/when-one-logic-not-enough-integrating-first-order-annotations-owl-ontologies
9. Glauer, M., Memariani, A., Neuhaus, F., Mossakowski, T., Hastings, J.: Interpretable Ontology Extension in Chemistry. Semant. Web J. (2022). https://doi.org/10.5281/ZENODO.6023497. https://zenodo.org/record/6023497
10. Hastings, J.: Primer on ontologies. In: Dessimoz, C., Škunca, N. (eds.) The Gene Ontology Handbook. MMB, vol. 1446, pp. 3–13. Springer, New York (2017). https://doi.org/10.1007/978-1-4939-3743-1_1
11. Hastings, J., Glauer, M., Memariani, A., Neuhaus, F., Mossakowski, T.: Learning chemistry: exploring the suitability of machine learning for the task of structure-based chemical ontology classification. J. Cheminform. **13**(23) (2021). https://doi.org/10.21203/rs.3.rs-107431/v1
12. Hastings, J., et al.: ChEBI in 2016: improved services and an expanding collection of metabolites. Nucleic Acids Res. **44**(D1), D1214–D1219 (2016). https://doi.org/10.1093/nar/gkv1031
13. Huang, R., et al.: Tox21Challenge to build predictive models of nuclear receptor and stress response pathways as mediated by exposure to environmental chemicals and drugs. Front. Environ. Sci. **3** (2016). https://www.frontiersin.org/articles/10.3389/fenvs.2015.00085
14. Idakwo, G.: Structure-activity relationship-based chemical classification of highly imbalanced Tox21 datasets. J. Cheminform. **12**(1), 1–19 (2020)
15. Jiang, J., Wang, R., Wei, G.W.: GGL-Tox: geometric graph learning for toxicity prediction. J. Chem. Inf. Model. **61**(4), 1691–1700 (2021). https://doi.org/10.1021/acs.jcim.0c01294
16. Jumper, J., et al.: Highly accurate protein structure prediction with AlphaFold. Nature 1–11 (2021). https://doi.org/10.1038/s41586-021-03819-2. https://www.nature.com/articles/s41586-021-03819-2
17. Kulmanov, M., Hoehndorf, R.: DeepPheno: predicting single gene loss-of-function phenotypes using an ontology-aware hierarchical classifier. PLoS Comput. Biol. **16**(11) (2020). https://doi.org/10.1371/journal.pcbi.1008453
18. Mayr, A., Klambauer, G., Unterthiner, T., Hochreiter, S.: DeepTox: toxicity prediction using deep learning. Front. Environ. Sci. **3** (2016). https://www.frontiersin.org/articles/10.3389/fenvs.2015.00080
19. Neuhaus, F., Hastings, J.: Ontology development is consensus creation, not (merely) representation. Appl. Ontol. **17**(4), 495–513 (2022). https://doi.org/10.3233/AO-220273

20. Peng, Y., Zhang, Z., Jiang, Q., Guan, J., Zhou, S.: TOP: a deep mixture representation learning method for boosting molecular toxicity prediction. Methods **179**, 55–64 (2020). https://doi.org/10.1016/j.ymeth.2020.05.013. https://www.sciencedirect.com/science/article/pii/S1046202320300888
21. Riegel, R., et al.: Logical neural networks. arXiv preprint arXiv:2006.13155 (2020)
22. Rombach, R., Blattmann, A., Lorenz, D., Esser, P., Ommer, B.: High-resolution image synthesis with latent diffusion models (2022)
23. von Rueden, L., et al.: Informed machine learning - a taxonomy and survey of integrating prior knowledge into learning systems. IEEE Trans. Knowl. Data Eng. **35**(1), 614–633 (2021). https://doi.org/10.1109/TKDE.2021.3079836
24. Sahoo, S.S., et al.: Ontology-based feature engineering in machine learning workflows for heterogeneous epilepsy patient records. Sci. Rep. **12**(1), 19430 (2022). https://doi.org/10.1038/s41598-022-23101-3. https://www.nature.com/articles/s41598-022-23101-3
25. Sayers, E.: PubChem: An Entrez Database of Small Molecules. NLM Tech. Bull. **2005 Jan-Feb**(342:e2) (2005)
26. Smaili, F.Z., Gao, X., Hoehndorf, R.: OPA2Vec: combining formal and informal content of biomedical ontologies to improve similarity-based prediction. Bioinformatics **35**(12), 2133–2140 (2019). https://doi.org/10.1093/bioinformatics/bty933. https://academic.oup.com/bioinformatics/article/35/12/2133/5165380
27. Vig, J., Madani, A., Varshney, L.R., Xiong, C., Socher, R., Rajani, N.F.: BERTology Meets Biology: Interpreting Attention in Protein Language Models. arXiv:2006.15222 (2021). http://arxiv.org/abs/2006.15222
28. Weininger, D.: SMILES, a chemical language and information system. 1. Introduction to methodology and encoding rules. J. Chem. Inf. Comput. Sci. **28**(1), 31–36 (1988). https://doi.org/10.1021/ci00057a005
29. Wu, Z., et al.: Moleculenet: a benchmark for molecular machine learning. Chem. Sci. **9**(2), 513–530 (2018)
30. Yang, H., Sun, L., Li, W., Liu, G., Tang, Y.: In silico prediction of chemical toxicity for drug design using machine learning methods and structural alerts. Front. Chem. **6** (2018). https://www.frontiersin.org/articles/10.3389/fchem.2018.00030
31. Zha, Y., et al.: Ontology-aware deep learning enables ultrafast and interpretable source tracking among sub-million microbial community samples from hundreds of niches. Genome Med. **14**(1), 43 (2022). https://doi.org/10.1186/s13073-022-01047-5
32. Zhang, N., et al.: Ontoprotein: protein pretraining with gene ontology embedding. In: The Tenth International Conference on Learning Representations, ICLR 2022, Virtual Event, 25–29 April 2022. OpenReview.net (2022). https://openreview.net/forum?id=yfe1VMYAXa4

A Novel Incremental Learning Strategy Based on Synthetic Data Generated from a Random Forest

Jordan Gonzalez(✉), Fatoumata Dama, and Laurent Cervoni

Talan's Research and Innovation Center, Paris, France
{jordan.gonzalez,fatoumata.dama,laurent.cervoni}@talan.com

Abstract. Accessing previous data when updating the model with new data is a common problem in some incremental learning applications. This prevents, for example, neural networks from suffering catastrophic forgetting. In this paper, we focus on the incrementing of NCMFs for which access to old data is required with classical incrementing strategies such as IGT. We propose a new incrementing strategy, named IGTLGSS, that allows these kind of random forests to continue to increment without relying on old data. For this purpose, the old data are replaced by synthetic data that are generated from the pre-trained NCMF which has to be incremented. Experimental studies are performed on UCI benchmarks. The results show that, for the used datasets, NCMFs are able to generate realistic synthetic data. Moreover, the first results obtained following the assessment of our incrementing strategy are encouraging.

Keywords: Incremental Learning · Synthetic data · RF · NCMF · GMM

1 Introduction

In many applications, it is common for datasets to be available only in small batches over time [5]. The question arises as to what to do with the model trained on a sub-dataset. The strategy that consists to re-train the model *from scratch* (from the beginning) is inefficient [6,9]. On the one hand, this method is very costly, on the other hand, it prevents the integration of new data in real time and may not be feasible if the original data are no longer available or their accumulation is impossible. Thus, it appears necessary to update the existing model incrementally.

It has been shown that Random Forests (RF) [1], in addition to their multi-class nature and ability to generalize, also have the ability to increment in data and classes [4,10]. Nearest Class Mean Forests (NCMFs) propose a different separation function from classical random forests as well as various incrementing strategies [13]. Nevertheless, the proposed strategies to continue growing the tree during the incremental phase require access to the previous data. In this paper, we propose a strategy to allow NCMFs to continue their incrementation without

D. Seipel and A. Steen (Eds.): KI 2023, LNAI 14236, pp. 46–59, 2023.
https://doi.org/10.1007/978-3-031-42608-7_5

accessing the actual old data, but rather generating it on demand in a synthetic manner when needed.

The structure of the paper is described as follows. Section 2 presents the related work. In Sect. 3, we describe the incrementing strategy proposed in this work. Section 4 presents the procedures for generating synthetic data from the NCMF model as well as the Gaussian mixture model (GMM) used for comparison. The experimental part of this study, described in Sect. 5, evaluates the generative capacity of NCMFs and the efficiency of the proposed incrementing strategy. In the same section, the results are presented and analyzed. The last section concludes our paper by giving some directions for future work.

2 Related Work: Incremental Learning

Incremental learning is a topic of interest for many machine learning approaches, in particular neural networks. A major challenge for the incremental learning of these networks is that they face the problem of "catastrophic forgetting". When learning new data, if a certain amount of old data is not present at the time of learning, the model will fail to recognize it and its performance may collapse. A number of approaches have been proposed to tackle this problem [11,12,14]. A comprehensive study on incremental learning of neural networks, has been proposed in [9] and [16] and is outside the scope of this article. We focus here on random forests, which are naturally incremental and require little data, and for which much work has also been proposed.

Hoeffding trees were proposed in [4] to be able to train trees continuously from data streams, as opposed to so-called offline trees, i.e. training on all available data. The idea of the authors was to provide a method to train trees on very large datasets (potentially of "infinite" size in the sense that they keep arriving) while taking a stable and short time to process each observation. Nevertheless, C4.5 trees seem to perform better for smaller datasets: less than 25,000 observations. Moreover, the authors assume that the observations are generated from a stationary stochastic process, whose distribution does not change over time, i.e. free from any concept drift.

Denil et al. describe in 2013 so-called "online" random forests [3]. Similar to [4], each tree starts as an empty root and grows incrementally. Then, each leaf maintains a list of k candidate *splits* and their quality scores. When a new observation is added, the scores of the corresponding node are updated. To reduce the risk of choosing a suboptimal split due to noise in the data, additional hyperparameters, such as the minimum number of examples required for a split (n_{stop}) and the minimum information gain to allow the split, are used. Once these criteria are satisfied in a node, the best split is chosen (making that leaf an internal node) and its two children are the new leaf nodes (with their own candidate splits). These methods may be inefficient for deep trees due to the high cost of maintaining candidate quality scores. Due to the large amount of information stored in the leaves, the authors require 1.6 to 10 GB of storage space when varying the parameters of their incremental forest [3].

Lakshminarayanan et al. propose in 2014 the Mondrian Forests (MF), a class of RFs that can do incremental learning [7]. Their method allows adding a node or continuing the construction from a leaf under certain conditions. MFs achieve competitive results with classical RFs trained on the same datasets.

The same year, Ristin et al. introduce the Nearest Class Mean Forests (NCMFs), a type of RFs based on the NCM principle [13]. The experimental studies show that NCMF model (without even using an incrementing phase) obtains better performances than classical RFs for large scale image classification. Furthermore, the authors propose several strategies to continue growing the tree during the incremental phase, which require access to the previous data, except the **Update Leaf Statistics** strategy (ULS). The latter is the most straightforward strategy: the structure of the pre-trained forest is not modified, only the leaf statistics are updated. The problem with this method is that during the incrementation phase, a majority label change may occur in a leaf. The model can therefore end up making different predictions than the model without incrementation and performance drops can occur on the old data. The authors have therefore proposed, among others, the **Incremental Growing Tree** strategy (IGT), which consists in enlarging the trees in depth and creating a split in the leaves where a majority label change occurs.

Because of their multi-class nature and their capacity of generalization, a very fast learning, a possible incrementation in data but also in classes and a better interpretability, we focused our study on the use of an NCMF. We have chosen to use the IGT strategy which we consider to be a good compromise in terms of accuracy and complexity among the incremental strategies proposed in [13]. Our proposal offers an extension of the IGT strategy that does not require access to the previous training data.

3 The Proposed Incrementing Strategy: IGTLGSS

This section starts with a brief presentation of NCMF model (in Subsect. 3.1). Then, the IGT strategy which is the baseline of our proposal is introduced (in Subsect. 3.2). To finish, our proposal is described in Subsect. 3.3.

3.1 Nearest Class Mean Forest (NCMF)

The NCMF model, proposed in [13], results from the combination of two classifiers: the Random Forest (RF) classifier and the Nearest Class Mean (NCM) classifier. In the latter classifier, the classification of an observation x is done by searching for the class $k \in \mathcal{K}$ whose centroid c_k is the closest to x. The centroids c_k's are defined as below:

$$c_k = \frac{1}{|X_k|} \sum_{i=1}^{|X_k|} x_k^i, \tag{1}$$

where $X_k = \{x_k^i, 1 \leq i \leq |X_k|\}$ corresponds to the set of observations belonging to class k.

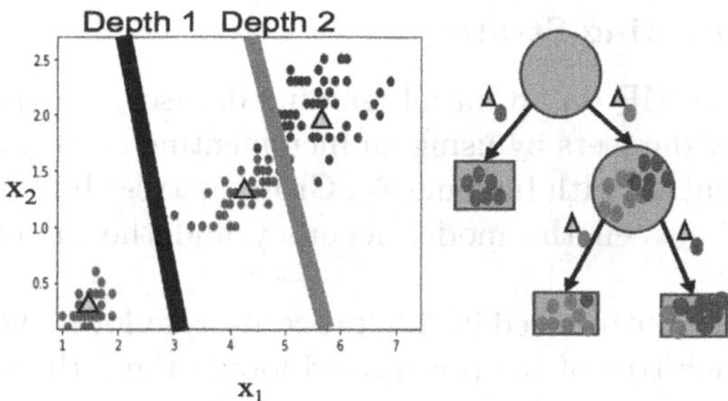

Fig. 1. Example of decision boundaries for an NCM decision tree.

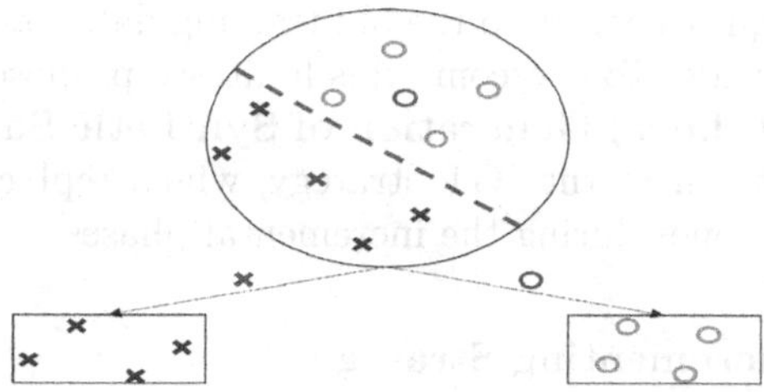

Fig. 2. Splitting decision in NCMF - The samples are directed to the child node associated with the nearest centroid (in red). (Color figure online)

NCMF model retains many of the characteristics of the "classic" forest: a set of trees trained on boostrap samples from the training base, and random selection of a subset of features when a node is created.

Training. The main difference with RF is that each node contains a binary NCM classifier. The training of a node begins with the random drawing of a subset of classes which defines the classes that the node will have to separate. For each pair of classes $\{k_i, k_j\}$ in the subset, we will train an NCM classifier, in other words, estimate the centroids of these classes c_i and c_j from the observations of these classes present in the node. Using the nearest centroid rule, the node's data set is separated into two subsets. Finally, the algorithm selects the pair of centroids that maximizes the information gain I; the separation achieved is then considered optimal. The subsets are passed on to the node's left and right children, and recursive tree construction continues. The stopping criteria are the same as for the "classic" random forest. Figure 1 illustrates the decision boundaries for an NCM decision tree.

Inference. The observation x goes through each tree, from its root to a single leaf. In each node, the NCM classifier orients x to the left or right, depending on its distance from the two centroids, as shown in Fig. 2. Once inside the leaf, x is assigned the majority class (among the classes present in the leaf). Finally, a majority vote is taken on all the tree predictions, to classify the observation x.

3.2 IGT Incrementing Strategy

For a pre-trained NCMF on an initial training dataset, it is possible to update the model with new datasets by using an incrementing strategy. As presented in Sect. 2, the incrementing with Incremental Growing Tree (IGT) strategy provides the best tradeoff between the model accuracy and the incrementing strategy complexity.

The IGT strategy, introduced in [13], proceeds as follows: the new observation is propagated in each tree of the pre-trained forest; then, the class distributions in each leaf it reaches are updated. In the case where the increment produces a majority class change within a leaf, this leaf is transformed into a node, and the recursive construction of a local subtree begins. An inherent limitation of the IGT strategy is that it requires access to the old training datasets in order to be able to compute the centroids. To overcome this limit, we propose the **Incremental Growing Tree with Local Generation of Synthetic Samples** (IGTLGSS) strategy, as an extension to the IGT strategy, which replaces the old training datasets by synthetic ones during the incremental phases.

3.3 IGTLGSS Incrementing Strategy

The Incremental Growing Tree with Local Generation of Synthetic Samples (IGTLGSS) method consists in generating local synthetic data using multivariate normal distributions. To make this possible, we have stored the covariance matrices conditioned on the classes at the node level, in addition to the centroids. In the following, we denote $\mathcal{K} = \{k_1, ..., k_l\}$ the set of l possible labels and $(X^{[INCR]}, y^{[INCR]})$ the dataset from which the pre-trained forest should increment, where $X^{[INCR]}$ represents the set of feature vectors and $y^{[INCR]}$ the associated labels.

For each observation (x, y) in $(X^{[INCR]}, y^{[INCR]})$, the incrementation proceeds as follows:

1. Each tree in the forest propagates the observation to a leaf to predict a label $k_i \in \mathcal{K}$;
2. Each tree t, for which the prediction k_i does not match y, is incremented. The idea here is to update only the trees in difficulty by executing the following steps.
   (i) (x, y) is propagated to a leaf, then the class repartition $S^t(x) = [S^t(k_1|x), ..., S^t(k_l|x)]$ stored in it is extracted, where $S^t(k_i|x)$ represents the number of occurrences of the class k_i in the leaf.
   (ii) For each class k_i, we use the corresponding local multivariate normal distribution in order to generate $S^t(k_i|x)$ observations. Note that the normal distribution parameters (the centroid and the covariance matrix corresponding to the class k_i) are stored within the parent node (*i.e.*, the old leaf). In this way, we generate a local synthetic set $(X^{[SYNTH]}, y^{[SYNTH]})$.
   (iii) The observation (x, y) is added to $(X^{[SYNTH]}, y^{[SYNTH]})$ and, from this position, the recursive construction of the subtree continues as in a classical learning phase.

This way, the NCMF model becomes able to increment from new data without needing to access the old datasets.

4 Synthetic Data Generation

In this section, we describe how synthetic data can be generated from NCMF and GMM models. The resulting synthetic data are considered in our experimentations (see Sect. 5).

4.1 Nearest Class Mean Forest (NCMF)

In order to generate synthetic data from a trained NCMF (introduced in Subsect. 3.1), we propose the procedure described as follows. For each class $k_i \in \mathcal{K}$, an observation is generated in three steps:

(i) A tree t is randomly drawn from the forest.
(ii) Then, a leaf is randomly chosen so that the more a leaf contains occurrences of the class k_i, the more likely it is to be chosen.
(iii) Finally, an observation is generated from the local multivariate normal distribution associated with k_i, whose parameters are stored in the parent node of the chosen leaf.

The procedure is repeated to obtain the desired number of observations per class.

4.2 Gaussian Mixture Model (GMM)

The GMM is a probabilistic model expressed as a mixture of Gaussian distributions (multivariate normal distributions) [18]. It allows to parametrically estimate the distribution of a set of dependent random variables by modeling it as the sum of l Gaussian distributions.

The GMM is typically used in unsupervised classification (clustering) [15,20]. In this case, each Gaussian distribution also called emission distribution is associated to a specific class or label. Then, once the model parameters are estimated, the data partitioning is performed by assigning each observation (vector of d features) to the class whose emission law has the highest probability of having generated the observation.

The parameters of the GMM model (the mean and the covariance matrix of the Gaussian distributions) are usually estimated in an unsupervised manner (*i.e.*, class labels are unknown) using the Expectation-Maximization (EM) algorithm [2,19]. However, when the data are labeled, as is the case in this study, a Gaussian distribution can be fitted to the data in each class in a supervised manner following the Maximum Likelihood method.

Synthetic Data Generation. As a probabilistic model, GMM allows for easy generation of synthetic data from a model trained on a real dataset. The data simulation is performed as follows:

(i) For each observation to be generated, its label noted y is randomly generated from the label space $\mathcal{K} = \{k_1, \ldots, k_l\}$ following a weighted sampling. The used weights are the prevalence of classes in the training set. In this way, the generated synthetic data will respect the class distribution of the training set.
(ii) Then, the observation or vector of *features* x is generated from the Gaussian distribution associated with the label y previously obtained.
(iii) The two first steps are repeated till to obtain the right number of observations.

5 Experiments

In this section, we assess the generative capacity of NCMF and the efficiency of our incrementing strategy. The dataset used in this evaluation are presented in Subsect. 5.1. Then, our experimental protocol is described in Subsect. 5.2. Finally, results are presented and analyzed in Subsect. 5.3.

5.1 Dataset Description

In our experimentations, five benchmark datasets from the *UCI*[1] machine learning repository were considered:

- *Breast Cancer Coimbra* dataset: diagnosis of breast cancer based on quantitative attributes such as patient age and parameters gathered in routine blood analysis (*e.g.*, glucose, insulin, resistin and leptin).
- *Banknote authentification* dataset: identification of falcifying banknote through features extracted from images that were taken from genuine and forged banknotes.
- *Avila* dataset: writer identification in medieval manuscripts using page layout features (*e.g.*, intercolumnar distance, upper/lower margin, interlinear spacing and peak number).
- *Speaker Accent Recognition* dataset: accent detection and recognition through a set of features extracted from English words read by speakers from different countries.
- *Optical Recognition of Handwritten Digits* dataset: handwritten digits recognition by means of normalized bitmaps of handwritten digits extracted from a preprinted form.

The datasets information (number of observations, number of feature and number of classes) are provided in Table 1. Each dataset was partitioned into a training set (80%) and a test set (20%), excepted *A.V.* dataset for which a split is already provided.

[1] https://archive.ics.uci.edu/ml/datasets/.

Table 1. Dataset desciption. From left to right: number of classes, number of observations and number of features.

Dataset	$\lvert\mathcal{K}\rvert$	n_{obs}	n_{feat}
Breast Cancer Coimbra (B.C.)	2	116	9
Banknote authentication (B.K.)	2	1372	4
Avila (A.V.)	12	20867	10
Speaker Accent Recognition (A.R.)	6	329	12
Optical Recognition of Handwritten Digits (H.G.)	10	5620	64

5.2 Experimental Protocol

Evaluation of the Generative Capacity of NCMF Forests. The first four datasets have been considered in this experimentation (see Table 1). For each dataset, NCMF composed of 50 trees and GMM models were trained on training dataset. Then, the resulting models were used to generate synthetic datasets $(X_{GMM}^{[SYNTH]}, y_{GMM}^{[SYNTH]})$ et $(X_{NCMF}^{[SYNTH]}, y_{NCMF}^{[SYNTH]})$, having the same length and the same class repartition as the training datasets. In order to assess the quality of our synthetic data, we use the following three metrics.

1. **Classification accuracy preservation**. This metric is inspired from the method proposed in [17]. It consists in training several classification models on both real and synthetic data. Then, the resulting models are evaluated on a test dataset only composed of real data. When the performance loss resulting from the use of synthetic data is low, we can conclude that synthetic data have a similar predictive capacity to real data. We have considered four classification algorithms: Random Forest (RF), Naive Bayes (NB), Support Vector Machine (SVM) and Nearest Class Mean Forest (NCMF).
2. **Feature-wise distribution**. With this metric, we test the ability for the generative models to reproduce the distribution of the real data. For this purpose, we use Kolmogorov-Smirnov test of homegeneity to compare the distribution (more precisely the cumulative density function) of the synthetic data to that of the real data. The null hypothesis of this statistical test assumes that the two compared samples have been drawn from the same distribution. A specific test is carried out for each feature of a given dataset. When the test p-value is greater than the tolerated risk level α (fixed at 5% as usual), the null hypothesis is not rejected and we can conclude there are no significant differences between the distributions of the real and synthetic data.
3. **Feature dependency preservation**. This metric aims at evaluating whether the synthetic data preserve the relationship or dependency between the features of a dataset. To this end, we compare the covariance matrices of the real and synthetic data with each other.

Incremental Learning. In order to compare IGTLGSS and IGT strategies, we have considered the H.G. dataset (see Tables 1). The protocol used here is

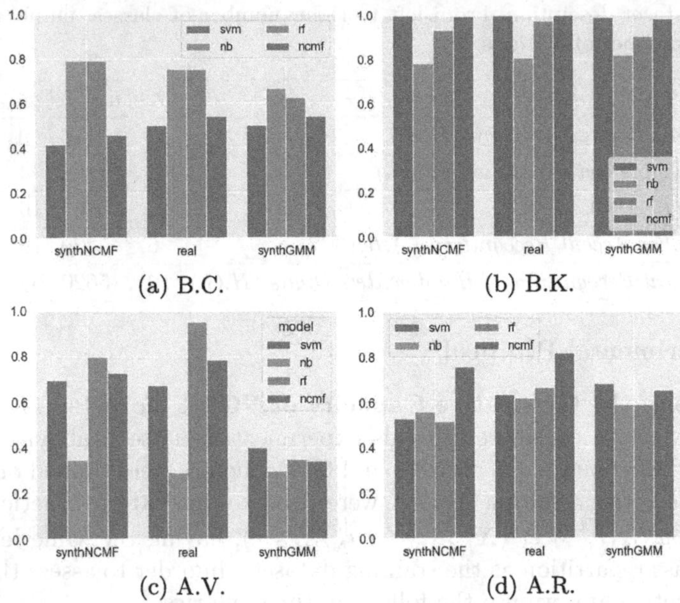

Fig. 3. Classification accuracy for different models (RF, NCMF, NB and SVM) and different data type (real data, synthetic data generated from GMM and NCMF).

similar to that described in [10]. The training dataset (80% of the total dataset) is randomly split into 50 subsets (*batches*) of identical size and identical class distributions (the 23 extra observations are skipped). The forest is trained on the first batch, then incremented on the others. We measure the accuracy of the model at the end of each increment.

5.3 Results and Analysis

Evaluation of the Generative Capacity of the NCMF. In Fig. 3, we compare the accuracy of the classification models (RF, NCMF, NB and SVM), for three types of data: the real data (*B.C.*, *B.K.*, *A.V.* and *A.R.*), the synthetic data generated from GMM (denoted *synthGMM*) and the synthetic data generated from NCMF (denoted *synthNCMF*). For each data type and each dataset, Table 2 presents the mean and the dispersion of the accuracies obtained for all classification models. As expected, the models trained on the real data provide the best performances. So, we essentially compare the performance obtained with synthetic datasets *synthNCMF* and *synthGMM*. The results show that the model that have been trained on *synthNCMF* globaly outperform those trained on *synthGMM*. Moreover, the use of *synthNCMF* datasets generates, on average, only 4% of performance loss. We can conclude that *synthNCMF* has a predictive ability similar to the real data.

We applied the feature-wise distribution comparison to real and *synthNCMF* datasets. The p-values of Kolmogorov-Smirnov test of homogeneity (applied to

Table 2. Average of accuracy metrics obtained for different classification models (RF, NCMF, NB and SVM). For each dataset type (real data, synthetic data generated from GMM and NCMF), the best score and the second best score are respectively displayed in bold and underlined.

Dataset	Real	*synthGMM*	*synthNCMF*
B.C.	**0.635** ± 0.133	0.583 ± 0.076	<u>0.615</u> ± 0.205
B.K.	**0.945** ± 0.093	0.924 ± 0.080	<u>0.928</u> ± 0.102
A.V.	**0.677** ± 0.280	0.348 ± 0.041	<u>0.642</u> ± 0.201
A.R.	**0.686** ± 0.090	<u>0.629</u> ± 0.045	0.591 ± 0.113

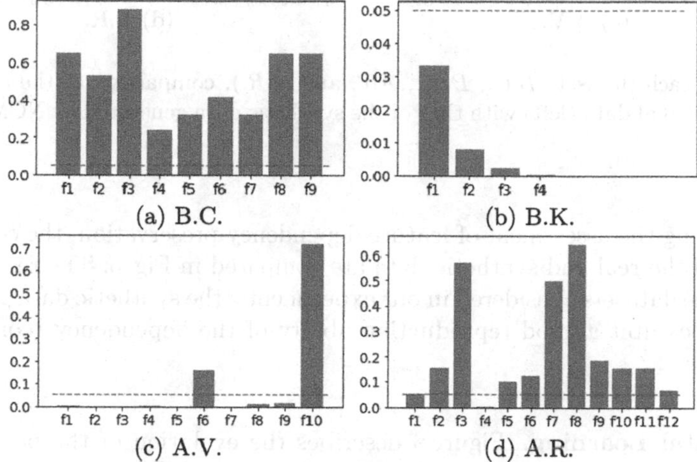

Fig. 4. p-values of Kolmogorov-Smirnov test of homogeneity. One test is executed for each feature of each dataset. The horizontal dashed line represents the risk level α which is set at 5%. When the p-value is greater than α, the null hypothesis of homogeneity is not rejected.

each single feature of each dataset) are depicted in Fig. 4. The results show that there are no significant differences between the feature-wise distribution of the real and synthetic data for the B.C. dataset (as the p-value is greater than α for all features, see Fig. 4a). The same analysis can be done for the A.R. dataset at the exception of the feature f4 for which the homogeneity test fails (see Fig. 4d). However, the results obtained for the datasets B.K. and A.V. are not satisfying, since the NCMF model correctly reproduce the distribution of zero features over 4 (resp. 2 features over 10) for the B.C. (resp. A.V.) dataset (see Figs. 4b and 4c). A further analysis of the feature-wise distributions of these two datasets shows that they are particularly skewed. However, the NCMF model generates data using the normal distribution which is symetric. For these two datasets, it would be better to consider a skewed distribution such as the lognormal or Weibull distributions.

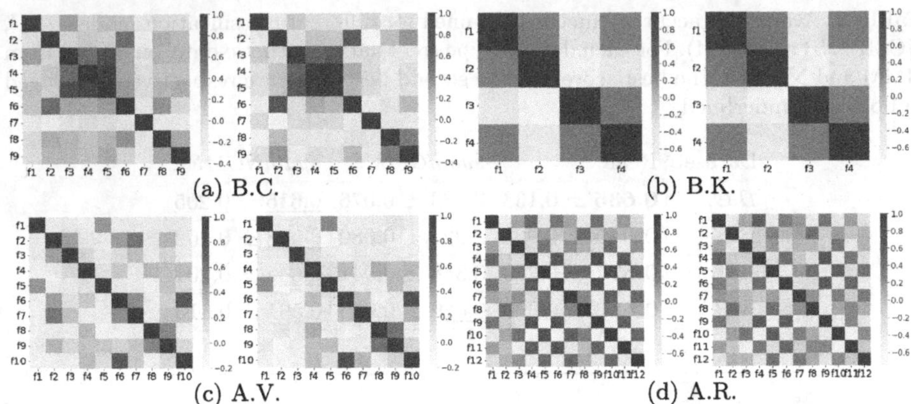

Fig. 5. For each dataset (*B.C.*, *B.K.*, *A.V.* and *A.R.*), comparison of the covariance matrix of the real data (left) with that of the synthetic data generated by NCMF model (right).

Regarding the assessment of feature dependency preservation, the covariance matrices of the real and synthetic data are compared in Fig. 5. The results show that, for the datasets considered in our experiments, the synthetic data generated by NCMF exhibit a good reproduction ability of the dependency (correlation) of features.

Incremental Learning. Figure 6 describes the evolution of the performance of the incremented NCMFs following the IGT or IGTLGSS strategies. We can observe that the performances are very satisfactory: only 0.01 point less on average for the synthetic data. The curve (displays in Fig. 6) shows that despite the fact that the data is synthetic, the NCMF is able to improve its performance up to the *batch* number 35. Beyond that, the IGTLGSS strategy seems to reach a ceiling, where the IGT method seems to allow a performance gain. This latter result can be analized as follows. When a split occurs during the incremental phase in a leaf, the construction generally continues until pure leaves are obtained. We distinguish between the old data and the new data. With the IGT strategy, we would gain in accuracy because there is no drop in performance on the old data, while we would gain in accuracy on the new data. Where the IGTLGSS method can be limited because the old data are generated around the mean of the distribution, discarding the particular cases for which the forest can have difficulty in classifying correctly. Hence the fact that we observe this stagnation in performance. It would be interesting to observe the behavior of the IGTLGSS strategy on larger datasets allowing to consider for example a hundred of *batches*.

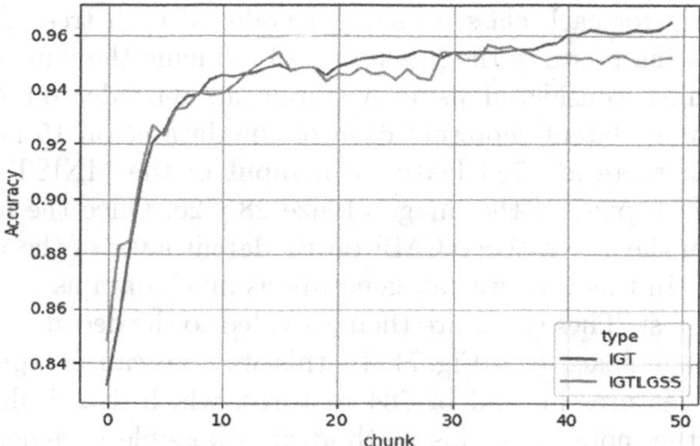

Fig. 6. Evolution of the accuracy according to the number of increments.

6 Conclusion and Perspectives

In this work, we propose a new incrementing strategy, named IGTLGSS, that extends the IGT strategy known to provide a good compromise accuracy/complexity. In contrast to the IGT strategy, our proposal allows to continue to increment the NCMF without requiring access to the old training data. To this end, the old data are replaced by synthetic data generated from the pre-trained NCMF to be incremented with new data. The result of our first experiment shows the similarity between the synthetic data generated from the NCMF model and the real data. This result is promising because, to our knowledge, this is the first time a random forest is used as a data generator. We have also compared IGTLGSS and IGT incrementing strategies. The results show that, on the used dataset, our proposal IGTLGSS obtains competitive performance.

It should be emphasized that the assumption made in this work is that the environment is stationary (no concept drift, data are assumed to belong to the same distribution). Furthermore, the method proposed here is specific to the architecture of an NCMF and is therefore not directly transposable to other machine learning models, such as neural networks.

Aside from these points, we have identified two main limitations to our incrementing strategy that can be addressed in future work. The first limitation is encountered with highly imbalanced datasets such as Avila dataset. For example, taking class C in the Avila dataset, which contains only 103 data, one per batch if 100 batches are used. This would make it impossible to compute a covariance matrix. Reducing the number of batches to 50, would imply having only two samples per batch, which makes the calculation of the covariance matrix unreliable. It could be interesting to continue the work on the handling of incremental learning when there is a class imbalance.

The second limitation observed in our experiments is the number of features in the dataset. Indeed, the more this number increases, the more difficult it becomes to keep the model in memory because it is necessary to calculate a covariance matrix for each class, in several nodes of each tree of the forest. It would be interesting to study the possibility of reducing the number of features. Currently, we have considered using a simple autoencoder on MNIST [8] to train it to learn the latent representation of the dataset on 15 nodes/features. For information, there are 784 features in input of the MNIST dataset, they correspond to each pixel of the image of size 28×28. Once the autoencoder is trained, we train the generative NCMF on the latent layer of the whole training MNIST dataset. In this way, we can generate as much data as we want for each class from the forest. These data are then provided to the decoder which projects them in the image space (see Fig. 7). In this way we can compute covariance matrices on 15 features instead of 784 features, which drastically reduces the model size. Furthermore, with this method, it is possible to generate images of the desired class on command, which is not possible in a standard autoencoder. In fact, in a standard autoencoder, one must normally generate random data without knowing the distribution in the latent layer, and without knowing the class that will be generated at the decoder output.

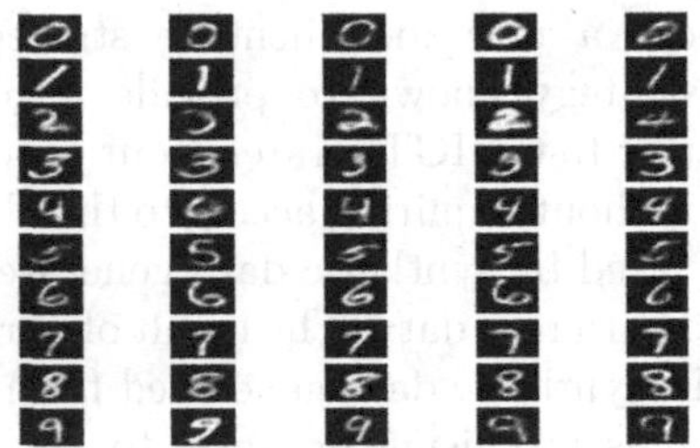

Fig. 7. Digits generated using our generative NCMF.

References

1. Breiman, L.: Random forests. Mach. Learn. **45**(1), 5–32 (2001)
2. Dempster, A.P., Laird, N.M., Rubin, D.B.: Maximum likelihood from incomplete data via the EM algorithm. J. Roy. Stat. Soc. Ser. B (Methodol.) **39**(1), 1–22 (1977)
3. Denil, M., Matheson, D., Freitas, N.: Consistency of online random forests. In: International Conference on Machine Learning (ICML), pp. 1256–1264 (2013)
4. Domingos, P., Hulten, G.: Mining high-speed data streams. In: Proceedings of the sixth ACM SIGKDD - International Conference on Knowledge Discovery and Data Mining, pp. 71–80 (2000)
5. Gepperth, A., Hammer, B.: Incremental learning algorithms and applications. In: European Symposium on Artificial Neural Networks (ESANN) (2016)
6. He, J., Mao, R., Shao, Z., Zhu, F.: Incremental learning in online scenario. In: Proceedings of the IEEE/CVF Conference on Computer Vision and Pattern Recognition, pp. 13926–13935 (2020)

7. Lakshminarayanan, B., Roy, D.M., Teh, Y.W.: Mondrian forests: efficient online random forests. Adv. Neural. Inf. Process. Syst. **27**, 3140–3148 (2014)
8. LeCun, Y.: The MNIST database of handwritten digits (1998). http://yann.lecun.com/exdb/mnist/
9. Parisi, G.I., Kemker, R., Part, J.L., Kanan, C., Wermter, S.: Continual lifelong learning with neural networks: a review. Neural Netw. **113**, 54–71 (2019)
10. Pecori, R., Ducange, P., Marcelloni, F.: Incremental learning of fuzzy decision trees for streaming data classification. In: 11th Conference of the European Society for Fuzzy Logic and Technology (EUSFLAT 2019), pp. 748–755 (2019)
11. Ratcliff, R.: Connectionist models of recognition memory: constraints imposed by learning and forgetting functions. Psychol. Rev. **97**(2), 285 (1990)
12. Rebuffi, S.A., Kolesnikov, A., Sperl, G., Lampert, C.H.: iCaRL: incremental classifier and representation learning. In: Proceedings of the IEEE Conference on Computer Vision and Pattern Recognition, pp. 2001–2010 (2017)
13. Ristin, M., Guillaumin, M., Gall, J., Van Gool, L.: Incremental learning of NCM forests for large-scale image classification. In: Proceedings of the IEEE Conference on Computer Vision and Pattern Recognition, pp. 3654–3661 (2014)
14. Robins, A.: Catastrophic forgetting, rehearsal and pseudorehearsal. Connect. Sci. **7**(2), 123–146 (1995)
15. Shen, X., Zhang, Y., Sata, K., Shen, T.: Gaussian mixture model clustering-based knock threshold learning in automotive engines. IEEE/ASME Trans. Mechatron. **25**(6), 2981–2991 (2020)
16. Shmelkov, K.: Approaches for incremental learning and image generation. Ph.D. thesis, Université Grenoble Alpes (2019)
17. Torfi, A., Fox, E.A.: CorGAN: correlation-capturing convolutional generative adversarial networks for generating synthetic healthcare records. arXiv preprint arXiv:2001.09346 (2020)
18. Wan, H., Wang, H., Scotney, B., Liu, J.: A novel gaussian mixture model for classification. In: 2019 IEEE International Conference on Systems, Man and Cybernetics (SMC), pp. 3298–3303. IEEE (2019)
19. Yang, M., Lai, C., Lin, C.: A robust EM clustering algorithm for gaussian mixture models. Pattern Recogn. **45**(11), 3950–3961 (2012)
20. Zhang, Y., et al.: Gaussian mixture model clustering with incomplete data. ACM Trans. Multimedia Comput. Commun. Appl. (TOMM) **17**(1s), 1–14 (2021)

RECol: Reconstruction Error Columns for Outlier Detection

Dayananda Herurkar[1,2(✉)], Mario Meier[3], and Jörn Hees[1,4]

[1] German Research Center for Artificial Intelligence (DFKI), Kaiserslautern, Germany
dayananda.herurkar@dfki.de
[2] RPTU Kaiserslautern-Landau, Kaiserslautern, Germany
[3] Deutsche Bundesbank, Frankfurt, Germany
mario.meier@bundesbank.de
[4] Bonn-Rhein-Sieg University of Applied Sciences, St. Augustin, Germany
joern.hees@h-brs.de

Abstract. Detecting outliers or anomalies is a common data analysis task. As a sub-field of unsupervised machine learning, a large variety of approaches exist, but the vast majority treats the input features as independent and often fails to recognize even simple (linear) relationships in the input feature space. Hence, we introduce RECol, a generic data pre-processing approach to generate additional columns (features) in a leave-one-out fashion: For each column, we try to predict its values based on the other columns, generating reconstruction error columns. We run experiments across a large variety of common baseline approaches and benchmark datasets with and without our RECol pre-processing method. From the more than 88k experiments, we conclude that the generated reconstruction error feature space generally seems to support common outlier detection methods and often considerably improves their ROC-AUC and PR-AUC values. Further, we provide parameter recommendations, such as starting with a simple squared error based random forest regression to generate RECols for new practical use-cases.

Keywords: Outlier Detection · Features · Column · Reconstruction Error

1 Introduction

Detecting outliers in data is an often occurring exercise in various domains. For example, the identification of certain diseases of patients [2] or fraudulent

This paper represents the authors' personal opinions and does not necessarily reflect the views of the Deutsche Bundesbank, the Eurosystem or their staff.

Supplementary Information The online version contains supplementary material available at https://doi.org/10.1007/978-3-031-42608-7_6.

D. Seipel and A. Steen (Eds.): KI 2023, LNAI 14236, pp. 60–74, 2023.
https://doi.org/10.1007/978-3-031-42608-7_6

behavior of individuals [25] can often be supported by applying outlier detection (OD) algorithms to data. Many different algorithms can be found in the literature to identify these outliers in unsupervised settings where no labels are available [19].

However, the existing approaches apply a certain definition of outliers that might not be appropriate for detecting outliers in all relevant contexts: outliers are typically defined as data points that are significantly different from the remaining data [1]. In contrast, one might actually be interested in data points deviating from underlying relations in the data. Such deviations might coincide with points far apart from others, but this need not always be the case.

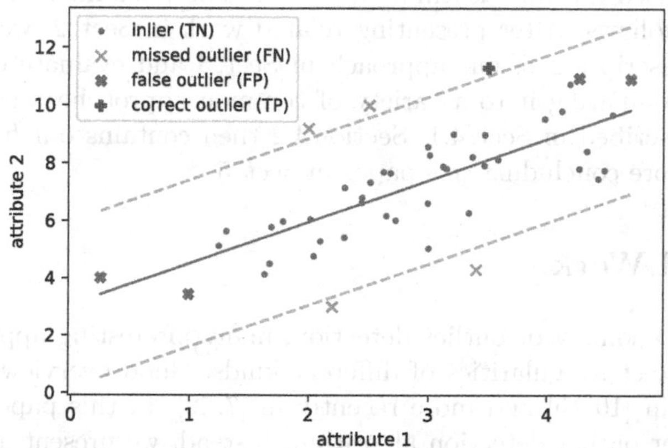

Fig. 1. Example dataset with a linear relation between 2 attributes. The dashed lines show the 2 σ interval. We can see that traditional OD methods (an isolation forest [18] in this case) incorrectly classify points at the far ends as outliers while missing outliers closer to the center.

We illustrate this issue in Fig. 1 with a toy example in two dimensions. The points in the scatter plot constitute some data. Applying a common outlier detection algorithm to this data flags the red and green points as outliers. However, notice that in our example there is a simple linear relationship between the two attributes. Hence, we argue that an outlier could also be defined by points deviating markedly from the underlying (often latent) relationship in the data. In our example, these are the points outside the 2σ interval of the linear relationship. As one can see, traditional approaches often flag points inside such bandwidth as outliers and miss points outside of it.

To tackle this problem, we present an approach to map the data to a reconstruction error space and then apply the outlier detection algorithm to a combination of the previous and this novel space. We do so by estimating a supervised model for each input feature using all other features as inputs. Consequently, for each input feature in the original dataset, one receives an additional (derived) feature containing the reconstruction errors. We can then use these new features as (additional) inputs for existing outlier detection algorithms. This enables them to easily identify data points that are hard to reconstruct by the relationships in the data. In Fig. 1 these would be the green and orange points, because their distance to the bandwidth is larger and therefore their reconstruction error is higher. In reconstruction error space they lie far apart from the other data points, making them suspicious to the outlier detection algorithm.

Therefore, we claim that by calculating additional reconstruction error columns (RECols) for each feature in the dataset, one can improve the detection of outliers in unsupervised contexts. Despite the simple example, one should note that our approach can also be beneficial in case of more complex and non-linear relationships within the data: For the generation of RECols, we can use any kind of supervised machine learning approach. The remainder of this paper is structured as follows. After presenting related work in Sect. 2, we will provide an in-depth description of our approach in Sect. 3 and evaluate our approach in Sect. 4 by comparing it to a variety of common approaches and benchmark datasets as described in Sect. 4.1. Section 4.2 then contains our findings and a discussion, before concluding this paper in Sect. 5.

2 Related Work

In the field of anomaly or outlier detection, many interesting approaches exist that try to detect irregularities of different kinds. Good overviews of the field can be found in [10,15] and more recently in [7,22]. In this paper, we do not present another outlier detection algorithm. Instead, we present a general pre-processing idea and argue that our idea is beneficial across use-cases and existing approaches. Hence, we base our work on Goldstein et al. [12]. They conduct a comprehensive study of unsupervised anomaly detection algorithms for multivariate data, analysing 19 different unsupervised anomaly detection algorithms (cluster, distance, density, and statistics based) using 10 different standard datasets, yielding a broad analysis of the advantages and disadvantages of each algorithm. In this paper, we hence treat their models and results as baseline results and compare our approach with these baselines to analyze the advantages generated by our pre-processing approach for unsupervised anomaly detection in general.

There are other methods studied in the outlier detection domain that are related to our approach. In regression-based models [8] for time series data, the models are first fitted to the data and then the residuals for test instances are computed. Anomaly scores are determined using the residuals. This method is related to our approach since it uses residuals or errors to determine anomaly scores but the RECol approach is more advanced as [8] computes one residual error per sample to detect anomalies, while RECol does it for every column in a leave-one-out manner. Additionally, in RECol we are providing the results to subsequent OD approaches while [8] uses the residual directly to compute anomaly scores. Furthermore, the Cross Prediction (CP) approach in [28] and the self-prediction methods in [16,21] are related to our overall approach in that they use a leave-one-out training approach for the columns. In contrast to our extensive evaluation and proposal of several supervised learning approaches, [28] barely discusses CP and lacks analysis, detailed study. Similarly, [16,21] differs from our approach in the sense that it tries to predict the individual column values, but then directly (count-based [16] or histogram/renormalized [21]) aggregates the incorrect predictions to generate an outlier score. All of the mentioned approaches are independent OD approaches however, we did not develop yet another OD approach, but propose RECol as a general pre-processing step for subsequent OD approaches. To summarize, with this research we are aiming to fill the gap from previous related studies by providing an approach that uses the combination of the common leave-one-out approach with reconstruction errors as a general purpose pre-processing step in OD and also conducting an extensive study showing that such an approach is applicable to most of the widely used unsupervised OD algorithms on a variety of available standard datasets.

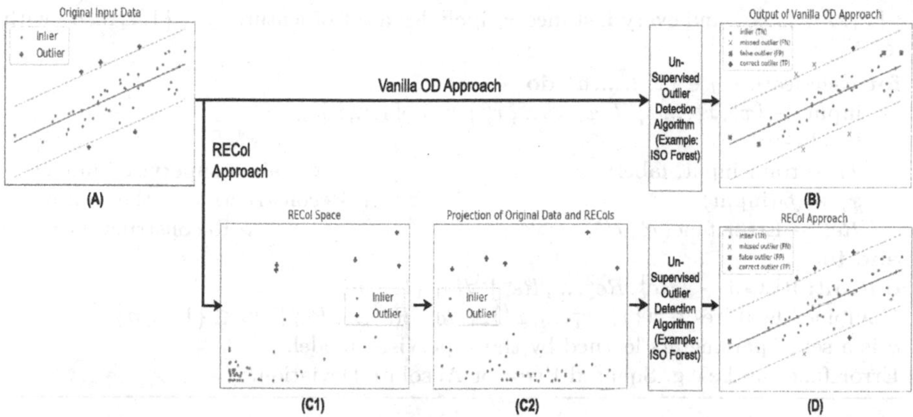

Fig. 2. Reconstruction Error Columns Approach. (A) Input data in 2D space with blue points as inliers and green points as outliers. (B) Result of directly applying a standard outlier detection algorithm to the input data missing obvious relations between attributes. (C1) The 2D Reconstruction Error space. (C2) A 2D projection of the 4D concatenation of the original and RECol space. We can see that in (C1) and (C2) the true outliers are easier separable from the inliers. (D) Results of the same standard outlier detection algorithm using our (RECol) pre-processing approach, resulting in a better prediction of true outliers and fewer misclassifications of outliers compared to (B). (Color figure online)

3 Approach

The standard approach for detecting outliers in unsupervised settings follows this outline: (1) pre-processing, (2) outlier detection algorithm, and (3) result evaluation. We extend the standard approach by expanding step (1): For each column in the dataset, we use a supervised machine learning approach to predict this column based on all other features in the dataset (so excluding the one to be predicted, hence also called a leave-one-out approach). After training the supervised model for a specific column we calculate the reconstruction (or prediction) error for each data point (e.g., the squared error or some other metric of the prediction error) in the train and test set. Repeating this process, we arrive at a corresponding reconstruction error feature column (RECol) for each of the original columns, effectively doubling the number of features. Algorithm 1 explains this process in step by step. After this adjusted first step, we can continue with (2), the outlier detection algorithm. However, one might also consider pre-processing the RECols first, e.g. scaling them appropriately.

Algorithm 1. Procedure for RECol calculation

data: Training dataset X consists of set of n instances (data points) represented as X=$\{x_1, x_2, ..., x_n\}$ and every instance x_i includes a set of features $j \in \{1, 2, ..., d\}$ with $x_i^j \in \mathbb{R}$.

for each feature $j \in \{1, 2, ..., d\}$ **do**
  input $\leftarrow \{x_i^1, x_i^2, .., x_i^{j-1}, x_i^{j+1}, ..., x_i^n\}\ \forall i \in \{1, .., n\}$
  label $\leftarrow x_i^j$
  $f_\theta \leftarrow$ train(input, label) ▷ train a supervised model f_θ
  $\hat{x}_i^j \leftarrow f_\theta$(input) ▷ Reconstruction of the sample x_i^j
  $Re_i^j \leftarrow$ Error_func($x_i^j, \hat{x}_i^j$) ▷ Reconstruction Error
end for
output: RECol $\leftarrow \{Re_i^1, Re_i^2, ..., Re_i^d\}\ \forall i \in \{1, .., n\}$
Combined Features $\leftarrow (\{x_i^1, x_i^2, ..., x_i^d\}, \{Re_i^1, Re_i^2, ..., Re_i^d\})\ \forall i \in \{1, ..., n\}$
θ is a set of parameters learned by the supervised model.
Error_func can be e.g. Squared Error or Absolute Deviation.

Figure 2 illustrates our extended approach for outlier detection. In (A) we plot a dataset with two features, where outliers are marked in green and inliers in blue. In this dataset, an outlier is a point that markedly deviates from the underlying relationship between the two features. By directly applying an outlier detection algorithm to the data usually receives predictions as highlighted by the red and green points in (B), missing the orange outliers and delivering a dissatisfying performance. This is caused by the outlier detection algorithms focusing on labeling data points that are far away from others in the data space. By applying our approach where we create RECols and apply the outlier detection algorithm to these features, we are able to classify points as outliers that deviate from the underlying relationship, as can be seen in (C1) and (C2). (C1) shows the 2D reconstruction error space and (C2) a 2D projection of the 4D space arising from a concatenation of the original data space with the RECols. One can see, that the true outliers are easier to separate in these new spaces introduced by our approach. Given these new representations, the outlier detection model is better able to classify outliers correctly, as can be seen in (D).

3.1 Possible Design Decisions

Aside from whether our approach leads to improvements in detecting outliers in data, our approach opens up a few questions regarding the specification of the input space for the outlier detection algorithm:

(a) Should one use only the RECols for estimation or the combined feature space with the original features and additional RECols? If one is only concerned about outliers that are hard to reconstruct from the underlying relationship in the data one can in principle estimate the outlier detection model abolishing the original data features. However, they might still contain valuable information for estimation. Therefore, in our evaluation, we compare the results of the outlier detection algorithm on the original data features only, the RECols only, and the combined (concatenated) feature space.
(b) Can the RECols be used as an outlier detection algorithm directly by fusing the reconstruction errors appropriately? In principle, one could directly use the RECols and aggregate them to calculate an outlier score for each data point. To test this, we normalize the RECols and average them to an outlier score directly. In the next section, we compare our results from this exercise with standard outlier detection algorithms.
(c) How should the supervised algorithm to calculate RECols be specified? Ideally, we want to estimate the true hypothesis in the data. In practice, however, there is noise in the data and the dataset is incomplete in the sense that relevant features are missing. Therefore, we want to estimate a good approximation of the true underlying relationship. In our experiments, we use a wide range of algorithms to check whether the choice of the supervised model affects the results. In the following, we call the supervised model used for computing RECols the "RECol model".
(d) Finally, how should RECols be calculated and pre-processed, and should all newly created columns be used for estimating output of the outlier detection algorithm? That is, which reconstruction error metric should be applied, e.g. squared error, absolute deviation, or some other prediction error metric? In terms of pre-processing, one can think about scaling the attributes, dropping uninformative RECols, or clipping RECol values.

To evaluate if our RECol approach is beneficial and to answer these additional questions, we have run several experiments, as described in the following.

4 Evaluation

4.1 Experimental Setup

In this section, we evaluate whether using RECols leads to improvements when looking for outliers in data. We measure the performance of an OD algorithm by either the area under the receiver operating characteristic (ROC-AUC) or the area under the precision-recall curve (PR-AUC), which are standard metrics in the outlier detection domain. Before estimating RECols or applying an outlier detection model we split the dataset into a train set containing 70% of the data and a test set with the remaining 30% of data. Then we estimate the RECols and outlier detection model on the train set and evaluate them on the test set to avoid data dredging. We calculate RECols in the test data using the trained RECol model and evaluate the outlier detection model with these data. Finally, we perform three types of experiments for estimating the outlier detection: (1) Estimating baseline results using only original data columns, (2) using only RECols, and (3) using both types of columns combined (concatenated).

Baselines. We compare our results to those of [12] using the same set of datasets and the most common outlier detection algorithms as in their paper. Similar to [12] we optimize hyper-parameters using a grid search to reproduce their results. In the end, we find very similar results and pick the best outlier detection model for each algorithm and each dataset to compare to our RECol approach. We apply the following algorithms: HBOS [11], ISO [18], KNN [23], KthNN [23], LDCOF [3], LOF [6], LOoP [17], Nu-OCSVM [4], OCSVM [26], CBLOF [14], uCBLOF [14]. These are typical outlier detection algorithms from the literature [12].

In addition to the aforementioned baseline results, we initially also experimented with several auto-encoder based approaches for outlier detection, such as memory-augmented auto-encoder [13], deep auto-encoder [24,29], adversarial auto-encoder [27], and robust deep auto-encoder [30]. However, despite experimenting with different architectures, varying the number of hidden layers, and the size of the smallest compression layer we found that our auto-encoder based attempts were only ever among the best models (with negligible improvement) on the kdd99 dataset. Due to this and the large number of additional design decisions, we decided to leave an in-depth analysis of a combination of auto-encoders with our RECol approach on multiple larger datasets for future work.

Datasets. As in [12] we apply all of our experiments to the following ten datasets: aloi [9], annthyroid [12], breast-cancer [5], kdd99 [5], letter [20], pen-global [5], pen-local [5], satellite [20], shuttle [5], and speech [12]. The datasets are very different in their number of observations, their dimensionality, and their fraction of outliers. This allows us to analyse the effect of RECols on a wide range of different types of datasets. A more detailed description of the datasets themselves can be found in [12].

Parameters. To give an answer to question (a) whether one should use RECols for outlier detection we decided to run many different experiments. By applying different experiments we can also learn about question (c) how one should specify the supervised algorithm for our leave-one-out approach. Finally, to answer (d) we need to try out different approaches to generate and process RECols, like the distance metric and whether all RECol columns should be used for training the outlier detection model.

Therefore, we experiment with various specifications of training and generating the RECols. For each fixed best parameter baseline OD model and dataset, we experiment with the following parameters for the RECol models:

1. Supervised algorithms: Decision Tree Regressor, Random Forest Regressor, Gradient Boosting Regressor, Linear Regression, Support Vector Regression with RBF kernel
2. Reconstruction error calculation: Squared Error, Absolute Deviation
3. Scaling of input features for the supervised algorithm: Min-max Scaler, Standard Scaler
4. Clipping reconstruction errors at twice the standard deviation of the RECol
5. Using only subsets of RECols depending on the R^2 metric of the RECol model: dropping RECols with R^2 values below (0.05, 0.10) or above (0.95, 0.90) certain thresholds

In total, we evaluated 895 different ways of creating RECols for each fixed best parameter baseline OD model and dataset. We evaluated them against the corresponding fixed best parameter baseline OD model for all datasets, expect kdd99. For kdd99 we ran only eight experiments due to the size of the dataset and the time each experiment consumes.[1] We will discuss our results in the following subsection.

4.2 Results and Discussion

Table 1. RECols vs. Baseline Results. Performance is measured in ROC-AUC and PR-AUC percent on the test set (Δ in p.p.). Depending on the dataset, we notice how RECols can have a strong positive effect (especially in the PR-AUC metric).

Dataset	Size	Columns	Outlier %	ROC-AUC			PR-AUC		
				Baseline	RECols	Δ	Baseline	RECols	Δ
pen-local	6724	16	0.15	99.47	99.65	**0.18**	17.57	50.08	**32.51**
pen-global	809	16	11.11	98.89	98.11	−0.78	93.83	86.99	−6.84
breast-cancer	367	30	2.72	100.00	98.46	−1.54	85.00	79.37	−5.63
speech	3686	400	1.65	66.14	61.79	−4.35	2.24	50.81	**48.57**
aloi	50000	27	3.02	78.52	80.37	**1.85**	11.65	9.34	−2.31
shuttle	46464	9	1.89	99.85	99.94	**0.09**	98.31	96.36	−1.95
letter	1600	32	6.25	89.50	92.78	**3.28**	40.26	53.13	**12.87**
satellite	5100	36	1.47	96.90	95.51	−1.39	55.83	68.07	**12.24**
annthyroid	6916	21	3.61	72.09	89.58	**17.49**	15.67	53.73	**38.06**
kdd99	620098	29	0.17	99.96	99.83	−0.13	71.17	69.42	−1.75

To evaluate whether RECols help to better detect outliers we first focus on the often-used ROC-AUC metric for evaluation. We define the best baseline model as

[1] This can only harm us, because we might miss many possible experiments that could outperform the baseline.

the maximum ROC-AUC value across all standard outlier detection algorithms where we also optimized hyper-parameters using a grid search as in [12]. The best model is picked based on the train set while we then compare their performance on the test set. In Table 1 these values are shown in the fifth column. Similarly, we select the best RECol model wrt. ROC-AUC value based on the train set across our experiments. However, we do not optimize hyper-parameters of the underlying outlier detection model for the RECol approach but take the same hyper-parameter specification as for the baseline.[2] The sixth column of Table 1 shows our results on the test set. In total, we improve in 5 out of 10 datasets. In light of the fact that for half of the datasets the baseline is very close to 100 % (due to the low contamination in the dataset), this is a relatively strong result. We have also compared the best baseline against our RECol approach for each dataset and outlier detection algorithm separately. The average improvement in this setup is around 6% points in terms of ROC-AUC and RECols improve the results in over 2/3 of cases. In over 1/3 of cases the improvement is even above 5% points in the ROC-AUC metric. When we only look at baselines with a baseline metric below 95% ROC-AUC we see in almost 60% of settings an improvement of over 5% points in the ROC-AUC metric by just adding RECols as additional input to the outlier detection model. In principle, one could further improve the results by optimizing hyper-parameters of each of the underlying OD models, after choosing one of the 895 ways to generate RECols. However, due to the combinatorial explosion of experiments, we decided to instead treat the OD models' parameters as fixed (to our disadvantage).

The ROC-AUC metric, however, has its weaknesses: If outlier contamination in the dataset is low (as often is the case in OD scenarios), the ROC-AUC value is high by construction. In particular, many OD datasets have very low contamination, leading to very high baseline ROC-AUC values (close to 100%).

An alternative to the ROC-AUC metric that does not suffer from the above mentioned weakness is the PR-AUC metric. We hence repeat our experiments to test whether using RECols improve the results compared to standard outlier detection algorithms using the PR-AUC metric. The results are shown in Table 1. The RECol approach leads to stunning improvements in five out of ten datasets while there is no improvement in the other five datasets. The average improvement in this setup is around 8.6% points in terms of PR-AUC and RECols improve the results in over 75% of cases. In over 55% of cases the improvement is even above 5% points in the PR-AUC metric. Hence, there are considerable improvements in the PR-AUC metric when RECols are added to the outlier detection algorithm[3].

4.3 Possible Design Decisions

We will now attempt to answer the questions raised in Sect. 3.1.

[2] These choices can only harm us in that we restrict ourselves to fewer options compared to the baseline results.

[3] Result and code available at https://github.com/DayanandVH/RECol.

(a) Should one replace original columns with RECols or use both columns for training? In Table 2 we compare the results of outlier detection algorithms using the RECols only and a combination of RECols and original features on different datasets. The result is close: In 4 out of 10 datasets the RECols only approach slightly outperforms the combination of RECols and original features. In 5 others the combination wins, with bigger margins especially on the letter and annthyroid datasets. Hence, overall the combination of original features and RECols seems beneficial. We reason that this is the case because original features might still contain valuable information for outlier detection. The larger dimensionality of the space with RECols has not led to problems in our experiments, but if such problems arise in high-dimensional use-cases, one can likely fail over to RECols only.

Table 2. Combined (RECols + original) features versus RECols only Results. Performance of OD-model is measured in ROC-AUC percent. RECols only results are promising, but due to larger upside, we recommend the combined approach.

Dataset	RECols + original	RECols Only
pen-local	99.65	99.65
pen-global	**98.11**	98.05
breast-cancer	98.46	**99.69**
speech	61.79	**63.16**
aloi	80.37	**81.10**
shuttle	**99.94**	99.65
letter	**92.78**	86.60
satellite	95.51	**97.15**
annthyroid	**89.58**	85.02
kdd99	**99.83**	99.15

(b) Can RECols directly be used as an outlier detection algorithm? To test this we have applied the following approach: We normalize the RECols using a min-max scaler and take the average across all columns to receive an outlier score for each observation. When we take this outlier score and evaluate them using the ROC-AUC metric we can compare this to other outlier detection algorithms. The results for the ROC-AUC metric are summarized in Table 3. We find that for 2 out of 10 datasets such a simple fusion of the RECols from our approach, called RECol-OD in the following, outperforms all other baselines. When we compare RECol-OD to the average performance of other algorithms we see that in 9 out of 10 datasets the RECol-OD algorithm performs better. Overall, RECol-OD outperforms many standard algorithms by a significant amount. This is somewhat surprising in light of the simple fusion method we applied. As before, we do the same comparison using the PR-AUC metric for evaluation as can be seen in Table 4. In 4 out of 10 datasets the RECol-OD algorithm is the most

Table 3. RECol-OD vs. Baseline Results measured in ROC-AUC percent (Δ in p.p.). RECol-OD here is the direct fusion of the RECols from our approach into an outlier score. Despite its simplicity, RECol-OD outperforms the avg. in many and the best baseline models in some cases.

Dataset	Best Baseline	Avg. Baseline	RECol-OD	Δ to Best	Δ to Avg
pen-local	99.47	95.03	95.93	−3.54	0.90
pen-global	98.89	90.54	94.86	−4.03	**4.32**
breast-cancer	100.00	91.69	98.77	−1.23	**7.07**
speech	66.14	58.69	57.19	−8.95	−1.50
aloi	78.52	60.48	63.39	−15.13	**2.91**
shuttle	99.85	90.16	99.43	−0.42	**9.27**
letter	89.50	75.82	87.93	−1.57	**12.10**
satellite	96.90	91.12	97.54	0.64	**6.41**
annthyroid	72.09	60.18	88.90	**16.81**	**28.71**
kdd99	99.96	86.17	99.34	−0.62	**13.17**

successful algorithm. In 7 out of 10 datasets the performance is considerably higher than the average algorithm.

(c) How should the supervised algorithm to calculate RECols be specified? One has many degrees of freedom to specify the RECol model, in particular, the choice of algorithm could be important. Hence, we have tried various algorithms and found that random forests or gradient boosting are relatively robust and good benchmarks. Depending on the dataset at hand other RECol algorithms might be more beneficial though. We have also reasoned about whether the performance of the RECol model impacts the performance of the outlier detection, in particular, one could think of an inverse u-shape here. This is because a RECol model with an accuracy of 100% delivers a RECol column full of zeros while a RECol model with an accuracy of 0% delivers the square of the feature to be predicted (in case of a squared error RECol). In both cases adding RECols should not improve outlier detection. Hence a model that is good but not perfect might be ideal.

(d) How should RECols be calculated and pre-processed? In addition to the choice of the RECol model, one has many options to calculate and pre-process the created RECols. In terms of calculating we have tried squared errors and absolute deviations. We advise to try out both metrics because depending on the dataset and algorithm there were performance differences between the two methods. In principle, one could think of other metrics too, like exponential distances. After having calculated the RECols it is important to scale them, for example by using a min-max scaler. Clipping RECol values or dropping RECols with high or low R-squared values have no big effects on outcomes, but might be relevant depending on the dataset at hand.

Table 4. RECol-OD vs. Baseline Results measured in PR-AUC percent (Δ in p.p.). The best baseline is the best OD Algorithm using original features only (picked on training split, which allows for the best value in test to be lower than the average, as can be seen in the speech dataset). Average baseline is the average PR-AUC value across all baseline OD models. RECol-OD here is the direct fusion of the RECols from our approach into an outlier score. Despite its simplicity, RECol-OD outperforms the avg. in many and the best baseline models in some cases.

Dataset	Best Baseline	Avg. Baseline	RECol-OD	Δ to Best	Δ to Avg
pen-local	17.57	11.66	2.08	−15.49	−9.59
pen-global	93.83	71.61	73.84	−19.99	**2.24**
breast-cancer	85.00	71.57	81.67	−3.33	**10.09**
speech	2.24	10.81	7.10	**4.86**	−3.72
aloi	11.65	5.98	9.55	−2.10	**3.56**
shuttle	98.31	62.45	86.87	−11.44	**24.42**
letter	40.26	19.60	41.96	**1.70**	**22.36**
satellite	55.83	49.16	68.89	**13.06**	**19.72**
annthyroid	15.67	5.98	34.50	**18.83**	**28.53**
kdd99	71.17	31.65	25.91	−45.26	−5.74

4.4 Parameter Recommendations

Based on the 88,693 experiments we ran to evaluate how to generate RECols across 10 standard datasets and 11 standard OD approaches for this study, we would now like to share some of our observations for practical use-cases, in which one might not want to run an equally comprehensive study.

In general, we would recommend starting with a simple random forest regressor as a supervised algorithm to create the RECols based on the MSE metric. This is due to our observation that this combination seems to be quite robust, often among the top performers, and additionally has the advantage of only a few parameters to tune. Wrt. scaling of input features and clipping the RECols the results are less conclusive, so we would recommend experimenting depending on the use case. Wrt. dropping RECols depending on R^2, we did not observe a clear pattern, but the differences were negligible in most cases. In the case of many dimensions, we would hence suggest dropping the most uninformative RECols based on R^2 to reduce the load of downstream components.

5 Conclusion and Future Work

In this paper, we introduced our RECol pre-processing approach. We showed how to calculate RECols and added them to the outlier detection algorithm as additional features. We find that adding RECols can often significantly improve model performance of unsupervised outlier detection models in terms of the

ROC-AUC and PR-AUC metric and using them does rarely harm model performance. Based on more than 88k experiments, we also provide parameter recommendations to quickly try our approach in practical use-cases. Further, simply fusing RECols to an outlier score (RECol-OD) delivers a surprisingly simple algorithm that outperforms standard algorithms in many cases.

Our results also open up many further avenues for future research. So far we have not optimized hyper-parameters of the OD algorithms when using RECols. This could further improve our results. In addition, one can experiment further on how to add RECols and can also try more complex approaches to fuse RECols for the RECol-OD algorithm. This is especially interesting as doubling the space of columns might be problematic in high-dimensional data. Finally, applying our approach to additional datasets, combining it with auto-encoders and also in more real-world applications might help to strengthen our findings further.

Acknowledgements. This work was supported by the BMWK project EuroDaT (Grant 68GX21010K) and XAINES (Grant 01IW20005).

References

1. Aggarwal, C.C.: Outlier Analysis. In: Aggarwal, C.C. (ed.) Data Mining, pp. 237–263. Springer, Cham (2015). https://doi.org/10.1007/978-3-319-14142-8_8
2. Amarbayasgalan, T., Park, K.H., Lee, J.Y., Ryu, K.H.: Reconstruction error based deep neural networks for coronary heart disease risk prediction. PLoS ONE **14**(12), 1–17 (2019). https://doi.org/10.1371/journal.pone.0225991
3. Amer, M., Goldstein, M.: Nearest-neighbor and clustering based anomaly detection algorithms for rapidminer. In: Fischer, S., Mierswa, I. (eds.) Proceedings of the 3rd RapidMiner Community Meeting and Conferernce (RCOMM 2012). RapidMiner Community Meeting and Conference (RCOMM-2012), 28–31 August, Budapest, Hungary, pp. 1–12. Shaker Verlag GmbH (2012)
4. Amer, M., Goldstein, M., Abdennadher, S.: Enhancing one-class support vector machines for unsupervised anomaly detection. In: Proceedings of the ACM SIGKDD Workshop on Outlier Detection and Description, ODD 2013, pp. 8–15. Association for Computing Machinery, New York (2013). https://doi.org/10.1145/2500853.2500857
5. Bache, K., Lichman, M.: UCI machine learning repository (2013). http://archive.ics.uci.edu/ml
6. Breunig, M.M., Kriegel, H.P., Ng, R.T., Sander, J.: LOF: identifying density-based local outliers. SIGMOD Rec. **29**(2), 93–104 (2000). https://doi.org/10.1145/335191.335388
7. Chalapathy, R., Chawla, S.: Deep learning for anomaly detection: a survey. CoRR abs/1901.03407 (2019). http://arxiv.org/abs/1901.03407
8. Chandola, V., Banerjee, A., Kumar, V.: Anomaly detection: a survey. ACM Comput. Surv. **41**(3) (2009). https://doi.org/10.1145/1541880.1541882
9. Geusebroek, J.M., Burghouts, G.J., Smeulders, A.W.M.: The Amsterdam library of object images. Int. J. Comput. Vision **61**(1), 103–112 (2005). https://doi.org/10.1023/B:VISI.0000042993.50813.60
10. Gogoi, P., Bhattacharyya, D., Borah, B., Kalita, J.K.: A survey of outlier detection methods in network anomaly identification. Comput. J. **54**(4), 570–588 (2011). https://doi.org/10.1093/comjnl/bxr026

11. Goldstein, M., Dengel, A.: Histogram-based outlier score (HBOS): a fast unsupervised anomaly detection algorithm. In: Wölfl, S. (ed.) KI-2012: Poster and Demo Track. German Conference on Artificial Intelligence (KI-2012), 24–27 September, Saarbrücken, Germany, pp. 59–63. Online (2012)
12. Goldstein, M., Uchida, S.: A comparative evaluation of unsupervised anomaly detection algorithms for multivariate data. PLoS ONE **11**(4), 1–31 (2016). https://doi.org/10.1371/journal.pone.0152173
13. Gong, D., et al.: Memorizing Normality to Detect Anomaly: Memory-augmented Deep Autoencoder for Unsupervised Anomaly Detection. arXiv e-prints arXiv:1904.02639 (2019)
14. He, Z., Xu, X., Deng, S.: Discovering cluster-based local outliers. Pattern Recogn. Lett. **24**(9–10), 1641–1650 (2003). https://doi.org/10.1016/S0167-8655(03)00003-5
15. Hodge, V., Austin, J.: A survey of outlier detection methodologies. Artif. Intell. Rev. **22**, 85–126 (2004). https://doi.org/10.1023/B:AIRE.0000045502.10941.a9
16. Huang, Y.A., Fan, W., Lee, W., Yu, P.S.: Cross-feature analysis for detecting ad-hoc routing anomalies. In: Proceedings of the 23rd International Conference on Distributed Computing Systems, ICDCS 2003, p. 478. IEEE Computer Society, USA (2003). https://doi.org/10.1109/ICDCS.2003.1203498
17. Kriegel, H.P., Kröger, P., Schubert, E., Zimek, A.: Loop: local outlier probabilities. In: Proceedings of the 18th ACM Conference on Information and Knowledge Management, CIKM 2009, pp. 1649–1652. Association for Computing Machinery, New York (2009). https://doi.org/10.1145/1645953.1646195
18. Liu, F.T., Ting, K.M., Zhou, Z.H.: Isolation forest. In: Proceedings of the 2008 Eighth IEEE International Conference on Data Mining, ICDM 2008, pp. 413–422. IEEE Computer Society, USA (2008). https://doi.org/10.1109/ICDM.2008.17
19. Ma, M.Q., Zhao, Y., Zhang, X., Akoglu, L.: The need for unsupervised outlier model selection: a review and evaluation of internal evaluation strategies. ACM SIGKDD Explor. Newsl. **25**(1) (2023)
20. Micenková, B., McWilliams, B., Assent, I.: Learning outlier ensembles: the best of both worlds-supervised and unsupervised. In: ACM SIGKDD 2014 Workshop ODD (2014)
21. Noto, K., Brodley, C., Slonim, D.: FRaC: a feature-modeling approach for semi-supervised and unsupervised anomaly detection. Data Min. Knowl. Discov. **25**(1), 109–133 (2012). https://doi.org/10.1007/s10618-011-0234-x
22. Pang, G., Shen, C., Cao, L., Hengel, A.V.D.: Deep learning for anomaly detection: a review. ACM Comput. Surv. **54**(2) (2021). https://doi.org/10.1145/3439950
23. Ramaswamy, S., Rastogi, R., Shim, K.: Efficient algorithms for mining outliers from large data sets. SIGMOD Rec. **29**(2), 427–438 (2000). https://doi.org/10.1145/335191.335437
24. Sakurada, M., Yairi, T.: Anomaly detection using autoencoders with nonlinear dimensionality reduction. In: Proceedings of the MLSDA 2014 2nd Workshop on Machine Learning for Sensory Data Analysis, MLSDA 2014, pp. 4–11. Association for Computing Machinery, New York (2014). https://doi.org/10.1145/2689746.2689747
25. Sattarov, T., Herurkar, D., Hees, J.: Explaining anomalies using denoising autoencoders for financial tabular data (2022)
26. Schölkopf, B., Platt, J.C., Shawe-Taylor, J.C., Smola, A.J., Williamson, R.C.: Estimating the support of a high-dimensional distribution. Neural Comput. **13**(7), 1443–1471 (2001). https://doi.org/10.1162/089976601750264965

27. Schreyer, M., Sattarov, T., Schulze, C., Reimer, B., Borth, D.: Detection of accounting anomalies in the latent space using adversarial autoencoder neural networks (2019). https://doi.org/10.48550/ARXIV.1908.00734. https://arxiv.org/abs/1908.00734
28. Ted, E., et al.: Detecting insider threats in a real corporate database of computer usage activity. In: Proceedings of the 19th ACM SIGKDD International Conference on Knowledge Discovery and Data Mining, pp. 1393–1401. Association for Computing Machinery (2013). https://doi.org/10.1145/2487575.2488213
29. Xia, Y., Cao, X., Wen, F., Hua, G., Sun, J.: Learning discriminative reconstructions for unsupervised outlier removal. In: 2015 IEEE International Conference on Computer Vision, ICCV 2015, Santiago, Chile, 7–13 December 2015, pp. 1511–1519. IEEE Computer Society (2015). https://doi.org/10.1109/ICCV.2015.177
30. Zhou, C., Paffenroth, R.C.: Anomaly detection with robust deep autoencoders. In: Proceedings of the 23rd ACM SIGKDD International Conference on Knowledge Discovery and Data Mining, KDD 2017, pp. 665–674. Association for Computing Machinery, New York (2017). https://doi.org/10.1145/3097983.3098052

Interactive Link Prediction as a Downstream Task for Foundational GUI Understanding Models

Christoph Albert Johns[1,2](✉), Michael Barz[2,3], and Daniel Sonntag[2,3]

[1] Aarhus University, Aarhus, Denmark
cajohns@cs.au.dk

[2] German Research Center for Artificial Intelligence (DFKI), Saarbrücken, Germany
{michael.barz,daniel.sonntag}@dfki.de

[3] Applied Artificial Intelligence, Oldenburg University, Oldenburg, Germany

Abstract. AI models that can recognize and understand the semantics of graphical user interfaces (GUIs) enable a variety of use cases ranging from accessibility to automation. Recent efforts in this domain have pursued the development of a set of foundation models: generic GUI understanding models that can be used off-the-shelf to solve a variety of GUI-related tasks, including ones that they were not trained on. In order to develop such foundation models, meaningful downstream tasks and baselines for GUI-related use cases will be required. In this paper, we present *interactive link prediction* as a downstream task for GUI understanding models and provide baselines as well as testing tools to effectively and efficiently evaluate predictive GUI understanding models. In interactive link prediction, the task is to predict whether tapping on an element on one screen of a mobile application (source element) navigates the user to a second screen (target screen). If this task is solved sufficiently, it can demonstrate an understanding of the relationship between elements and components across screens and enable various applications in GUI design automation and assistance. To encourage and support research on interactive link prediction, this paper contributes (1) a preprocessed large-scale dataset of links in mobile applications (18,830 links from 5,362 applications) derived from the popular RICO dataset, (2) performance baselines from five heuristic-based and two learning-based GUI understanding models, (3) a small-scale dataset of links in mobile GUI prototypes including ratings from an online study with 36 end-users for out-of-sample testing, and (4) a Figma plugin that can leverage link prediction models to automate and assist mobile GUI prototyping.

Keywords: Link prediction · GUI prototyping · Mobile GUIs · Online machine learning

---

Supplementary Information The online version contains supplementary material available at https://doi.org/10.1007/978-3-031-42608-7_7.

D. Seipel and A. Steen (Eds.): KI 2023, LNAI 14236, pp. 75–89, 2023.
https://doi.org/10.1007/978-3-031-42608-7_7

1 Introduction

The goal of graphical user interface (GUI) understanding is to model the semantics of a GUI by learning effective representations of user interface screens and their components. These semantic representations can then be used to enable various assistive technologies related to GUI accessibility such as more effective screen readers and voice assistants (e.g., [3,29]) or related to GUI design and prototyping such as GUI search or recommendation engines and design diagnostic tools (e.g., [23,25]). Since GUIs are inherently multimodal, layered, and complex, they offer a particular challenge to any single embedding or modeling technique [18]. Recent efforts within the field of GUI understanding have thus moved away from the development of narrow single-purpose models designed to solve a specific GUI understanding task (e.g., icon labeling [3] or tappability prediction [23]) in pursuit of a set of foundation models that can be used off-the-shelf to solve a variety of GUI-related downstream tasks without requiring any or while only requiring minimal finetuning (see, for example, [15,18,26]).

The development of such general GUI understanding models requires their evaluation against a range of varied and meaningful downstream tasks. Previous efforts in this domain have mainly focused on single-screen tasks (e.g., summarization [26,27], captioning [17], or labeling [3]). In this paper, we present *interactive link prediction*, a downstream task that requires an understanding of the relationship between two GUI screens and their components and complements existing downstream tasks related to screen similarity [8] and click prediction [12]. To facilitate the development of GUI understanding models capable of interactive link prediction, we provide a preprocessed large-scale dataset of links in Android applications (18,830 links from 5,362 applications) derived from the popular RICO dataset [7] as well as a custom small dataset of links in mobile GUI prototypes including ratings from an online study with 36 end-users that can be used to test the generalization of link prediction models to out-of-sample data[1]. We further provide performance baselines from five heuristic-based and two learning-based GUI understanding models inspired by recent GUI understanding model architectures. Finally, we present a custom plugin to the GUI design software Figma that can be used to leverage and evaluate link prediction models for design assistance in real-world applications (see footnote 1).

2 Related Work

GUI understanding techniques and datasets are tools and resources used to analyze and understand the structure and functionality of screens and components of GUI-based applications. These techniques commonly leverage supervised machine learning models trained on large-scale mobile GUI datasets and have proven successful across a wide range of tasks related to link prediction

[1] The datasets and source code for the models are available at https://github.com/christophajohns/link-prediction. The source code for the Figma plugin prototype is available at https://github.com/christophajohns/suggested-links.

like tappability prediction [23,25,29] or screen transition prediction [8]. Schoop *et al.* [23], for example, use the large-scale mobile GUI dataset RICO [7] to train a visual recognition model that predicts the tappability of GUI elements in Android applications. Feiz *et al.* [8], similarly, utilize the RICO dataset to train a model that predicts whether a pair of screenshots assumed to be part of the same interaction trace represents the same underlying screen (e.g., representing different scroll positions or showing a notification). Finally, Wang *et al.* [26] utilize prompting of a pre-trained large language model to solve a variety of single-screen GUI understanding tasks including screen summarization and screen question answering. These efforts are emblematic of a recent development in the field of GUI understanding: the pursuit of a set of foundational GUI understanding models that can be used off-the-shelf to solve a variety of interaction tasks in GUI-based applications (see, for instance, [15,18,26]). The downstream task presented in this paper, *interactive link prediction*, supports the development of such foundational GUI understanding models by offering a new challenge to evaluate models' multi-screen generalization.

3 Methodology

Interactive link prediction refers to a binary two-screen GUI classification task that is designed to facilitate and test a model's understanding of the relationship between screens and components of a GUI-based application. Given a 3-tuple (S, T, e) of a source screen S, a target screen T, and a source element e from the same application as input, the task is to predict whether a tap gesture (i.e., an interaction with a single touch point) performed on the source element e triggers a transition to the target screen T (see Fig. 1). The source screen S and the target screen T are defined as element hierarchies; the source element e is a unique identifier for an element on the source screen S. The goal of the interactive link prediction task is to learn a function f that can accurately predict whether a transition occurs between the source and target screens given the source element:

$$f(S, T, e) = \begin{cases} 1 & \text{if tapping on element e on screen S triggers a} \\ & \text{transition to screen T} \\ 0 & \text{otherwise} \end{cases} \tag{1}$$

A model f that is able to recognize and predict interactive links should be able to identify semantic relationships between elements and screens across source element types (e.g., icons or buttons) and application contexts (e.g., banking or entertainment) from the screens' visual, structural, and textual information alone. To provide benchmarks for such understanding, we implement, train, and test seven link prediction models, five based on heuristics and two based on supervised learning, on a set of links in Android applications derived from the popular RICO dataset [7] and report accuracy, recall, and F1 scores for all models across the preprocessed dataset.

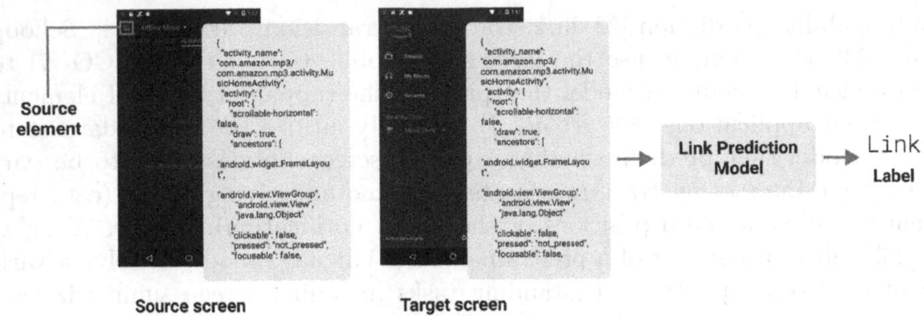

Fig. 1. Given a 3-tuple of a source screen, a target screen and a source element, the task in *interactive link prediction* is to predict whether tapping on the source element triggers a transition to the target screen. The source and target screen are represented as view hierarchies (GUI element trees); the source element is specified by its identifier.

3.1 Datasets

Our baselines utilize the RICO [7] dataset: a large-scale collection of element hierarchies and user interactions of free Android applications from 2017. Using RICO's [7] interaction traces as basis, we derive a sub-dataset called $\mathbf{RICO_{links}}$ that contains data on links in mobile GUIs using heuristic selection. In a first step all traces with at least three screens and at least one tap gesture are filtered, resulting in 5,726 remaining traces. Then, for each tap gesture in those traces, three samples are constructed: (1) a positive sample representing a link from the touched element (i.e., the node lowest in the view hierarchy whose bounds contain the touch point; source element) on a given screen (source screen) to the next screen in the trace (target screen), (2) a negative 'non-source' sample originating from a randomly selected visible[2], static[3] element to the true target screen and (3) a negative 'non-target' sample representing a link from the true source element to a randomly selected screen from the same trace. In total, 18,830 samples are constructed for each of the three sample types, resulting in 56,790 screen pairs across 5,362 applications.

We additionally create a custom small-scale dataset of links in high-fidelity interactive mobile GUI prototypes called $\mathbf{Figma_{links}}$ and conduct an online study with 36 end-users to rate a set of potential links across these prototypes. These data can facilitate the evaluation of a model's generalization to recent visual design styles and from real applications to operating system-independent high-fidelity prototypes. We, first, create a custom set of mock-ups for 4 small applications in the GUI design software Figma spanning 4 screens each and representing common visual styles and application categories, which are chosen

[2] A node within a view hierarchy from the RICO dataset is considered visible if its `visibility` or `visible-to-user` attribute is set to `True` and it has positive width and height as well as at least part of its bounding area is within the screen bounds.

[3] A node with the `clickable` attribute set to `False`.

following the procedure described in [7]. The view hierarchies of these mock-ups are then extracted using a custom Figma plugin. Next, three trained designers with an academic degree in human-computer interaction who use GUI design and prototyping tools at least once a week (all male, age ranging from 25 to 28 years) are recruited and individually tasked to add links to these previously created static application mock-ups using Figma while thinking out loud. Finally, we use the intersection of their three individual link sets (Fleiss' $\kappa = 0.978$) to form the basis for our rating task. In total, 30 designer-created links are selected across the 4 mobile application prototypes and 16 GUI screens. These links are then extended with a random sample of 70 additional potential links across the 4 example applications. This resulting set of 100 potential links is then rated by end-users in an asynchronous online study. 36 adult smartphone users—19 female, 17 male—aged 21 to 64 (mean: 35.3; std: 15.3) are recruited via convenience sampling and instructed to complete a rating task that asks participants to indicate on a five-point Likert scale how strongly they expect that tapping on a given link's source element will open that link's target screen. This data adds further nuance to prediction evaluation and may be used to validate positive predictions that lie outside the set of designer-created links. In total, 100 links are rated by the 36 end-users—including the 30 designer-created links—, across all 4 mobile GUI prototypes and 16 screens (Fleiss' $\kappa = 0.305$).

3.2 Heuristic Models

To provide meaningful baselines for model evaluation, we implement five heuristic models inspired by related work on information foraging theory and information scent (see, for instance, [4]) based on rules about the text in and around the source element and about the text on the target screen. If a model includes a free hyperparameter, its value is tuned using mean F1 score across threefold cross-validation via twenty iterations of Bayesian optimization on a training set derived from the RICO$_{links}$ dataset. Our baseline model *PageContainsLabel* is built upon the assumption that a target screen is accurately described by the label of the source elements that link to it [4]. The model suggests adding a link, if the label[4] of the source element contains at least one word from the target screen. All other heuristic models are based on a similar principle. The *LargestTextElementsContainLabel* model suggests adding a link, if the label of the source element contains at least one word of the n largest text elements on the target screen as measured by the elements' area on screen ($n = 20$). It assumes that prominent text elements accurately describe the content of their screen. The *LabelTextSimilarity* model is based on the assumption that semantic text embeddings (i.e., vector representations that aim to encode the meaning of a word or phrase) better capture the semantic similarity of two texts than simple bag-of-words approaches, which treat texts as arbitrarily ordered collections of character sequences and typically rely on co-occurrence as a measure of

[4] A label in the RICO$_{links}$ dataset is given by any non-empty value of a node's `text` attribute.

relatedness. This heuristic model uses the natural language model all-MiniLM-L6-v2 [21], a small-footprint successor to the Sentence-BERT model that was successfully used in the Screen2Vec GUI embedding technique [16]. The model suggests adding a link if the cosine similarity between the text embeddings of the source element's label and of text on the target screen exceeds a tuned threshold t (here, $t = 0.122$). The *TextSimilarity* model addresses the issue that models relying on the source element's label are unable to predict links for common GUI elements that do not have a label (e.g., a menu item consisting of an icon and a text element). It suggests adding a link if the cosine similarity between the text embeddings of the source element and of the target screen exceeds a tuned threshold t (here, $t = 0.263$). The text of an element here is defined as its own label or the labels of all its descendants if it does not have a label of its own. The *TextSimilarityNeighbors* model is designed to consider non-text elements (i.e., GUI elements without own or descendant labels; e.g., images or icons) as well as elements whose label is not descriptive of their target's content (e.g., buttons labeled "Click here" or "Read more"). In addition to the source element's own text, its embedding includes the text of the source element's n closest neighboring leaf text elements (here, $n = 2$) as measured by Euclidean distance between their bounding boxes. The model suggests adding a link if the cosine similarity between the embeddings of the text in and around the source element and of the target screen's text exceeds a tuned threshold t (here, $t = 0.292$). This context-based approach is mirrored in both learning-based GUI understanding techniques such as Screen2Vec [16] and in non-learning based techniques like information scent models (e.g., [4]).

3.3 Supervised Learning Models

In addition to the five heuristic models, we implement two supervised learning-based models using binary online learning classifiers. The selection of binary classifier included in each of the two models is detailed in Sect. 3.4 below. The *TextOnly* model is similar in architecture and implementation to the *TextOnly* model included in the Screen2Vec GUI embedding technique described in [16]. Using the same preprocessing of the *TextSimilarityNeighbors* model, extracting the text in and around the source element and on the target screen and encoding them using the all-MiniLM-L6-v2 language model, the *TextOnly* model labels the concatenated embeddings using a binary random forest classifier trained with random under-sampling. In addition to the number of neighboring leaf text elements to consider n (here, $n = 6$), this model includes two hyperparameters of the random forest classifier: the number of decision trees t (here, $t = 10$) and the size of the random subsets of features to consider f (here, $f = 85$). The *LayoutOnly* model utilizes the layout autoencoder from the original RICO publication [7] to predict links based on the concatenated layout embeddings of the source and target screens as well as a vector denoting the relative position and size of the source element. As was the case with the *TextOnly* model above, a random forest classifier using random under-sampling is used as the classification model with a tuned number of decision trees t (here, $t = 24$) and size of the random subsets of features to consider f (here, $f = 52$). The models are implemented using Python 3.10 and the scikit-learn [20], imbalanced-learn [14], and PyTorch packages [19]. They are trained, tuned and tested using a MacBook Pro with a 2.3 GHz 8-Core Intel Core i9 CPU on a random split of the RICO_{links} dataset into a training (80%) and test (20%) set, ensuring similar distribution of source element types (i.e., containing or not containing text), application categories and number of data points per application. The models' hyperparameters are tuned using mean F1 score across threefold cross-validation via twenty iterations of Bayesian optimization on the training set.

3.4 Model Selection

We explore a selection of common binary classifiers from supervised learning literature for the task of link prediction as defined above: random forest, logistic regression, and naive Bayes. These models are chosen due to the beneficial properties of both probabilistic—being able to produce meaningful confidence estimates—and online learning classifiers—being able to continue learning during deployment in interactive systems. To further address the issue of class imbalance in the training set, balanced variants of the classification models and imbalanced learning techniques (i.e., random under-sampling, random over-sampling, SMOTE [2] and ADASYN [11]) are included. We estimate the performance of all models based on their mean F1 score from a threefold cross-validation. We found a random forest classifier with a random under-sampler to be the best performing model for both the *TextOnly* and *LayoutOnly* models, and thus selected it for the final implementation of these two models.

4 Experiments

Using the test set constructed from the RICO_{links} dataset, the benchmark models are compared to an additional baseline of a random classifier predicting a positive label with a probability reflecting the frequency of true links in the RICO_{links} dataset (i.e., 0.333). While all but the *TextOnly* model outperform this baseline random classifier (see Table 1), our benchmark models' overall level of performance is rather low, indicating the challenge of the prediction task. Among our models, the *LabelTextSimilarity* model is able to most accurately predict the links in the given test set.

Further analyzing the performance of the models by source element type (i.e., text or non-text), all models achieve better prediction performance using only text elements. Comparing the best model's (i.e., the *LabelTextSimilarity* model's) performance across the ten most common application categories in the test set, its performance generalizes well across categories with F1 scores ranging from 0.496 (Entertainment) to 0.603 (Weather). To further illustrate some of the limitations of the *LabelTextSimilarity* model, exemplary data points from the true and false positives with highest heuristic decision score and false negatives with lowest score are displayed in Fig. 2. These cases demonstrate

Table 1. Prediction performance of heuristic and learning-based models for links in Android applications using the RICO$_{links}$ test set.

Model	Precision	Recall	F1
LabelTextSimilarity	0.476	**0.629**	**0.542**
PageContainsLabel	0.543	0.450	0.492
LargestTextElementsContainLabel	**0.571**	0.407	0.475
LayoutOnly	0.381	0.584	0.461
TextSimilarity	0.436	0.397	0.416
TextSimilarityNeighbors	0.445	0.358	0.397
TextOnly	0.223	0.301	0.256
(Random (stratified))	0.339	0.337	0.338

some of the shortcomings of the RICO$_{links}$ dataset that is used to train and tune the link prediction models. Both misclassifications can be traced back to mislabeled or mismatched data from the original RICO (see false negative) or derived RICO$_{links}$ datasets (see false positive). For further discussion of the shortcomings of the RICO dataset, we refer to recent reviews in [6,13], and [28].

When applying the benchmark models to the out-of-sample data in the Figma$_{links}$ dataset, the models are found to predict sets of links that strongly differ from those created by human designers. For each of the seven benchmark models, we thus compare the ratings for the predicted links and predicted non-links across the dataset using descriptive statistics. Additionally, we construct a baseline of human performance by comparing the ratings for the designer-created links and the additional random set of links included in the sample.

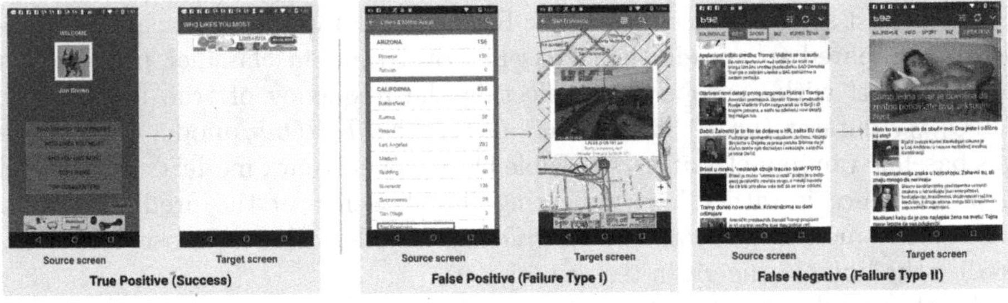

Fig. 2. Selected true positive (left), false positive (middle) and false negative (right) samples with highest/low decision score as determined by the best-performing model in this evaluation. The source element is highlighted with a red frame. While the true positive sample shows the general capability of the model to identify meaningful relations, the false negative (result of a full-screen advertisement interrupting the screen transition) and false positive (mismatched view hierarchy or misreported touch point) samples illustrate limitations of the RICO$_{links}$ dataset. (Color figure online)

The designer-created links received a mean rating of 4.281 (std: 1.211) from the end-users, indicating a high-quality of the chosen links. The random sample of additional links not chosen by the designers, by comparison, received a mean rating of 1.938 (std: 1.415), resulting in a considerable difference in mean rating of Δ Mean = 2.342 between the two sets. If a foundational GUI understanding models achieved similar discriminative performance to the human designers, the difference in mean rating between their predicted links and predicted non-links should be expected to fall into a similar range of this delta. All benchmark models included in this study fail to achieve a comparable result. The *TextSimilarityNeighbors* model is able to most closely match the designers' difference in mean rating, but still performs considerably worse with Δ Mean = 1.355. Both supervised learning-based models *LayoutOnly* (Δ Mean = −0.263) and *TextOnly* (Δ Mean = −0.357) as well as the heuristic *LabelTextSimilarity* (Δ Mean = −0.508) even produce negative differences where the predicted non-links achieve a higher mean rating than the predicted links.

We further analyze the relationship between the models' decision score (heuristic models) or confidence estimate (supervised learning-based models) and the end-user ratings by comparing the ratings for the ten predicted links with highest ("Top 10") and the ten predicted non-links with lowest decision score ("Bottom 10"). Only the *TextSimilarity* and *TextSimilarityNeighbors* models achieve a comparable difference in mean rating to the designers' sets, further suggesting a limited predictive performance of the benchmark models included in this evaluation. An additional Wilcoxon signed-rank test shows a significant difference in rating between the top ten and bottom ten predictions for all but the *LabelTextSimilarity* and *TextOnly* models (see Table 2). Overall, our baselines illustrate the challenge in interactive link prediction to generalize across element types, application contexts, and available modalities.

Table 2. Comparison between the mean ratings of the ten potential links with highest and lowest decision score (heuristic models) or confidence estimate (supervised learning-based models) per model, including the results of a Wilcoxon signed-rank test. The difference in mean rating between the designer-created and random additional links in the dataset was Δ Mean = 2.342. Note: *: $p < 0.05$; **: $p < 0.01$; ***: $p < 0.001$

Model	Δ Mean Rating	Top 10 (std)	Bottom 10 (std)
TextSimilarity	**2.556*** **	**4.333 (1.143)**	**1.778 (1.371)**
TextSimilarityNeighbors	**2.556*** **	**4.333 (1.143)**	**1.778 (1.371)**
LayoutOnly	0.567***	2.428 (1.663)	1.861 (1.351)
PageContainsLabel	0.336**	3.097 (1.792)	2.761 (1.715)
LargestTextElementsContainLabel	0.336**	3.097 (1.792)	2.761 (1.715)
LabelTextSimilarity	0.111	2.264 (1.604)	2.153 (1.628)
TextOnly	−0.403	2.411 (1.663)	2.814 (1.664)

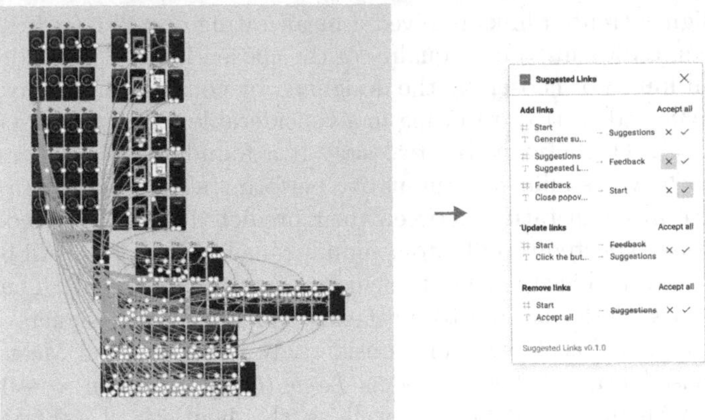

Fig. 3. Screenshot of a prototype in the GUI design application Figma (left). To address the issue of system state visibility, an assistive plugin suggesting links to add, update or remove for a given Figma prototype is implemented (right).

5 Application: 'Suggested Links' Figma Plugin

To illustrate and facilitate potential uses of GUI understanding models capable of interactive link prediction, we present a prototype for an assistive GUI prototyping tool called 'Suggested Links' that can leverage GUI understanding models to support designers of interactive GUI prototypes. A common problem in the development and maintenance of interactive GUI prototypes is that actively developing static GUI designs often leads to cluttered, outdated, and inappropriate links in their corresponding prototypes [9,10]. An example of such a GUI prototype is shown in Fig. 3 (left). We envision a system that utilizes capable link prediction models to automatically infer meaningful connections for a given prototype, suggesting them to a designer, and reducing the required effort to develop or maintain that GUI design. Within a research-oriented setting, the same tool could be used to rapidly and interactively diagnose the ability of link prediction models to accurately predict appropriate links in custom prototypes.

To demonstrate the feasibility of creating such a real-time link suggestion system and to furthermore illustrate the promise of GUI understanding models that can learn from user feedback, we develop a plugin to the well-known interface design tool Figma, based on our presented models. The plugin computes the set of all potential links in a given prototype and uses these potential connections as input for a pre-trained link prediction model. The resulting predictions are then compared to the current set of links in the design and grouped into connections that can be added, updated, or removed by the user (see Fig. 3, right). If a user decides to accept one of these suggestions, this supervisory signal can be used to update the underlying the link prediction model. We used online learning for our benchmark models in this work to enable such seamless model updates. However, several techniques including Reinforcement Learning from Human Feedback (RLHF) [5] can be imagined that fulfill a similar purpose and enable users to train a shared global model or create a personal one.

The presented prototype is a first encounter in the direction of an AI-supported GUI design process that saves human time and cost when developing new mobile interfaces. However, our predictive benchmark models do not perform well enough to enable suitable link suggestions and, hence, the utility of our prototype is currently very limited. Still, we could demonstrate the feasibility of integrating link suggestions with interactive feedback in a state-of-the-art GUI design tool. To leverage the full potential such a tool, strong GUI understanding models will be required that are capable of accurate link prediction and which can adapt models to individual or project-specific needs. One option to achieve this is to use interactive machine learning techniques as described above.

6 Discussion

We reflect on the pros and cons of our task definition and our provided tools for the development and evaluation of foundational GUI understanding models.

Challenges in Interactive Link Prediction. Interactive link prediction is a challenging task within GUI understanding as it not only requires an understanding of individual screens and their components but of the relationship between components across screens, including non-text elements like images or icons. This is difficult for many of the state-of-the-art GUI understanding models as they commonly focus on either single screens (e.g., [23]) or simple components (e.g., [26]) and thus are likely to struggle with this task by nature of their architecture alone. Link prediction, similarly to other challenging tasks like screen summarization [26,27] and tappability prediction [23,25], is furthermore subject to individual differences in perception, requiring effective GUI understanding models to handle ambiguity and contradictions in the training and testing data. The out-of-sample GUI prototype data and its associated end-user ratings provided in this paper may aid research in this issue.

Baseline Model Limitations. The baseline models provided here are limited due to their reliance on unimodal text and layout-based GUI embedding techniques, which may not be as effective in handling non-text source elements or capturing the relationships between multiple GUI elements. They are furthermore trained on a dataset where true links are defined to originate from the lowest element in the screen's element hierarchy that contains a screen transition's touch point. While following convention (cf. [12]), this is in conflict with common GUI design patterns and alternative approaches that have been shown to be effective in tappability prediction (e.g., using the element *highest* in the view hierarchy [25]). As a result, the baseline models' performance appears to generalize poorly to data from high-fidelity prototypes. Future research could explore this issue of generalization and investigate alternative definitions for true links in the RICO dataset. Further, we could investigate the use of computer vision-based techniques (cf. [8,23]) for improved performance in dealing with non-text elements, as well as of multi-GUI pre-training tasks (cf. [12]) to develop more effective semantic representations for both real applications and high-fidelity prototypes. Outside the domain of GUI understanding, interactive link prediction models may benefit from taking inspiration from factual link prediction in knowledge graphs (see, for example, [22]). While this direction is not further explored in this work, this body of literature provides relevant insights into alternative problem formulations for multi-screen and multi-component relationship prediction and may aid in the development of new approaches to GUI understanding.

GUI Design Automation and Assistance. Models that can understand and effectively predict semantic relationships across components and screens as required by interactive link prediction hold great promise for design automation and assistance. As our prototypical Figma plugin illustrates, such models could empower designers to spend more time on their high-fidelity prototypes and automate costly maintenance tasks. Beyond this illustration, several use cases can be imagined that enhance existing design automation tools (e.g., [1,24]) where rough GUI sketches are automatically translated into high-fidelity interactive prototypes, enabling more rapid iterations in early stages of the design process. When link prediction is accurate and efficient, it can lead to time and cost savings in the development process, ultimately resulting in better user experiences.

Towards Foundational GUI Understanding Models. Interactive link prediction points to future research directions in developing more accurate and tailored models that can recognize and understand the relationships between components and screens across parts of an application. As our Figma plugin suggests, adapting models to specific design requirements and personal preferences of designers may be an effective way to create more targeted and efficient design tools. Employing interactive machine learning techniques or other personalization techniques such as RLHF [5] and leveraging additional input modalities may guide the development of foundational GUI understanding models in this vain and further improve the efficiency and effectiveness of the design process.

7 Conclusion

In this work, we proposed a new downstream task for foundational GUI understanding models called *interactive link prediction* and implemented and evaluated seven benchmark models using interactive links from real mobile applications and in high-fidelity GUI prototypes. Our analysis revealed that link prediction is a challenging task that requires understanding not only of various GUI components such as texts, images, and icons but also of their relationship on a single screen and across multiple screens of an application. We discussed the limitations of our presented datasets and benchmark models and further considered interactive machine learning as a methodology to overcome them. Furthermore, we showcased how interactive link suggestions can be integrated in the GUI design process by implementing a plugin for the popular GUI design application

Figma. We envision that foundational GUI understanding models that are capable of effective and adaptive link prediction can help to significantly improve the future GUI design process. We hope that the tools we provide here can encourage and support researchers who want to take up this challenge and develop the next generation of adaptive and flexible GUI understanding models.

Acknowledgements. This work was partially funded by the German Federal Ministry of Education and Research (BMBF) under grant number 01IW23002 (No-IDLE), by the Lower Saxony Ministry of Science and Culture, and the Endowed Chair of Applied Artificial Intelligence of the University of Oldenburg.

References

1. Baulé, D., Hauck, J.C.R., Júnior, E.C.V.: Automatic code generation from sketches of mobile applications in end-user development using Deep Learning, p. 18 (2021). https://arxiv.org/abs/2103.05704
2. Chawla, N.V., Bowyer, K.W., Hall, L.O., Kegelmeyer, W.P.: SMOTE: synthetic minority over-sampling technique. J. Artif. Intell. Res. **16**, 321–357 (2002). https://doi.org/10.1613/jair.953. https://www.jair.org/index.php/jair/article/view/10302
3. Chen, J., Swearngin, A., Wu, J., Barik, T., Nichols, J., Zhang, X.: Towards complete icon labeling in mobile applications. In: CHI Conference on Human Factors in Computing Systems, CHI 2022, pp. 1–14. Association for Computing Machinery, New York (2022). https://doi.org/10.1145/3491102.3502073
4. Chi, E.H., Pirolli, P., Chen, K., Pitkow, J.: Using information scent to model user information needs and actions and the web. In: Proceedings of the SIGCHI Conference on Human Factors in Computing Systems, CHI 2001, pp. 490–497. Association for Computing Machinery, New York (2001). https://doi.org/10.1145/365024.365325
5. Christiano, P.F., Leike, J., Brown, T.B., Martic, M., Legg, S., Amodei, D.: Deep reinforcement learning from human preferences. In: Proceedings of the 31st International Conference on Neural Information Processing Systems, NIPS 2017, pp. 4302–4310. Curran Associates Inc., Red Hook (2017)
6. Deka, B., et al.: An early rico retrospective: three years of uses for a mobile app dataset. In: Li, Y., Hilliges, O. (eds.) Artificial Intelligence for Human Computer Interaction: A Modern Approach. HIS, pp. 229–256. Springer, Cham (2021). https://doi.org/10.1007/978-3-030-82681-9_8
7. Deka, B., et al.: Rico: a mobile app dataset for building data-driven design applications. In: Proceedings of the 30th Annual ACM Symposium on User Interface Software and Technology, UIST 2017, pp. 845–854. Association for Computing Machinery, New York (2017). https://doi.org/10.1145/3126594.3126651
8. Feiz, S., Wu, J., Zhang, X., Swearngin, A., Barik, T., Nichols, J.: Understanding screen relationships from screenshots of smartphone applications. In: 27th International Conference on Intelligent User Interfaces, IUI 2022, pp. 447–458. Association for Computing Machinery, New York (2022). https://doi.org/10.1145/3490099.3511109
9. Figma Community Forum: Header nav and prototype spaghetti - Ask the community (2021). https://forum.figma.com/t/header-nav-and-prototype-spaghetti/1534/4

10. Figma Community Forum: How to change several interactions at once? - Ask the community (2022). https://forum.figma.com/t/how-to-change-several-interactions-at-once/33856
11. He, H., Bai, Y., Garcia, E.A., Li, S.: ADASYN: adaptive synthetic sampling approach for imbalanced learning. In: 2008 IEEE International Joint Conference on Neural Networks (IEEE World Congress on Computational Intelligence), pp. 1322–1328 (2008). https://doi.org/10.1109/IJCNN.2008.4633969. ISSN 2161-4407
12. He, Z., et al.: ActionBert: leveraging user actions for semantic understanding of user interfaces. In: Proceedings of the AAAI Conference on Artificial Intelligence, vol. 35, no. 7, pp. 5931–5938 (2021). https://ojs.aaai.org/index.php/AAAI/article/view/16741
13. Leiva, L.A., Hota, A., Oulasvirta, A.: Enrico: a dataset for topic modeling of mobile UI designs. In: 22nd International Conference on Human-Computer Interaction with Mobile Devices and Services, MobileHCI 2020, pp. 1–4. Association for Computing Machinery, New York (2020). https://doi.org/10.1145/3406324.3410710
14. Lemaître, G., Nogueira, F., Aridas, C.K.: Imbalanced-learn: a python toolbox to tackle the curse of imbalanced datasets in machine learning. J. Mach. Learn. Res. **18**(17), 1–5 (2017). http://jmlr.org/papers/v18/16-365.html
15. Li, G., Li, Y.: Spotlight: Mobile UI Understanding using Vision-Language Models with a Focus (2023). https://doi.org/10.48550/arXiv.2209.14927. http://arxiv.org/abs/2209.14927. arXiv:2209.14927
16. Li, T.J.J., Popowski, L., Mitchell, T., Myers, B.A.: Screen2Vec: semantic embedding of GUI screens and GUI components. In: Proceedings of the 2021 CHI Conference on Human Factors in Computing Systems, CHI 2021, pp. 1–15. Association for Computing Machinery, New York (2021). https://doi.org/10.1145/3411764.3445049
17. Li, Y., Li, G., He, L., Zheng, J., Li, H., Guan, Z.: Widget Captioning: Generating Natural Language Description for Mobile User Interface Elements. arXiv:2010.04295 (2020). http://arxiv.org/abs/2010.04295. arXiv:2010.04295
18. Li, Y., Li, G., Zhou, X., Dehghani, M., Gritsenko, A.: VUT: Versatile UI Transformer for Multi-Modal Multi-Task User Interface Modeling (2021). https://doi.org/10.48550/arXiv.2112.05692. http://arxiv.org/abs/2112.05692. arXiv:2112.05692
19. Paszke, A., et al.: PyTorch: an imperative style, high-performance deep learning library. In: Wallach, H., Larochelle, H., Beygelzimer, A., Alché-Buc, F.D., Fox, E., Garnett, R. (eds.) Advances in Neural Information Processing Systems, vol. 32. Curran Associates, Inc. (2019). https://proceedings.neurips.cc/paper/2019/file/bdbca288fee7f92f2bfa9f7012727740-Paper.pdf
20. Pedregosa, F., et al.: Scikit-learn: machine learning in python. J. Mach. Learn. Res. **12**, 2825–2830 (2011)
21. Reimers, N.: Hugging Face - sentence-transformers/all-MiniLM-L6-v2 (2021). https://huggingface.co/sentence-transformers/all-MiniLM-L6-v2
22. Rossi, A., Barbosa, D., Firmani, D., Matinata, A., Merialdo, P.: Knowledge graph embedding for link prediction: a comparative analysis. ACM Trans. Knowl. Discov. Data **15**(2), 1–49 (2021). https://doi.org/10.1145/3424672
23. Schoop, E., Zhou, X., Li, G., Chen, Z., Hartmann, B., Li, Y.: Predicting and explaining mobile UI tappability with vision modeling and saliency analysis. In: CHI Conference on Human Factors in Computing Systems, New Orleans, LA, USA, pp. 1–21. ACM (2022). https://doi.org/10.1145/3491102.3517497. https://dl.acm.org/doi/10.1145/3491102.3517497

24. Suleri, S., Sermuga Pandian, V.P., Shishkovets, S., Jarke, M.: Eve: a sketch-based software prototyping workbench. In: Extended Abstracts of the 2019 CHI Conference on Human Factors in Computing Systems, CHI EA 2019, pp. 1–6. Association for Computing Machinery, New York (2019). https://doi.org/10.1145/3290607.3312994
25. Swearngin, A., Li, Y.: Modeling mobile interface tappability using crowdsourcing and deep learning. In: Proceedings of the 2019 CHI Conference on Human Factors in Computing Systems, CHI 2019, pp. 1–11. Association for Computing Machinery, New York (2019). https://doi.org/10.1145/3290605.3300305
26. Wang, B., Li, G., Li, Y.: Enabling conversational interaction with mobile UI using large language models. In: Proceedings of the 2023 CHI Conference on Human Factors in Computing Systems, Hamburg, Germany, pp. 1–17. ACM (2023). https://doi.org/10.1145/3544548.3580895. https://dl.acm.org/doi/10.1145/3544548.3580895
27. Wang, B., Li, G., Zhou, X., Chen, Z., Grossman, T., Li, Y.: Screen2Words: automatic mobile UI summarization with multimodal learning. In: The 34th Annual ACM Symposium on User Interface Software and Technology, Virtual Event USA, pp. 498–510. ACM (2021). https://doi.org/10.1145/3472749.3474765. https://dl.acm.org/doi/10.1145/3472749.3474765
28. Wu, J., Wang, S., Shen, S., Peng, Y.H., Nichols, J., Bigham, J.P.: WebUI: a dataset for enhancing visual UI understanding with web semantics. In: Proceedings of the 2023 CHI Conference on Human Factors in Computing Systems, Hamburg, Germany, pp. 1–14. ACM (2023). https://doi.org/10.1145/3544548.3581158. https://dl.acm.org/doi/10.1145/3544548.3581158
29. Zhang, X., et al.: Screen recognition: creating accessibility metadata for mobile applications from pixels. In: Proceedings of the 2021 CHI Conference on Human Factors in Computing Systems, Yokohama, Japan, pp. 1–15. ACM (2021). https://doi.org/10.1145/3411764.3445186. https://dl.acm.org/doi/10.1145/3411764.3445186

Harmonizing Feature Attributions Across Deep Learning Architectures: Enhancing Interpretability and Consistency

Md Abdul Kadir[1](✉), GowthamKrishna Addluri[1], and Daniel Sonntag[1,2]

[1] German Research Center for Artificial Intelligence (DFKI), Saarbrücken, Germany
{abdul.kadir,Gowthamkrishna.Addluri,daniel.sonntag}@dfki.de
[2] University of Oldenburg, Oldenburg, Germany

Abstract. Enhancing the interpretability and consistency of machine learning models is critical to their deployment in real-world applications. Feature attribution methods have gained significant attention, which provide local explanations of model predictions by attributing importance to individual input features. This study examines the generalization of feature attributions across various deep learning architectures, such as convolutional neural networks (CNNs) and vision transformers. We aim to assess the feasibility of utilizing a feature attribution method as a future detector and examine how these features can be harmonized across multiple models employing distinct architectures but trained on the same data distribution. By exploring this harmonization, we aim to develop a more coherent and optimistic understanding of feature attributions, enhancing the consistency of local explanations across diverse deep-learning models. Our findings highlight the potential for harmonized feature attribution methods to improve interpretability and foster trust in machine learning applications, regardless of the underlying architecture.

Keywords: Explainability · Trustworthiness · XAI · Interpretability

1 Introduction

Deep learning models have revolutionized various domains, but their complex nature often hampers our ability to understand their decision-making processes [1,8]. Interpretability techniques have emerged, with local and global explanations being two significant categories [7]. Local explanations focus on understanding individual predictions, highlighting the most influential features for a specific instance. This method is valuable for understanding model behavior at a granular level and providing intuitive explanations for specific predictions. On the other hand, global explanations aim to capture overall model behavior and identify patterns and trends across the entire dataset. They offer a broader perspective and help uncover essential relationships between input features and model prediction. This paper delves into interpretability in deep learning models, particularly model-agnostic feature attribution, a subset of local explanation techniques.

D. Seipel and A. Steen (Eds.): KI 2023, LNAI 14236, pp. 90–97, 2023.
https://doi.org/10.1007/978-3-031-42608-7_8

Feature attribution refers to assigning importance or relevance to input features in a machine learning model's decision-making process [2]. It aims to understand which features have the most significant influence on the model's predictions or outputs. Feature attribution techniques provide insights into the relationship between input features and the model's decision, shedding light on the factors that drive specific outcomes. These techniques are precious for interpreting complex models like deep learning, where the learned representations may be abstract and difficult to interpret directly [18]. By quantifying the contribution of individual features, feature attribution allows us to identify the most influential factors, validate the model's behavior, detect biases, and gain a deeper understanding of the decision-making process.

Feature attribution methods can be evaluated through various approaches and metrics [18]. Qualitative evaluation involves visually inspecting the attributions and assessing their alignment with domain knowledge. Perturbation analysis tests the sensitivity of attributions to changes in input features [17]. Sanity checks ensure the reasonableness of attributions, especially in classification problems. From a human perspective, we identify objects in images by recognizing distinct features [14]. Similarly, deep learning models are trained to detect features from input data and make predictions based on these characteristics [19]. The primary objective of deep learning models, irrespective of the specific architecture, is to learn the underlying data distribution and capture unique identifying features for each class in the dataset.

Various deep learning architectures have proven proficient in capturing essential data characteristics within the training distribution [23]. We assume that if a set of features demonstrates discriminative qualities for one architecture, it should likewise exhibit discriminative properties for a different architecture, provided both architectures are trained on the same data. This assumption forms the foundation for the consistency and transferability of feature attributions across various deep learning architectures.

Our experiments aim to explore the generalizability of features selected by a feature attribution method for one deep learning architecture compared to other architectures trained on the same data distribution. We refer to this process as harmonizing feature attributions across different architectures. Our experimental results also support our assumption and indicate that different architectures trained on the same data have a joint feature identification capability.

2 Related Work

Various explanation algorithms have been developed better to understand the internal mechanisms of deep learning models. These algorithms, such as feature attribution maps, have gained significant popularity in deep learning research. They offer valuable insights into the rationale behind specific predictions made by deep learning models [4]. Notable examples of these explanation methods include layer-wise relevance propagation [12], Grad-CAM [20], integrated gradient [22], guided back-propagation [21], pixel-wise decomposition [3], and contrastive explanations [11].

Various methods have been developed to evaluate feature attribution maps. Ground truth data, such as object-localization or masks, has been used for evaluation [5,15]. Another approach focuses on the faithfulness of explanations, measuring how well they reflect the model's attention [17]. The IROF technique divides images into segments and evaluates explanations based on segment relevance [16]. Pixel-wise evaluations involve flipping pixels or assessing attribution quality using pixel-based metrics [3,17].

Chen et al. [6] has demonstrated the utility of feature attribution methods for feature selection. Additionally, research conducted by Morcos et al. [13] and Kornblith et al. [10] has explored the internal representation similarity between different architectures. However, to the best of our knowledge, the generalization of feature attributions across diverse neural architectures still needs to be explored.

Motivated by the goal of evaluating feature attributions, we are investigating a novel approach that involves assessing feature attributions across multiple models belonging to different architectural designs. This method aims to provide a more comprehensive understanding of feature attribution in various contexts, thereby enhancing the overall explainability of deep neural networks.

3 Methodology

This experiment investigates the generalizability and transferability of feature attributions across different deep learning architectures trained on the same data distribution. The experimental process involves generating feature attribution maps for a pretrained model, extracting features from input images, and passing them to two models with distinct architectures. The accuracy and output probability distribution are then calculated for each architecture.

In this experiment, we employ a modified version of the Soundness Saliency (SS) method [9] for generating explanations. The primary objective with a network f, for a specific input x (Fig. 1(a)), and label a, is to acquire a map or mask $M \in \{0,1\}^{hw}$.

This map aims to minimize the expectation $E_{\tilde{x}\sim(x,M)}[-\sum f_i(\tilde{x})log(f_i(\tilde{x}))]$, wherein the probability assigned by the network to a modified or composite input x is maximized.

$$\tilde{x} \sim \Gamma(x,a) \equiv \overline{x} \sim \mathcal{X}, \tilde{x} = M \odot x + (1-M) \odot \overline{x} \tag{1}$$

Here, M (Fig. 1(b)) represents the feature attribution map generated by the Soundness Saliency algorithm. The saliency map M provides information about the importance of each pixel and the extent of its contribution to the classification. If the value of M (Fig. 1(b)) for a specific pixel is 0, it implies that the

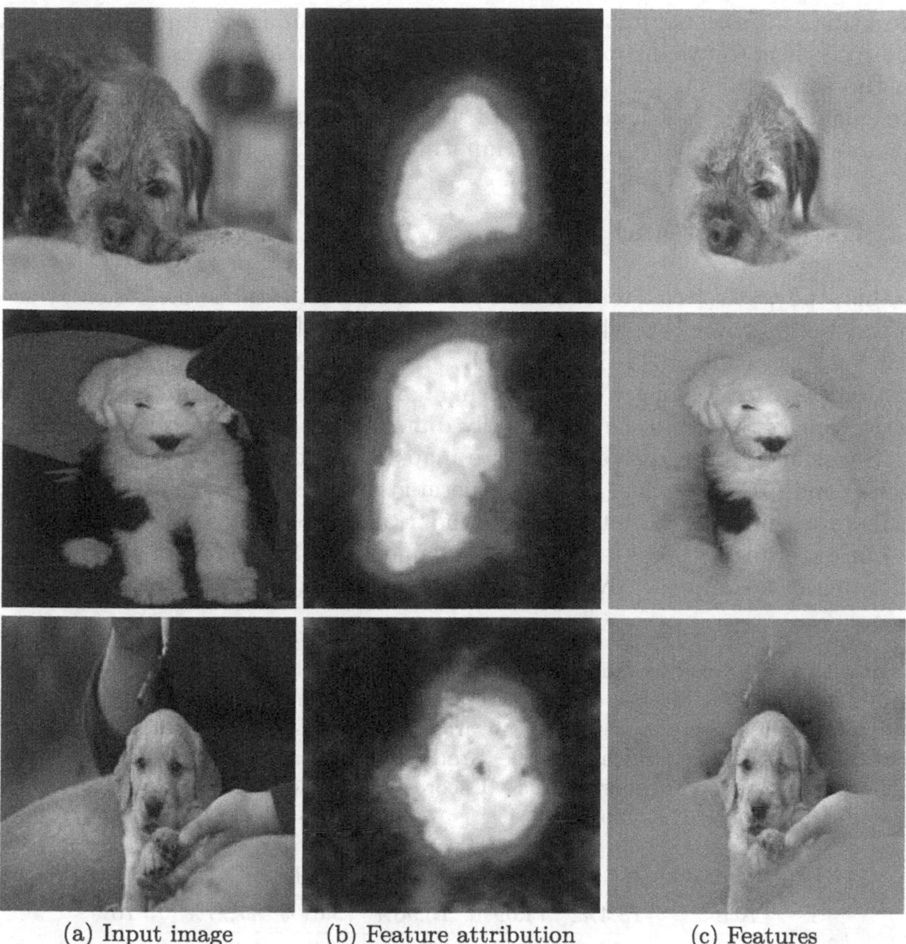

Fig. 1. Columns a, b, and c represent the input image, the feature attribution map generated by the soundness saliency algorithm, and the extracted features of the image based on the feature attribution map, respectively.

pixel has no significance in the classification process. Conversely, if the value of M for a particular pixel is high, it indicates that the pixel is highly important for the classification. We enhance the extraction of important features (Fig. 1(c)) from input by applying the Hadamard product between each input channel and the corresponding attribution map M. In addition to the Soundness Saliency (SS) algorithm, we also employ Grad-CAM [20] (GC) for feature extraction.

We utilize the selected features Fig. 1(c) extracted through feature attributions and feed them to two distinct models with different architectures, albeit trained on the same training data. Accuracy, F1 score, and output probability scores are calculated for these models. The focus is observing model prediction changes when only the selected features are inputted rather than the entire image.

4 Experiment and Results

In this study, we selected four distinct pretrained architectures: the Vision Transformer architecture (ViT) [8], EfficientNet-B7 (E-7) [23], EfficientNet-B6 (E-6) [23], and EfficientNet-B5 (E-5) [23]. To generate feature attribution maps, we first employed E-7 along with a challenging subset[1] of the ImageNet validation data, which is known to be particularly difficult for classifiers.

Subsequently, we generated feature maps for all test data and passed them to E-6 and E-5. In parallel, we also generated features for ViT and followed the same procedure. We chose to utilize both a transformer and a CNN architecture in our experiments because they are fundamentally different from one another, allowing for a comprehensive evaluation of the various architectures.

Table 1. The SS row examines model performance with image (I) and feature (F) inputs generated by the E-7 architecture and SS algorithm. It shows stable accuracy and F1 scores across the E-6 and E-5 architectures when using feature inputs. In contrast, the GC row, which uses the Grad-CAM algorithm for feature generation, demonstrates a drop in accuracy and F1 scores across the E-6, and E-5 architectures when features are used as input.

Exp.	Metric	E-7 (I)	E-7 (F)	E6 (I)	E-6 (F)	E-5 (I)	E-5 (F)
SS	Acc	78.4%	74.09%	75.90%	73.61%	77.27%	73.76%
	F1	0.87	0.84	0.85	0.84	0.86	0.84
GM	Accuracy	78.47%	58.87%	75.90%	55.74%	77.27%	57.24%
	F1	0.87	0.72	0.85	0.70	0.86	0.71

Our experimental results (Tables 1 and 2) indicate that features generated by a neural architecture can be detected by other architectures trained on the same data. This implies that feature attribution maps encapsulate sufficient data distribution information. Consequently, feature maps created using attribution maps on one architecture can be recognized by another architecture, provided that both are trained on the same data. As depicted in Fig. 2, when we feed only features to the model, the class probability increases (Fig. 2(b), (d), and (f)), particularly when using similar architectures for feature generation and evaluation. When employing different types of architectures (e.g., Transformer

[1] https://github.com/fastai/imagenette.

Table 2. The SS row evaluates model performance using image (I) and feature (F) inputs, generated through the ViT architecture and the SS algorithm. There's a slight decrease in accuracy and F1 score across architectures (E-6, E-5) with feature inputs. Similarly, the GC row, utilizing the Grad-CAM algorithm for feature generation, shows a comparable drop in accuracy and F1 scores across the E-6 and E-5 architectures when using feature inputs.

Exp.	Metric	ViT (I)	ViT (F)	E6 (I)	E-6 (F)	E-5 (I)	E-5 (F)
SS	Acc	89.92%	88.83%	75.90%	71.90%	77.27%	73.10%
	F1	0.94	0.94	0.85	0.83	0.86	0.83
GM	Ac	89.92%	58.56%	75.90%	44.21%	77.27%	43.47%
	F1	0.87	0.73	0.85	0.60	0.86	0.58

for generating feature maps and CNN for evaluating them), there is a slight drop in accuracy (Fig. 2(j) and (l)), but the performance remains consistent.

Accuracy decreases when features are extracted with Grad-CAM saliency maps, suggesting these maps might not capture crucial information on the data distribution. However, when examining row GC in Tables 1 and 2, it's observed that accuracy remains consistent across various architectural configurations when features are generated using Grad-CAM. This suggests that different architectures have harmony in detecting certain features from data.

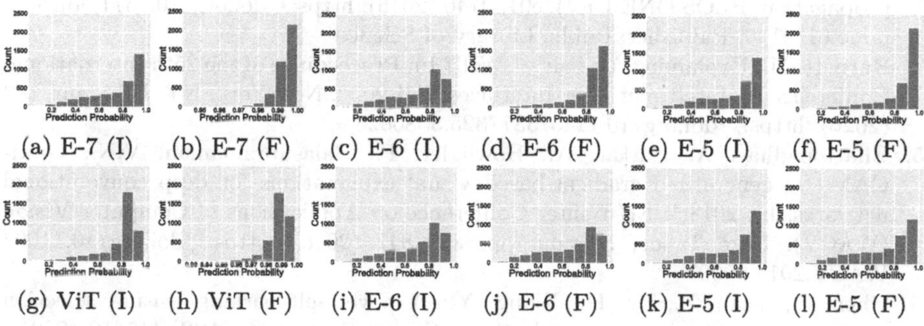

(a) E-7 (I) (b) E-7 (F) (c) E-6 (I) (d) E-6 (F) (e) E-5 (I) (f) E-5 (F)

(g) ViT (I) (h) ViT (F) (i) E-6 (I) (j) E-6 (F) (k) E-5 (I) (l) E-5 (F)

Fig. 2. Histogram of class prediction probability distribution across different architectures, comparing image-only (I) and feature-only (F) inputs for the entire test dataset.

5 Conclusion

The experiment validates our hypothesis that various architectures acquire shared features from a common data distribution. We noticed a notable rise in class prediction probabilities when utilizing selected features as inputs, particularly when employing similar neural architecture building blocks such as Convolution. Additionally, the consistency of predictions on future attribution maps across architectures demonstrates that different architectures are not randomly learning features from the data, thereby enhancing the reliability of the models. These findings underscore the potential to generalize features and emphasize the need for additional research to harmonize feature attribution maps, expanding their applicability in various domains.

Acknowledgements. This work was partially funded by the German Federal Ministry of Education and Research (BMBF) under grant number 16SV8639 (Ophthalmo-AI) and 2520DAT0P2 (XAINES) and German Federal Ministry of Health (BMG) under grant number 2520DAT0P2 (pAItient) and supported by the Lower Saxony Ministry of Science and Culture and the Endowed Chair of Applied Artificial Intelligence (AAI) of the University of Oldenburg.

References

1. Alirezazadeh, P., Schirrmann, M., Stolzenburg, F.: Improving deep learning-based plant disease classification with attention mechanism. Gesunde Pflanzen **75**(1), 49–59 (2022). https://doi.org/10.1007/s10343-022-00796-y
2. Ancona, M., Ceolini, E., Öztireli, A.C., Gross, M.H.: A unified view of gradient-based attribution methods for deep neural networks. CoRR abs/1711.06104 (2017). https://doi.org/10.48550/arXiv.1711.06104
3. Bach, S., Binder, A., Montavon, G., Klauschen, F., Müller, K.R., Samek, W.: On Pixel-wise explanations for non-linear classifier decisions by layer-wise relevance propagation. PLOS ONE **10**(7), e0130140 (2015). https://doi.org/10.1371/journal.pone.0130140. Publisher: Public Library of Science
4. Burnett, M.: Explaining AI: Fairly? Well? In: Proceedings of the 25th International Conference on Intelligent User Interfaces, IUI 2020, New York, NY, USA, pp. 1–2 (2020). https://doi.org/10.1145/3377325.3380623
5. Chattopadhay, A., Sarkar, A., Howlader, P., Balasubramanian, V.N.: Grad-CAM++: generalized gradient-based visual explanations for deep convolutional networks. In: 2018 IEEE Winter Conference on Applications of Computer Vision (WACV), Lake Tahoe, Nevada, pp. 839–847 (2018). https://doi.org/10.1109/WACV.2018.00097
6. Chen, J., Yuan, S., Lv, D., Xiang, Y.: A novel self-learning feature selection approach based on feature attributions. Expert Syst. Appl. **183**, 115219 (2021). https://doi.org/10.1016/j.eswa.2021.115219
7. Doshi-Velez, F., Kim, B.: Towards a rigorous science of interpretable machine learning (2017). https://doi.org/10.48550/arXiv.1702.08608
8. Dosovitskiy, A., et al.: An image is worth 16x16 words: transformers for image recognition at scale. CoRR abs/2010.11929 (2020). https://doi.org/10.48550/arXiv.2010.11929
9. Gupta, A., Saunshi, N., Yu, D., Lyu, K., Arora, S.: New definitions and evaluations for saliency methods: staying intrinsic, complete and sound. ArXiv abs/2211.02912 (2022). https://doi.org/10.48550/arXiv.2211.02912
10. Kornblith, S., Norouzi, M., Lee, H., Hinton, G.E.: Similarity of neural network representations revisited. CoRR abs/1905.00414 (2019). https://doi.org/10.48550/arXiv.1905.00414

11. Luss, R., et al.: Leveraging latent features for local explanations. In: Proceedings of the 27th ACM SIGKDD Conference on Knowledge Discovery & Data Mining, KDD 2021, New York, NY, USA, pp. 1139–1149, August 2021. https://doi.org/10.1145/3447548.3467265
12. Montavon, G., Samek, W., Müller, K.R.: Methods for interpreting and understanding deep neural networks. Digit. Signal Process. **73**, 1–15 (2018). https://doi.org/10.1016/j.dsp.2017.10.011
13. Morcos, A.S., Raghu, M., Bengio, S.: Insights on representational similarity in neural networks with canonical correlation. In: Proceedings of the 32nd International Conference on Neural Information Processing Systems, NIPS 2018, Red Hook, NY, USA, pp. 5732–5741 (2018). https://doi.org/10.48550/arXiv.1806.05759
14. Mozer, M.: Object recognition: theories. In: International Encyclopedia of the Social & Behavioral Sciences, Pergamon, Oxford, pp. 10781–10785 (2001). https://doi.org/10.1016/B0-08-043076-7/01459-5
15. Nunnari, F., Kadir, M.A., Sonntag, D.: On the overlap between grad-CAM saliency maps and explainable visual features in skin cancer images. In: Holzinger, A., Kieseberg, P., Tjoa, A.M., Weippl, E. (eds.) CD-MAKE 2021. LNCS, vol. 12844, pp. 241–253. Springer, Cham (2021). https://doi.org/10.1007/978-3-030-84060-0_16
16. Rieger, L., Hansen, L.K.: IROF: a low resource evaluation metric for explanation methods, March 2020. https://doi.org/10.48550/arXiv.2003.08747, arXiv:2003.08747 [cs]
17. Samek, W., Binder, A., Montavon, G., Lapuschkin, S., Müller, K.R.: Evaluating the visualization of what a deep neural network has learned. IEEE Trans. Neural Netw. Learn. Syst. **28**(11), 2660–2673 (2017). https://doi.org/10.1109/TNNLS.2016.2599820
18. Samek, W., Montavon, G., Lapuschkin, S., Anders, C.J., Müller, K.R.: Explaining deep neural networks and beyond: a review of methods and applications. Proc. IEEE **109**(3), 247–278 (2021). https://doi.org/10.1109/JPROC.2021.3060483
19. Schulz, H., Behnke, S.: Deep learning. KI - Künstliche Intelligenz **26**(4), 357–363 (2012). https://doi.org/10.1007/s13218-012-0198-z
20. Selvaraju, R.R., Cogswell, M., Das, A., Vedantam, R., Parikh, D., Batra, D.: Grad-CAM: visual explanations from deep networks via gradient-based localization. In: 2017 IEEE International Conference on Computer Vision (ICCV), Venice, Italy, pp. 618–626 (2017). https://doi.org/10.1109/ICCV.2017.74
21. Springenberg, J., Dosovitskiy, A., Brox, T., Riedmiller, M.: Striving for simplicity: the all convolutional net. In: ICLR (Workshop Track), San Diego, California, p. 10 (2015). https://doi.org/10.48550/arXiv.1412.6806
22. Sundararajan, M., Taly, A., Yan, Q.: Axiomatic attribution for deep networks. In: Proceedings of the 34th International Conference on Machine Learning, ICML 2017, Sydney, Australia, vol. 70, pp. 3319–3328 (2017). https://doi.org/10.48550/arXiv.1703.01365
23. Tan, M., Le, Q.V.: EfficientNet: rethinking model scaling for convolutional neural networks. CoRR abs/1905.11946 (2019). https://doi.org/10.48550/arXiv.1905.11946

Lost in Dialogue: A Review and Categorisation of Current Dialogue System Approaches and Technical Solutions

Hannes Kath[1,2(✉)], Bengt Lüers[1], Thiago S. Gouvêa[1(✉)], and Daniel Sonntag[1,2]

[1] German Research Center for Artificial Intelligence (DFKI), Oldenburg, Germany
{hannes_berthold.kath,bengt.lueers,thiago.gouvea,daniel.sonntag}@dfki.de
[2] University of Oldenburg, Applied Artificial Intelligence (AAI), Oldenburg, Germany

Abstract. Dialogue systems are an important and very active research area with many practical applications. However, researchers and practitioners new to the field may have difficulty with the categorisation, number and terminology of existing free and commercial systems. Our paper aims to achieve two main objectives. Firstly, based on our structured literature review, we provide a categorisation of dialogue systems according to the objective, modality, domain, architecture, and model, and provide information on the correlations among these categories. Secondly, we summarise and compare frameworks and applications of intelligent virtual assistants, commercial frameworks, research dialogue systems, and large language models according to these categories and provide system recommendations for researchers new to the field.

Keywords: Dialogue System · Conversational AI · Task-oriented · Natural Language Processing · Survey

1 Introduction

Major advances in natural language processing (NLP) through deep learning have tremendously strengthened research in dialogue systems [8,41]. However, the vast number of dialogue system descriptions and surveys lack standardised terminology and dialogue systems are mainly categorised by their objective [5,9,54], neglecting current research topics such as multi-modality [30]. That can be confusing and daunting for researchers and practitioners new to this field, and even experienced researchers could benefit from a structured review. The main goal of this paper is to facilitate researchers' entry into the field of dialogue systems. To achieve this, we first present a theoretical background that introduces the categories of modality, domain, architecture and model that we have derived from the literature, in addition to the objective. Then, we categorise applications and frameworks of intelligent virtual assistants, commercial frameworks, research dialogue systems, and large language models based on the derived categories and provide beginner-friendly system recommendations. The rest of the

D. Seipel and A. Steen (Eds.): KI 2023, LNAI 14236, pp. 98–113, 2023.
https://doi.org/10.1007/978-3-031-42608-7_9

paper is structured as follows: Sect. 2 outlines our structured literature review approach. Section 3 gives an overview of dialogue system categories derived from the literature and explains the relationships between them. Section 4 provides descriptions and recommendations for applications and frameworks suitable for different purposes.

2 Methods

The aim of our structured literature review is to provide a summary of categories, terminologies, applications and frameworks related to dialogue systems to assist researchers new to the field. We have manually reviewed contributions from the main technically oriented research and application venues for dialogue systems, in particular SIGdial[1] and Interspeech[2]. This resulted in a selection of 63 papers. To complement this selection with other technically sound survey papers, we extended the results by searching Scopus[3] (TITLE("DIALOG* SYSTEM*" AND ("SURVEY" OR "REVIEW")) AND PUBYEAR>2016) and ACM Digital Library[4] ([[TITLE:"DIALOG* SYSTEMS"] OR [TITLE:"DIALOG* SYSTEM"]] AND [[TITLE:SURVEY] OR [TITLE:REVIEW]] AND [PUBLICATION DATE: (01/01/2017 TO *)]). From 25 results we excluded duplicates (3), non-English articles (3), articles focusing on specific languages (2) and articles focusing on specific fields (medical domain) (3). The remaining 14 surveys cover the topics general knowledge [5,41], evaluation [7,9,14,29], deep learning [6,44], task-oriented dialogue system components (natural language understanding [31], dialogue state tracking [1]), empathy [36], corpora [35,54] and multi-modality [30]. While these surveys focus on specific topics in the field of dialogue systems, to our knowledge there is no elaboration that introduces newcomers to the topic and offers practical suggestions for applications and frameworks. The whole selection has 77 relevant papers. Acronyms used throughout the paper are listed in Table 1.

3 Dialogue Systems

A dialogue system, in literature also called conversational agent, virtual agent, (intelligent) virtual assistant, digital assistant, chat companion system, chatbot or chatterbot, is an interactive software system that engages in natural language conversations with humans. The communication is usually structured in *turns* (one or more utterances from one speaker), *exchanges* (two consecutive turns) and the *dialogue* (multiple exchanges) [9]. We present the criteria extracted from the structured literature review for categorising dialogue systems in the following subsections.

[1] https://www.sigdial.org/.
[2] https://www.interspeech2023.org/.
[3] https://www.scopus.com.
[4] https://dl.acm.org.

Table 1. List of acronyms

ASR	automatic speech recognition	LLM	large language model
CDS	conversational dialogue system	NLG	natural language generation
DM	dialogue manager	NLP	natural language processing
DRAS	dialogue response action selection	NLU	natural language understanding
DST	dialogue state tracking	QADS	question answering dialogue system
GUI	graphical user interface	TDS	task-oriented dialogue system
IVA	intelligent virtual assistant	TTS	text to speech

3.1 Objective

Most surveys divide dialogue systems into task-oriented dialogue systems (TDSs), conversational dialogue systems (CDSs) and question answering dialogue systems (QADSs) according to their objective. The terminology used in the literature is not consistent and occasionally ambiguous. A TDS is also referred to as task-specific, task-based, goal-driven or goal-oriented dialogue system and sometimes simply dialogue system [54] or conversational agent [41]. A CDS is also referred to as open-domain, chit-chat, non-goal-driven, social, non-task-orientated or chat-oriented dialogue system. The term chatbot is not clearly defined and used for TDSs [41], CDSs [7,36] or dialogue systems in general [44]. We use the term dialogue system as defined above and TDS, CDS and QADS as subcategories.

Task-oriented Dialogue Systems (TDSs) are designed to help users complete specific tasks as efficiently as possible [5,9], such as providing information (e.g. timetable information) or carrying out actions (e.g. ordering food, booking a hotel). Due to the clearly defined goal the dialogue is highly structured. The initiative (party that initiates an exchange) is shared, as the user defines the target (e.g. ordering food) and the TDS requests the necessary information about the constraints (e.g. type of food) [9]. Requesting information allows multiple turns, but the dialogue is kept as short as possible. Evaluation metrics for TDSs are accuracy (correct result) and efficiency (number of turns). Evaluation methods include *user satisfaction modelling*, which derives objective properties (e.g. accuracy) from subjective user impressions using frameworks such as PARADISE [64], and *user simulation*, which mimics humans to assess comprehensibility, relevance of responses, user satisfaction and task performance. Most TDSs use the semantic output for agenda-based user simulation (ABUS) [51], while newer ones use the system output for neural user simulation (NUS) [24].

Conversational Dialogue Systems (CDSs) are designed to have long-term social conversations without solving a specific task [9,54]. Social conversations require extensive analysis (of content, user personality, emotions, mood, and background), system consistency (no inconsistencies in personality, language style or content) and interactivity, resulting in complex systems [20]. The dialogue is unstructured and aims to emulate natural human conversations. A response generation engine computes the response $Y_t \in \Omega$ out of the response-space Ω from the current utterance X_t and the dialogue context C_t. In *retrieval-based methods*, Ω consists of a corpus of predefined utterances, and the response Y_t is produced by first ranking Ω based on X_t and C_t, and then selecting an element from some top subset. Ranking can be achieved with traditional learning-to-rank methods [32] or modern neural models [15,20,21,34] *Generation-based methods* use a corpus V of predefined words (dictionary). The response-space $\Omega = V^m$ is large, where m is the response length in words. These methods are mainly implemented by neural models [20] such as sequence-to-sequence models [55,57,58,63], conditional variational autoencoders [77], generative adversarial networks [28,72] or transformers [71,75]. *Hybrid methods* combine the advantages of retrieval-based methods (grammatically correct, well-structured responses of high quality) and generation-based methods (large response-space) [69,73]: a response selected from a corpus of predefined utterances is chosen and adapted using a corpus of predefined words [20].

As the aim of CDSs is to entertain the user, interactivity is required and the initiative is shared. Emulating social interactions leads to long dialogues [9].

There has been no agreement on how to evaluate CDSs due to the goal of user entertainment being vaguely defined [11]. Manual evaluation is time-consuming, costly and subjective [20]. Existing methods either use Turing test techniques [33] or evaluate the appropriateness of the generated responses [11].

Question Answering Dialogue Systems (QADSs) are designed to answer specific questions (e.g. extract information from an input sheet), often neglecting the naturalness of the answers generated [9]. The dialogue is unstructured but follows the question and answer style [9]. The initiative is user-centred, with the user posing a specific question.

The conversation is kept brief, with three distinct approaches being identified: *Single turn QADSs* respond without queries and are often used for simple tasks such as extracting information. While open QADSs use web pages or external corpora as knowledge sources [13], the more common reading comprehension extracts information from a document [9]. Modern approaches use pre-trained large language models such as BERT [10] or XLNet [74] (see Sect. 4.4). *Context QADSs* (also known as multi-turn or sequential QADSs) break down complex tasks into simple questions using follow-up questions and are used in reading comprehension to extract quotations. They often consist of single turn QADSs with extended input for dialogue flow [9]. *Interactive QADSs* combine context QADS with TDS and primarily coordinate constraints (e.g. pages to search). Few (many) constraints lead to many (few) results and are therefore added (removed) [9].

The evaluation metrics for QADS are accuracy and, less commonly, dialogue flow. Single turn QADSs are evaluated by mean average precision, mean reciprocal rank or F-Score, while context QADSs are mostly evaluated qualitatively by hand [9,53].

3.2 Modality

Modality refers to the channels of communication between humans and computers that can be text-based, speech-based, or multi-modal [30]. Multi-modal interfaces enable more expressive, efficient, and robust interaction by processing multimedia [45]. Dialogue systems can use different modalities for input and output.

3.3 Domain

A domain is the topic or specific area of knowledge that a dialogue system covers. *Single-domain* dialogue systems are restricted to a specific domain, such as a flight booking system. *Multi-domain* dialogue systems are restricted to several specific domains, e.g. weather, news, appointments and reminders. Single-domain and multi-domain dialogue systems are equipped with special knowledge bases tailored to the respective domains. *Open-domain* dialogue systems are not restricted to specific domains, but can in principle answer questions from any domain. On the other hand, they are usually unable to convey specific knowledge, tend to give ambiguous or incorrect answers due to incorrect semantic analysis, and are usually unable to handle complex, multi-step tasks such as booking a flight.

3.4 Architecture

The architecture of a dialogue system can be either modular (also called pipelined), where each task is performed by a separate module, or end-to-end, where intermediate results are not interpretable and outputs are generated directly from inputs [5,14,44]. In addition, the literature contains end-to-end modules, integrated systems that bypass interpretable intermediate results within the module.

Modular Dialogue Systems consist of different modules (see Fig. 1) [44,54], which are described in more detail below.

Automatic speech recognition (ASR) transcribes the audio signal. As ASR is not part of the dialogue system, we will not go into further detail, but refer to [37] for a comprehensive overview of the state of the art.

Natural language understanding (NLU) extracts information from written text by performing intent and semantic analysis (see Table 2), also known as slot-filling [31,54]. The specific slots that need to be filled are not predetermined.

The *dialogue manager (DM)* is responsible for controlling the flow of the conversation, selecting appropriate responses based on the current context and user input, and maintaining the overall coherence and relevance of the dialogue.

Dialogue state tracking (DST) is the first module of the DM and computes the current dialogue state using the dialogue history [5,44] and either the output of NLU [17,18,70] or of ASR [19,22,66]. States are computed by slot-filling, but

Table 2. Extracted information for the example sentence *show restaurant at New York tomorrow* by the module natural language understanding [5]

Sentence	show	restaurant	at	New	York	tomorrow
Slots	O	O	O	B-desti	I-desti	B-date
Intent	Find Restaurant					
Category	Order					

unlike NLU, the slots are known in advance. States consist of target constraints (possible values: unimportant, not named or user defined), a list of requested slots and the search method (possible values: by constraints (user specifies the information about the requested slots), by alternatives (user wants an alternative for the requested slot), finished (user ends the dialogue)) [44].

Dialogue response action selection (DRAS) is the second module of the DM and selects for the dialogue state S_t at time t either an action from the action-space $A = \{a_1, ..., a_n\}(f : S_t \rightarrow a_i \in A)$ [44] or a request for missing information about necessary constraints to perform an action [5]. Recent DRAS implementations largely use reinforcement learning [16,44].

Natural language generation (NLG) generates a human-readable utterance from the action selected by DRAS. Pipeline structures of content planning, sentence planning and concrete realisation [50] have largely been replaced by end-to-end approaches [60,68,78].

Text to speech (TTS) converts the text output of NLG into an audio output. As TTS is not part of the dialogue system, we will not go into detail, but refer to [59] for a comprehensive overview of the state of the art.

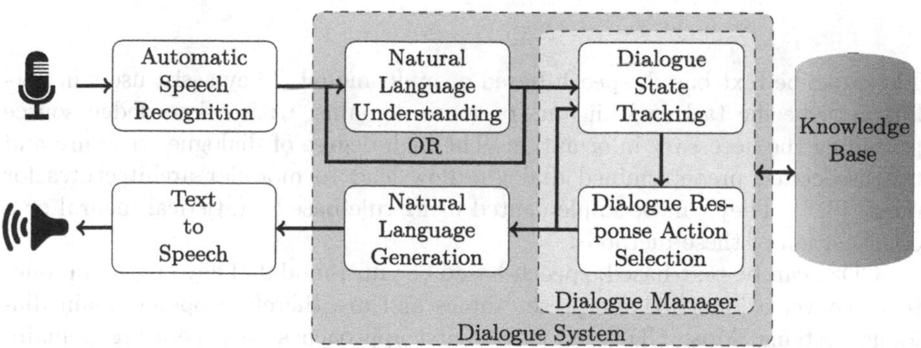

Fig. 1. Structure of a modular speech-based dialogue system [44,54]

End-to-end Dialogue Systems address the drawbacks of modular systems: modules must be well matched, improving individual modules may not improve the whole dialogue system [5,44], and training the system with backpropagation requires all modules to be differentiable (which requires a gradient computation) [2,76]. Hence, current designs use either differentiable modules [26], facing the problem that knowledge retrieval is not differentiable [44], or end-to-end architectures that generate answers without interfaces/intermediate results from a discrete action-space or by statistical means, making the system more domain independent [54].

3.5 Model

Dialogue systems are usually created using artificial intelligence techniques. Rule-based models use symbolic artificial intelligence, where fixed sets of rules are implemented [5,41]. Statistical models using machine learning and neural models using deep learning, on the other hand, are data-driven approaches [6,44] that are trained on dialogue or speech corpora [35,54].

3.6 Correlations Among Categories

We have divided dialogue systems into categories, although it should be noted that the categories are partially correlated. To illustrate their relationships, Table 3 provides an overview of the main correlations, grouped by objectives.

Table 3. Correlations of the categories of dialogue systems, grouped by objective

Objective	Modality	Domain	Architecture	Model
TDS	text/speech/multi	single/multi	modular	rule/statistical/neural
CDS	text/speech/multi	open	end-to-end	neural
QADS	text	multi/open	modular/end-to-end	rule/statistical/neural

TDSs can be text-based, speech-based or multi-modal. They assist users in performing specific tasks within one or more domains, with a knowledge source providing the necessary information. The high degree of dialogue structure and the associated predetermined dialogue flow lead to modular architectures for most TDSs. They can be implemented using rule-based, statistical, neural or a combination of these methods.

CDSs can be text-based, speech-based or multi-modal. They engage in long-term conversations without preset topics and are therefore open-domain dialogue systems. Most CDSs are end-to-end approaches to generating domain-independent responses, but modular systems similar to Fig. 1 are also possible [20]. End-to-end approaches are most often implemented using neural models because of their ability to handle the high complexity of such systems.

QADSs are mostly text-based, but can also be speech-based or multi-modal. They are either tailored to specific domains or designed to handle questions from open domains. The architecture and method of implementing the functions of QADSs vary according to the specific purpose of the dialogue system.

4 Applications and Frameworks

Dialogue systems have become ubiquitous, e.g. in the form of emotional care robots, website guides and telephone assistants [31]. While applications are ready-to-use systems, frameworks provide a development environment for creating applications. In this section, we provide a detailed description of the four main categories and recommend applications or frameworks for each. *Intelligent virtual assistants (IVAs)* are applications (either virtual embodied agents or voice assistants), typically developed by large companies, and provide personalised answers or actions for various problems in real time. *Commercial frameworks* refer to frameworks used by companies to develop their own applications and integrate them into their business. *Research dialogue systems* are applications or frameworks that are developed for the purpose of advancing technology, investigating new functionalities, and addressing existing challenges. Table 4 gives an overview of these three categories. *Large language models (LLMs)* are models, sometimes applications, that can represent the semantics and syntax of human language more accurately than traditional machine learning models. Besides dialogue systems, LLMs are used in NLP for a variety of other tasks. They do not fully fit into the categories outlined in Table 4 and have therefore been excluded.

4.1 Intelligent Virtual Assistants (IVAs)

IVAs, also called voice assistants, are usually freely available in multiple languages, not open source TDSs and designed to provide users with a quick and effective way to interact with technology through a text or speech interface. Challenges include recognising wake words (e.g. "Hey Siri"), assisting users with a variety of tasks and providing instant information retrieval with little effort [9], simplifying daily tasks and activities. The top part of Table 4 contains a selection of the most commonly used IVAs. Amazon's Alexa, Samsung's Bixby, Microsoft's Cortana, Google's Assistant and Apple's Siri differ mainly in the platforms and integrations they can be used on, which are based on each company's devices. Due to the similarity of these systems, we recommend using the IVA that best matches the researcher's existing devices to get a first insight into their functions. XiaoIce is different in that it is a Chinese CDS optimised for long-term user engagement [79].

4.2 Commercial Frameworks

Commercial frameworks (middle part of Table 4) are designed to easily integrate an interactive interface into business applications to elicit responses or perform actions. The diversity of customer companies leads to a low-code or no-code policy for all frameworks except Rasa. A free trial is available for all frameworks except Dragon and SemVox. The ontology used in commercial frameworks to create TDSs consists of *Intents* and *Entities* (also called *Concepts*), which are learned through examples. An intent is the goal or purpose behind a user's input, while entities are lists of specific information relevant to fulfilling that

Table 4. Selected intelligent virtual assistants, commercial frameworks and research dialogue systems categorised by objective, modality, domain, architecture and model. (•) means it applies, (-) not applicable, (x) no information available.

		Objective			Modality			Domain			Architecture		Model				
	System name	Task-oriented	Conversational	Question Answering	Text	Speech	Multi-modal	Single	Multi	Open	Modular	End-to-end	Rule-based	Statistical	Neural	Open source	Specifications
Intelligent Virtual Assistants	Amazon Alexa[a]	•	-	-	•	•	-	-	•	-	-	•	•	•	•	-	
	Bixby[b]	•	-	-	•	•	-	-	•	-	x	x	•	•	•	-	
	Cortana [48]	•	-	-	•	•	-	-	•	-	•	-	•	•	•	-	
	Google Assistant[c]	•	-	-	•	•	-	-	•	-	x	x	•	•	•	-	
	Siri[d]	•	-	-	•	•	-	-	•	-	x	x	•	•	•	-	
	XiaoIce [79]	-	•	-	•	•	-	-	-	•	•	-	•	-	•	-	Emotional module, Chinese
Commercial Frameworks	Cerence Studio[e]	•	-	-	•	•	-	-	•	-	•	-	-	•	•	-	
	Conversational AI[f]	•	-	-	•	•	-	-	•	-	-	•	x	x	x	-	
	Dialogflow[g]	•	-	-	•	•	-	-	•	-	-	•	•	-	•	-	
	Dragon[h]	-	-	-	-	•	-	-	•	•	x	x	-	-	•	-	ASR (dictation software)
	LUIS[i]	•	-	-	•	•	-	-	•	-	x	x	-	-	•	-	
	Nuance Mix[j]	•	-	•	•	•	-	-	•	•	•	-	•	-	•	-	
	Rasa[k]	•	-	-	•	-	-	-	•	-	•	-	-	•	•	•	NLU&DM, also research system
	SemVox [3]	•	-	-	-	•	•	-	•	-	•	-	x	x	x	-	also IVA
	Watson Assistant[l]	•	-	•	•	•	-	-	•	-	•	-	-	-	•	-	
Research Dialogue Systems	DenSPI [52]	-	-	•	•	-	-	-	-	•	-	•	-	-	•	•	
	DrQA [4]	-	-	•	•	-	-	-	-	•	•	-	•	-	•	•	
	R^3 [65]	-	-	•	•	-	-	-	-	•	•	-	-	-	•	•	
	SmartWeb [56]	-	-	•	•	•	•	-	•	-	•	-	•	-	-	•	
	ELIZA [67]	-	•	-	•	-	-	-	-	•	•	-	•	-	-	•	First dialogue system (1966)
	ConvLab [27]	•	-	-	•	-	-	-	•	-	•	•	-	-	•	•	GUI included
	DialogOS [23]	•	-	-	-	•	-	-	•	-	•	-	•	-	-	•	GUI included
	Nemo [25]	•	•	•	-	•	-	-	-	•	•	-	-	-	•	•	
	ParlAI [40]	•	•	•	•	-	•	-	-	•	-	•	-	-	•	•	
	Plato [47]	•	-	-	•	-	-	-	•	-	•	•	•	•	-	•	Speech interface integrable
	PyDial [61]	•	-	-	•	•	-	-	•	-	•	-	-	•	-	•	
	ReTiCo [38,39]	•	-	-	•	•	-	-	•	-	•	-	-	•	•	•	DM based on Rasa
	Siam-dp [42]	•	-	•	-	•	•	-	•	-	•	-	•	-	-	•	

[a] https://developer.amazon.com/alexa
[b] https://bixbydevelopers.com
[c] https://developers.google.com/assistant
[d] https://developer.apple.com/siri
[e] https://developer.cerence.com/landing
[f] https://cai.tools.sap
[g] https://cloud.google.com/dialogflow
[h] https://www.nuance.com/dragon.html
[i] https://www.luis.ai
[j] https://docs.nuance.com/mix
[k] https://rasa.com/
[l] https://www.ibm.com/products/watson-assistant

intent. The example sentence "I want a coffee" could be mapped to the intent ORDER_DRINK with the entity DRINK_TYPE and it's value (from a predefined list) set to COFFEE.

The nine commercial frameworks can be summarised as follows: SAP's Conversational AI, Google's Dialogflow, Microsoft's LUIS (next version: CLU) and IBM's Watson Assistant allow easy integration of dialogue systems into applications such as Twitter, Facebook, Slack, etc., do not offer separate access to intermediate results, and the last three frameworks require (even to use the free trial) further personal data such as phone number or payment option. Cerence Studio, Dragon and Nuance Mix were originally developed by Nuance Communications, a leading provider of speech processing solutions for businesses and consumers. Dragon is not a framework for dialogue systems, but one of the leading dictation software. Cerence Studio and Nuance Mix use similar graphical user interfaces (GUIs) and workflows, both providing separate access to each module of the dialogue system. SemVox allows multi-modal interfaces and also uses the ASR and TTS modules from Cerence Studio. Unlike the other commercial frameworks, Rasa can be used offline and is open source. Rasa is very popular in the research community and is used for many projects such as ReTiCo [38, 39].

To help developers and researchers get started with commercial frameworks, we recommend Cerence Studio because it offers a state-of-the-art demo version that allows the retrieval of intermediate results from any dialogue system module, has an intuitive GUI, and provides straightforward tutorials. It is also free to use and does not require any personal information other than name and email address. The workflow is divided into two steps, using pre-trained ASR and TTS modules. For each intent of the NLU module, the user creates example phrases with associated concepts in the *.nlu* tab to train the model. The DST, DRAS and NLG modules are trained in the *.dialog* tab. The user creates a table with example sentences, collected concepts, actions, and answers/requests. Deploying an application generates an app_id, an app_key and a context_tag, which are used to access the services via a WebSocket connection (a GUI-enabled JavaScript client is provided for testing).

4.3 Research Dialogue Systems

Research dialogue systems are usually open source and serve either as frameworks for research environments or as applications developed to present research results. The bottom part of Table 4 provides a brief summary of the research dialogue systems identified during our structured literature review. Table 5 compares these systems in terms of features relevant to implementation and use.

Large companies have become interested in researching dialogue systems as the technology has been integrated into various fields such as healthcare, education, transport and communication: DrQA and ParlAI were developed by Facebook, R^3 by IBM, ConvLab by Microsoft, Nemo by Nvidia and Plato by Uber. DialogOS was developed by the smaller company CLT Sprachtechnologie GmbH. The pioneering ELIZA dialogue system (1966) and ReTiCo were developed by

Table 5. Properties relevant for implementation of research dialogue systems: (•) means applicable, (-) not applicable, as of status 04/2023

System	Application	Framework	Code Availability	Programming Language	Windows	MacOS	Linux	Tutorial	Demonstration	Last Activity (Releases, Commits, Issues)	Purpose
DenSPI	•	-	•	Python	•	•	•	-	•	06/2022	Real-time, Wikipedia-based
DrQA	•	-	•	Python	-	•	•	-	•	11/2022	Machine reading at scale
R^3	•	-	•	Lua, Python	•	•	•	-	-	04/2018	Reinforcement learning
SmartWeb	•	-	-	Java	•	•	•	-	-	2014	Multi-modal access to Web
ELIZA	•	•	•	Python	•	•	•	-	•	1966	Simulate interlocutors
ConvLab	-	•	•	Python	-	•	•	•	-	04/2023	Reusable experimental setup
DialogOS	-	•	•	Java	•	•	•	-	-	12/2022	Student projects
Nemo	-	•	•	Python	•	•	•	•	-	04/2023	Reuse code and models
ParlAI	-	•	•	Python	-	•	•	•	•	04/2023	Share, train and test systems
Plato	-	•	•	Python	•	•	•	•	-	09/2020	Multi-agent setting
PyDial	-	•	•	HTML	•	•	•	•	•	04/2022	Research on modules
ReTiCo	-	•	•	Python	•	•	•	-	-	04/2023	Real time, incremental
Siam-dp	-	•	•	Java	•	•	•	-	-	02/2017	Multi-modal development

a single researcher, while DenSPI, SmartWeb, PyDial and Siam-dp were developed by research institutions. Typically, research focused on dialogue systems is designed to investigate specific research questions, often resulting in small, specialised projects that may still be in the development phase when published and become inactive shortly afterwards. For a first insight into research, we recommend using the DrQA application and the NeMo framework. Both are state of the art, fully functional, user friendly, actively used and provide tutorials or demonstrations.

DrQA is an open-domain QADS consisting of a *Document Retriever*, which selects relevant documents from a large unstructured knowledge source (e.g. Wikipedia), and a *Document Reader*, which searches them to answer the given question. The installation process includes the following steps: (1) clone repository, (2) set up virtual environment, (3) install requirements, (4) configure CoreNLP tokeniser, (5) install Document Retriever, (6) install Document Reader, (7) install Wikipedia snapshot, (8) download evaluation datasets. We provide a Docker image[5] to set up DrQA with a single command. Users can combine, train and test different tokenisers, Document Retrievers, and Document Readers on supplied datasets via the command line. DrQA supports the development of new Document Retriever and Document Reader models, accompanied by documentation.

[5] `docker exec -it bengt/drqa venv/bin/python scripts/pipeline/interactive.py`.

NeMo is a framework for building dialogue system applications through reusability, abstraction, and composition. Detailed installation instructions are provided with the software, and numerous Google Colab tutorials are available to help users get started. The question-answering tutorial for example provides step-by-step instructions from the installation to the usage of LLMs to extract or generate answers. Using the datasets SQuAD [49] and MS MARCO [43], the pre-trained LLMs are fine-tuned and evaluated. NeMo supports the creation of experiments and applications by using different models, varying hyperparameters, and implementing other components.

4.4 Large Language Models (LLMs)

While there is no formal definition, LLMs typically refer to pre-trained deep transformer models [62] with a vast number of parameters that can generate human-like language. They can perform a range of language-related tasks, including text summarisation, sentiment analysis and translation. Training LLMs from scratch is costly due to the vast number of parameters and the large text datasets, so it is primarily suitable for large companies. Figure 2 shows recently developed LLMs along with their model size and developer. To get a practical insight into the performance of LLMs in the field of dialogue systems, we suggest ChatGPT[6], which is free to use, although it requires users to provide an email address and a phone number. LLM research currently focuses on multi-modality: Google devel-

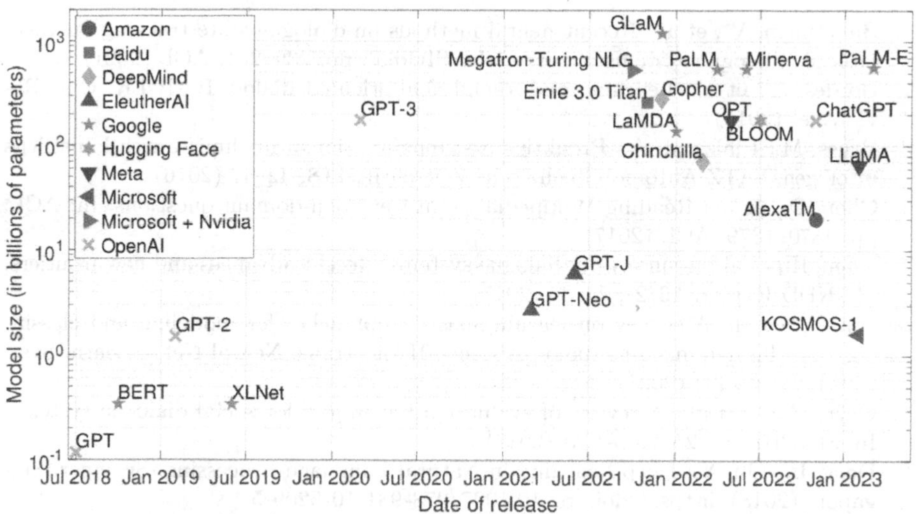

Fig. 2. Comparison of selected large language models developed since 2018, with their model size and developer. GPT-4, released in March 2023, was excluded due to an unknown number of parameters.

[6] https://openai.com/blog/chatgpt.

oped PaLM-E [12] by integrating vision transformers with PaLM, and OpenAI extended GPT-3 to include vision data processing in GPT-4 [46].

5 Conclusion and Future Work

The aim of this paper is to provide researchers with easy access to the field of dialogue systems. We presented a structured literature review and provided a list of relevant papers, articles and books covering all relevant technical topics. The main findings of our literature review have been presented through the clarification of terminology and the derived categories of objective, modality, domain, architecture and model and their main correlations. To facilitate practical entry into dialogue systems research, we have described the four main application and framework categories and recommended dialogue systems for each category: Cerence Studio as a commercial framework, DrQA and NeMo as research dialogue systems, and ChatGPT as an large language model. For intelligent virtual assistants, our recommendation depends on the device used.

Future work includes extending the structured literature review to other knowledge bases and extending the search terms to include synonyms identified in the literature. In addition, a performance comparison based on a benchmark dataset will allow a more accurate comparison of the dialogue systems.

References

1. Balaraman, V., et al.: Recent neural methods on dialogue state tracking for task-oriented dialogue systems: a survey. In: SIGdial. pp. 239–251. ACL (2021)
2. Bordes, A., et al.: Learning end-to-end goal-oriented dialog. In: ICLR. OpenReview.net (2017)
3. Bruss, M., Pfalzgraf, A.: Proaktive assistenzfunktionen für hmis durch künstliche intelligenz. ATZ Automobiltechnische Zeitschrift **118**, 42–47 (2016)
4. Chen, D., et al.: Reading Wikipedia to answer open-domain questions. In: ACL, pp. 1870–1879. ACL (2017)
5. Chen, H., et al.: A survey on dialogue systems: Recent advances and new frontiers. SIGKDD Explor. **19**(2), 25–35 (2017)
6. Cui, F., et al.: A survey on learning-based approaches for modeling and classification of human-machine dialog systems. IEEE Trans. Neural Netw. Learn. Syst. **32**(4), 1418–1432 (2021)
7. Curry, A.C., et al.: A review of evaluation techniques for social dialogue systems. In: SIGCHI, pp. 25–26. ACM (2017)
8. Deng, L., Liu, Y.: Deep Learning in Natural Language Processing. Springer, Singapore (2018). https://doi.org/10.1007/978-981-10-5209-5
9. Deriu, J., et al.: Survey on evaluation methods for dialogue systems. Artif. Intell. Rev. **54**(1), 755–810 (2021)
10. Devlin, J., et al.: BERT: pre-training of deep bidirectional transformers for language understanding. In: NAACL-HLT, pp. 4171–4186. ACL (2019)
11. Dinan, E., et al.: The second conversational intelligence challenge (convai2). CoRR abs/1902.00098 (2019)

12. Driess, D., et al.: PaLM-E: an embodied multimodal language model. CoRR abs/2303.03378 (2023)
13. Fader, A., et al.: Paraphrase-driven learning for open question answering. In: ACL, pp. 1608–1618. ACL (2013)
14. Fan, Y., Luo, X.: A survey of dialogue system evaluation. In: 32nd IEEE, ICTAI, pp. 1202–1209. IEEE (2020)
15. Fan, Y., et al.: MatchZoo: a toolkit for deep text matching. CoRR abs/1707.07270 (2017)
16. Henderson, J., et al.: Hybrid reinforcement/supervised learning of dialogue policies from fixed data sets. Comput. Linguist. **34**(4), 487–511 (2008)
17. Henderson, M., et al.: The second dialog state tracking challenge. In: SIGDIAL, pp. 263–272 (2014)
18. Henderson, M., et al.: The third dialog state tracking challenge. In: SLT, pp. 324–329. IEEE (2014)
19. Hu, J., et al.: SAS: dialogue state tracking via slot attention and slot information sharing. In: ACL, pp. 6366–6375. ACL (2020)
20. Huang, M., et al.: Challenges in building intelligent open-domain dialog systems. ACM Trans. Inf. Syst. **38**(3), 21:1-21:32 (2020)
21. Huang, P., et al.: Learning deep structured semantic models for web search using clickthrough data. In: ACM, pp. 2333–2338. ACM (2013)
22. Kim, S., et al.: Efficient dialogue state tracking by selectively overwriting memory. In: ACL, pp. 567–582. ACL (2020)
23. Koller, A., et al.: DialogOS: simple and extensible dialogue modeling. In: Interspeech, pp. 167–168. ISCA (2018)
24. Kreyssig, F., et al.: Neural user simulation for corpus-based policy optimisation of spoken dialogue systems. In: SIGdial, pp. 60–69. ACL (2018)
25. Kuchaiev, O., et al.: Nemo: a toolkit for building AI applications using neural modules. CoRR abs/1909.09577 (2019)
26. Le, H., et al.: Uniconv: a unified conversational neural architecture for multi-domain task-oriented dialogues. In: EMNLP, pp. 1860–1877. ACL (2020)
27. Lee, S., et al.: ConvLab: multi-domain end-to-end dialog system platform. In: ACL, pp. 64–69. ACL (2019)
28. Li, J., et al.: Adversarial learning for neural dialogue generation. In: EMNLP, pp. 2157–2169. ACL (2017)
29. Li, X., et al.: A review of quality assurance research of dialogue systems. In: AITest, pp. 87–94. IEEE (2022)
30. Liu, G., et al.: A survey on multimodal dialogue systems: recent advances and new frontiers. In: AEMCSE, pp. 845–853 (2022)
31. Liu, J., et al.: Review of intent detection methods in the human-machine dialogue system. J. Phys. Conf. Ser. **1267**(1), 012059 (2019)
32. Liu, T.: Learning to rank for information retrieval. In: SIGIR, p. 904. ACM (2010)
33. Lowe, R., et al.: Towards an automatic turing test: learning to evaluate dialogue responses. In: ACL, pp. 1116–1126. ACL (2017)
34. Lu, Z., Li, H.: A deep architecture for matching short texts. In: NeurIPS, pp. 1367–1375 (2013)
35. Ma, L., et al.: Unstructured text enhanced open-domain dialogue system: a systematic survey. ACM Trans. Inf. Syst. **40**(1), 9:1-9:44 (2022)
36. Ma, Y., et al.: A survey on empathetic dialogue systems. Inf. Fus. **64**, 50–70 (2020)
37. Malik, M., et al.: Automatic speech recognition: a survey. Multim. Tools Appl. **80**(6), 9411–9457 (2021)

38. Michael, T.: ReTiCo: an incremental framework for spoken dialogue systems. In: SIGdial, pp. 49–52. ACL (2020)
39. Michael, T., Möller, S.: ReTiCo: an open-source framework for modeling real-time conversations in spoken dialogue systems. In: ESSV, pp. 134–140 (2019)
40. Miller, A.H., et al.: ParlAI: a dialog research software platform. In: EMNLP, pp. 79–84. ACL (2017)
41. Motger, Q., et al.: Software-based dialogue systems: survey, taxonomy, and challenges. ACM Comput. Surv. **55**(5), 1–42 (2022)
42. Nesselrath, R., Feld, M.: SiAM-dp: a platform for the model-based development of context-aware multimodal dialogue applications. In: IE, pp. 162–169. IEEE (2014)
43. Nguyen, T., et al.: MS MARCO: a human generated machine reading comprehension dataset. In: NeurIPS, vol. 1773. CEUR-WS.org (2016)
44. Ni, J., et al.: Recent advances in deep learning based dialogue systems: a systematic survey. CoRR abs/2105.04387 (2021)
45. Obrenovic, Z., Starcevic, D.: Modeling multimodal human-computer interaction. Computer **37**(9), 65–72 (2004)
46. OpenAI: GPT-4 technical report. CoRR abs/2303.08774 (2023)
47. Papangelis, A., et al.: Plato dialogue system: a flexible conversational AI research platform. CoRR abs/2001.06463 (2020)
48. Paul, Z.: Cortana-intelligent personal digital assistant: a review. Int. J. Adv. Res. Comput. Sci. **8**, 55–57 (2017)
49. Rajpurkar, P., et al.: SQuAD: 100,000+ questions for machine comprehension of text. In: EMNLP, pp. 2383–2392. ACL (2016)
50. Reiter, E.: Has a consensus NL generation architecture appeared, and is it psycholinguistically plausible? In: INLG (1994)
51. Schatzmann, J., et al.: Agenda-based user simulation for bootstrapping a POMDP dialogue system. In: NAACL HLT, pp. 149–152. ACL (2007)
52. Seo, M.J., et al.: Real-time open-domain question answering with dense-sparse phrase index. In: ACL, pp. 4430–4441. ACL (2019)
53. Serban, I.V., et al.: A hierarchical latent variable encoder-decoder model for generating dialogues. In: AAAI, pp. 3295–3301. AAAI Press (2017)
54. Serban, I.V., et al.: A survey of available corpora for building data-driven dialogue systems: the journal version. Dialogue Discourse **9**(1), 1–49 (2018)
55. Shang, L., et al.: Neural responding machine for short-text conversation. In: ACL, pp. 1577–1586. ACL (2015)
56. Sonntag, D.: Ontologies and Adaptivity in Dialogue for Question Answering, Studies on the Semantic Web, vol. 4. IOS Press (2010)
57. Sordoni, A., et al.: A neural network approach to context-sensitive generation of conversational responses. In: NAACL HLT, pp. 196–205. ACL (2015)
58. Sutskever, I., et al.: Sequence to sequence learning with neural networks. In: NeurIPS, pp. 3104–3112 (2014)
59. Tan, X., et al.: A survey on neural speech synthesis. CoRR abs/2106.15561 (2021)
60. Tran, V.-K., Nguyen, L.-M.: Semantic refinement GRU-based neural language generation for spoken dialogue systems. In: Hasida, K., Pa, W.P. (eds.) PACLING 2017. CCIS, vol. 781, pp. 63–75. Springer, Singapore (2018). https://doi.org/10.1007/978-981-10-8438-6_6
61. Ultes, S., et al.: PyDial: a multi-domain statistical dialogue system toolkit. In: ACL, pp. 73–78. ACL (2017)
62. Vaswani, A., et al.: Attention is all you need. In: NeurIPS, pp. 5998–6008 (2017)
63. Vinyals, O., Le, Q.: A neural conversational model. CoRR abs/1506.05869 (2015)

64. Walker, M.A., et al.: PARADISE: a framework for evaluating spoken dialogue agents. In: ACL, pp. 271–280. ACL (1997)
65. Wang, S., et al.: R^3: reinforced ranker-reader for open-domain question answering. In: AAAI, pp. 5981–5988. AAAI Press (2018)
66. Wang, Y., et al.: Slot attention with value normalization for multi-domain dialogue state tracking. In: EMNLP, pp. 3019–3028. ACL (2020)
67. Weizenbaum, J.: ELIZA - a computer program for the study of natural language communication between man and machine. Commun. ACM **9**(1), 36–45 (1966)
68. Wen, T., et al.: Stochastic language generation in dialogue using recurrent neural networks with convolutional sentence reranking. In: SIGDIAL, pp. 275–284. ACL (2015)
69. Weston, J., et al.: Retrieve and refine: improved sequence generation models for dialogue. In: SCAI, pp. 87–92. ACL (2018)
70. Williams, J.D., et al.: The dialog state tracking challenge. In: SIGDIAL, pp. 404–413. ACL (2013)
71. Wolf, T., et al.: TransferTransfo: a transfer learning approach for neural network based conversational agents. CoRR abs/1901.08149 (2019)
72. Xu, J., et al.: Diversity-promoting GAN: a cross-entropy based generative adversarial network for diversified text generation. In: EMNLP, pp. 3940–3949. ACL (2018)
73. Yang, L., et al.: A hybrid retrieval-generation neural conversation model. In: CIKM, pp. 1341–1350. ACM (2019)
74. Yang, Z., et al.: XLNet: generalized autoregressive pretraining for language understanding. In: NeurIPS, pp. 5754–5764 (2019)
75. Zhang, Y., et al.: DIALOGPT: large-scale generative pre-training for conversational response generation. In: ACL, pp. 270–278. ACL (2020)
76. Zhao, T., Eskénazi, M.: Towards end-to-end learning for dialog state tracking and management using deep reinforcement learning. In: SIGDIAL, pp. 1–10. ACL (2016)
77. Zhao, T., et al.: Learning discourse-level diversity for neural dialog models using conditional variational autoencoders. In: ACL, pp. 654–664. ACL (2017)
78. Zhou, H., et al.: Context-aware natural language generation for spoken dialogue systems. In: COLING, pp. 2032–2041. ACL (2016)
79. Zhou, L., et al.: The design and implementation of xiaoice, an empathetic social chatbot. Comput. Linguist. **46**(1), 53–93 (2020)

Cost-Sensitive Best Subset Selection for Logistic Regression: A Mixed-Integer Conic Optimization Perspective

Ricardo Knauer(✉) and Erik Rodner

KI-Werkstatt, University of Applied Sciences Berlin, Berlin, Germany
ricardo.knauer@htw-berlin.de
https://kiwerkstatt.htw-berlin.de

Abstract. A key challenge in machine learning is to design interpretable models that can reduce their inputs to the best subset for making transparent predictions, especially in the clinical domain. In this work, we propose a certifiably optimal feature selection procedure for logistic regression from a mixed-integer conic optimization perspective that can take an auxiliary cost to obtain features into account. Based on an extensive review of the literature, we carefully create a synthetic dataset generator for clinical prognostic model research. This allows us to systematically evaluate different heuristic and optimal cardinality- and budget-constrained feature selection procedures. The analysis shows key limitations of the methods for the low-data regime and when confronted with label noise. Our paper not only provides empirical recommendations for suitable methods and dataset designs, but also paves the way for future research in the area of meta-learning.

Keywords: cost-sensitive · best subset selection · mixed-integer conic optimization · interpretable machine learning · meta-learning

1 Introduction

Transparency is of central importance in machine learning, especially in the clinical setting [16,19]. The easier it is to comprehend a predictive model, the more likely it is to be trusted and applied in practice. Large feature sets decrease the comprehensibility and can therefore be problematic, which is particularly relevant for intrinsically interpretable models like logistic regression. Furthermore, it is often not reasonable to collect large numbers of features for predictive models in the first place. Consider a situation in which a physician would like to apply a decision support system that relies on a test battery of 5 questionnaires, 5 physical examination tests and 10 other assessments - measurement time constraints would make it impossible to utilize such a system in practice. Feature selection procedures therefore typically aim to reduce the feature set to the k best predictors without sacrificing performance. The most intuitive way to select

D. Seipel and A. Steen (Eds.): KI 2023, LNAI 14236, pp. 114–129, 2023.
https://doi.org/10.1007/978-3-031-42608-7_10

the best subset of k features is to perform an exhaustive search over all possible feature combinations [10,23]. However, this problem is NP-hard [42], which is why exhaustive enumeration is computationally intractable except for small predictor (sub)set sizes. There is a need for feature selection procedures that can find the best subset among all possible subsets for moderate (sub)set sizes, while taking into account the cost to collect predictors in practice [23,39], such as the limited amount of time for measuring them [16,45]. In this work, we therefore focus on cost-sensitive best subset selection for an intrinsically interpretable machine learning model, logistic regression, which is often considered as a default for predictive modeling due to its simplicity [11,38,45].

The key contributions of our paper are as follows:

1. We propose a **cost-sensitive best subset selection** for logistic regression given a budget constraint, such as the available time to obtain predictors in practice. To the best of our knowledge and although there is a vast literature on feature selection (Sect. 2), we provide the first mixed-integer conic optimization formulation for this problem (Sect. 3.2).
2. We show how to generate **synthetic data for prognostic models** that can be used as a blueprint for future research, based on sensible parameter settings from the relevant clinical literature (Sect. 4).
3. This allows us to provide an **extensive evaluation of both heuristic and optimal cardinality- and budget-constrained feature selection procedures** with recommendations for the feature selection strategy, outcome events per variable, and label noise level, and is also an important step towards meta-learning and foundation models in this domain (Sect. 5).

2 Related Work and Preliminary Definitions

In the following, we provide a brief overview about heuristics that have been proposed over the years to tackle the best subset selection problem without certificates of optimality. We then present contemporary approaches that solve the feature selection problem to provable optimality given an explicit cardinality constraint k as well as approaches that consider the cost to measure predictors.

Heuristic Feature Selection. Heuristic feature selection procedures can be divided into three broad categories: filters, wrappers, and embedded methods [23]. Filters reduce the feature set without using the predictive model itself and are frequently based on the Pearson correlation coefficient [24] or univariable classifiers [20]. In comparison to filters, wrappers employ the predictive model as a black box to score feature subsets. Prominent examples are forward or backward selection procedures. In forward selection, the most significant features are added to the predictive model; in backward selection, the most insignificant features are removed from the model [1,21]. Both can be performed either once or stepwise, and are often combined in practice [3,17,31,32]. Finally, embedded methods incorporate some form of subset selection into their training process.

A well-known embedded method is L1-regularization which indirectly selects predictors by shrinking some of the model parameters to exactly zero [47,49]. A more direct way to induce sparsity is to penalize the number of nonzero parameters via L0-regularization [6,14].

Cardinality-Constrained Feature Selection. In practice, it is often desirable to explicitly constrain the number of nonzero parameters to a positive scalar k. Feature selection with such an explicit cardinality constraint has greatly benefited from recent advances in mathematical optimization [8,14,15,46]. Mathematical programming has not only made it possible to select the best subset of k features for moderate (sub)set sizes, but also to deliver certificates of optimality for the feature selection process, at least for some predictive models (see [48] for a survey).

We introduce the following notation. Let $\boldsymbol{X} \in \mathbb{R}^{M\times N}$ be the design matrix of M training examples and N features, $\boldsymbol{x}_m \in \mathbb{R}^N$ a single training example, $\boldsymbol{y} \in \{-1,1\}^M$ the labels, $y_m \in \{-1,1\}$ a single label, $\boldsymbol{\theta} \in \mathbb{R}^N$ the model parameters, $\theta_n \in \mathbb{R}$ a single parameter, $\theta_0 \in \mathbb{R}$ the intercept, $\|\cdot\|_0 \in \mathbb{R}_+$ the number of nonzero parameters, *i.e.*, the L0 "norm", $\lambda \in \mathbb{R}_+$ a regularization hyperparameter for the squared L2 norm, and $k \in \mathbb{R}_+$ a sparsity hyperparameter for the L0 "norm". In [8], the feature selection problem was solved to provable optimality for two predictive models, namely support vector classification with a hinge loss $\ell(y_m, \boldsymbol{\theta}^T\boldsymbol{x}_m + \theta_0) = \max(0, 1 - y_m(\boldsymbol{\theta}^T\boldsymbol{x}_m + \theta_0))$ and logistic regression with a softplus loss $\ell(y_m, \boldsymbol{\theta}^T\boldsymbol{x}_m + \theta_0) = \log(1 + \exp(-y_m(\boldsymbol{\theta}^T\boldsymbol{x}_m + \theta_0)))$, by means of the following optimization objective:

$$\begin{aligned} \boldsymbol{\theta}^*, \theta_0^* = \underset{\boldsymbol{\theta},\theta_0}{\operatorname{argmin}} \quad & \sum_{m=1}^{M} \ell(y_m, \boldsymbol{\theta}^T\boldsymbol{x}_m + \theta_0) + \frac{\lambda}{2}\sum_{n=1}^{N}\theta_n^2 \\ \text{subject to} \quad & \|\boldsymbol{\theta}\|_0 \leq k \quad . \end{aligned} \tag{1}$$

Using pure-integer programming with a cutting-plane algorithm, optimal solutions to eq. (1) could be obtained within minutes for feature set sizes up to 5000 and subset sizes up to 15. A drawback of the employed algorithm is its stochasticity, *i.e.*, it is not guaranteed that successive runs yield similar results. More recently, the cardinality-constrained feature selection problem (1) was approached from a mixed-integer conic optimization perspective for logistic regression, and it was shown that optimal solutions can be found within minutes for features set sizes as large as 1000 and subset sizes as large as 50 [15]. In contrast to [8], the mixed-integer conic solver in [15] is designed to be run-to-run deterministic [41], which makes the feature selection process more predictable. We extend the evaluation of [15] by assessing the cardinality-constrained conic optimization formulation with respect to the interpretability of the selection process (Sect. 5), therefore asking whether this method is worth to use in practice.

Budget-Constrained Feature Selection. In addition to heuristic and cardinality-constrained approaches to best subset selection, there have also been studies that considered the budget, especially the available time, to measure predictors in practice. In [33], easily obtainable predictors were included in a logistic regression model first, before controlling for them and adding significant predictors that were harder to measure. In a patient-clinician encounter, asking questions during history-taking often takes less time than performing tests during physical examination, which is why predictors derived from physical examination were considered last. This hierarchical forward selection was followed by a stepwise backward selection. Similarly, features from history-taking were included and controlled for before features from physical examination in [43].

Next to the aforementioned heuristics, a cost-sensitive best subset selection procedure for support vector classification was proposed in [2]. In particular, the summed costs associated with the selected features were constrained to not exceed a predefined budget. Generalized Benders decomposition was used to allow the mixed-integer program to be solved for moderate feature (sub)set sizes in a reasonable amount of time. The procedure was then assessed on a range of synthetic and real-world problems, with each cost set to one and the budget set to k, *i.e.*, with an explicit cardinality constraint as a special case of budget constraint (Sect. 3.2). Best subsets could be selected within minutes on many of the problems, for instance with $N = 2000$ and $k = 17$ where similar approaches have struggled [34]. More recently, budget-constrained best subset selection for support vector classification using mixed-integer optimization was also evaluated for moderate (sub)set sizes with individual feature costs [36]. No hyperparameter tuning was performed, which makes it difficult to put the experimental results into context with previous studies as predictive performance and computational time can vary significantly depending on the hyperparameter setting [2,8,15,34].

Overall, there is an extensive literature for heuristic and mathematical optimization approaches to feature selection. Mathematical programming [8,14,15,46] can deliver certifiably optimal solutions for budget-constrained best subset selections, with cardinality-constrained best subset selections as a special case. With respect to logistic regression, a mixed-integer conic optimization perspective seems particularly appealing since it can deal with moderate feature (sub)set sizes while still being run-to-run deterministic [15,41]. To the best of our knowledge, we present the first optimal budget-constrained feature selection for logistic regression from a mixed-integer conic programming perspective (Sect. 3.2) and systematically evaluate it in terms of the interpretability of the selection process (Sect. 5).

3 Best Subset Selection

In this section, we describe two ways to approach optimal feature selection for logistic regression in case of moderate feature (sub)set sizes. We first review how to arrive at the best feature subset given a cardinality constraint. Based on that, we propose a new way to solve the best subset selection problem given a budget constraint, such as the available time to measure predictors in practice.

3.1 Cardinality-Constrained Best Subset Selection

A cardinality-constrained formulation for best subset selection for logistic regression has already been introduced in Eq. (1):

$$\begin{aligned} \boldsymbol{\theta}^*, \theta_0^* = \underset{\boldsymbol{\theta},\theta_0}{\operatorname{argmin}} \quad & \sum_{m=1}^{M} \log\left(1+\exp\left(-y_m(\boldsymbol{\theta}^T \boldsymbol{x}_m + \theta_0)\right)\right) + \frac{\lambda}{2}\sum_{n=1}^{N} \theta_n^2 \\ \text{subject to} \quad & \|\boldsymbol{\theta}\|_0 \le k \quad . \end{aligned} \tag{2}$$

The first term of the optimization objective is known as the softplus (loss) function, the second term adds L2-regularization to prevent overfitting and make the model robust to perturbations in the data [5,7], and the constraint limits the L0 "norm". However, explicitly constraining the number of nonzero parameters is not easy (in fact, it is NP-hard). To model our L2-regularized L0-constrained logistic regression, we choose a conic formulation similar to [15]. This allows us to restructure our optimization problem so that it can be solved within a reasonable amount of time in practice, even for moderate feature (sub)set sizes.

In conic programming, $\boldsymbol{\theta}$ is bounded by a K-dimensional convex cone $\mathcal{K}^K$, which is typically the Cartesian product of several lower-dimensional cones. A large number of optimization problems can be framed as conic programs. With a linear objective and linear constraints, for example, the nonnegative orthant $\mathbb{R}_+^N$ would allow us to solve linear programs [4,9]. For our nonlinear objective, though, we use $2M$ exponential cones as well as N rotated quadratic (or second-order) cones. The exponential cone is defined as follows [40]:

$$\mathcal{K}_{exp} = \{[\vartheta_1, \vartheta_2, \vartheta_3] \,|\, \vartheta_1 \ge \vartheta_2 e^{\vartheta_3/\vartheta_2}, \vartheta_2 > 0\} \cup \{[\vartheta_1, 0, \vartheta_3] \,|\, \vartheta_1 \ge 0, \vartheta_3 \le 0\}\,. \tag{3}$$

With this cone, we can model our softplus function using the auxiliary vectors $\boldsymbol{t}, \boldsymbol{u}, \boldsymbol{v} \in \mathbb{R}^M$:

$$\begin{aligned} &[u_m, 1, -y_m(\boldsymbol{\theta}^T \boldsymbol{x}_m + \theta_0) - t_m], [v_m, 1, -t_m] \in \mathcal{K}_{exp}, u_m + v_m \le 1 \\ &\Leftrightarrow t_m \ge \log(1+\exp(-y_m(\boldsymbol{\theta}^T \boldsymbol{x}_m + \theta_0))) \;\; \forall 1 \le m \le M \;\; . \end{aligned} \tag{4}$$

The rotated quadratic cone is given by [40]:

$$\mathcal{Q}_R^K = \{\boldsymbol{\vartheta} \in \mathbb{R}^K \,|\, \vartheta_1\vartheta_2 \ge \frac{1}{2}\|\boldsymbol{\vartheta}_{3:K}\|_2^2, \vartheta_1, \vartheta_2 \ge 0\}. \tag{5}$$

We use the rotated quadratic cone to represent our L2-regularization with the auxiliary vectors $\boldsymbol{z} \in \{0,1\}^N$ and $\boldsymbol{r} \in \mathbb{R}_+^N$. Additionally, the cone naturally allows us to set $\theta_n = 0$ if $z_n = 0$:

$$[z_n, r_n, \theta_n] \in \mathcal{Q}_R^3 \Leftrightarrow z_n r_n \ge \frac{1}{2}\theta_n^2 \;\; \forall 1 \le n \le N. \tag{6}$$

Finally, we use an explicit cardinality constraint to limit the number of nonzero parameters:

$$\sum_{n=1}^{N} z_n \leq k. \tag{7}$$

Our reformulated optimization problem therefore becomes:

$$\begin{aligned}
\boldsymbol{\theta}^*, \theta_0^* = \underset{\boldsymbol{\theta}, \theta_0}{\text{argmin}} \quad & \sum_{m=1}^{M} t_m + \lambda \sum_{n=1}^{N} r_n \\
\text{subject to} \quad & [u_m, 1, -y_m(\boldsymbol{\theta}^T \boldsymbol{x}_m + \theta_0) - t_m] \in \mathcal{K}_{exp} \ \forall 1 \leq m \leq M \\
& [v_m, 1, -t_m] \in \mathcal{K}_{exp} \ \forall 1 \leq m \leq M \\
& u_m + v_m \leq 1 \ \forall 1 \leq m \leq M \\
& [z_n, r_n, \theta_n] \in \mathcal{Q}_R^3 \ \forall 1 \leq n \leq N \\
& \sum_{n=1}^{N} z_n \leq k \\
& \boldsymbol{\theta} \in \mathbb{R}^N, \boldsymbol{t}, \boldsymbol{u}, \boldsymbol{v} \in \mathbb{R}^M, \boldsymbol{r} \in \mathbb{R}_+^N, \boldsymbol{z} \in \{0,1\}^N.
\end{aligned} \tag{8}$$

3.2 Budget-Constrained Best Subset Selection

Cardinality-constrained best subset selection for logistic regression is already very useful to limit the number of candidate predictors. However, there are situations in which the cardinality is not the most suitable criterion to select features in practice, such as when two less costly predictors are preferred over one costly predictor. For instance, given a limited time budget to perform a diagnostic process, it may be more practical for a clinician to ask two quick questions rather than to perform one lengthy physical examination test. We therefore propose a cost-sensitive best subset selection procedure for logistic regression from a mixed-integer conic optimization perspective. To that end, we generalize eq. (7) by using a cost vector $\boldsymbol{c} \in \mathbb{R}^N$ and a budget $b \in \mathbb{R}_+$ [2,34,36]:

$$\sum_{n=1}^{N} c_n z_n \leq b. \tag{9}$$

The budget constraint forces the summed costs that are associated with the selected features to be less than or equal to our budget b. Note that this formulation can be regarded as a 0/1 knapsack problem; like a burglar who aims to find the best subset of items that maximize the profit such that the selected items fit within the knapsack, we aim to find the best subset of features that maximize the log likelihood of the data such that the selected features fit within our budget. If $\boldsymbol{c} = \boldsymbol{1}$ and $b = k$, we recover the explicit cardinality constraint. Our budget-constrained best subset selection procedure is therefore more general than its cardinality-constrained counterpart. This allows us to capture situations in which $\mathbf{c} \neq \mathbf{1}$, *i.e.*, individual feature costs can be different from one another, and increases the applicability of best subset selection.

4 Data Synthesis for Prognostic Models

Machine learning models can be broadly deployed in a variety of settings, so getting a comprehensive picture about their performance is essential. Synthetic dataset generators are useful tools to systematically investigate predictive models under a diverse range of conditions and thus give recommendations about key performance requirements. They also provide an opportunity to pretrain complex machine learning models so that small real-world datasets can be used more efficiently [30]. In the following, we describe our approach to synthetic data generation in the domain of prognostic model research. In particular, we derive parameter settings to generate sensible prognostic model data for multidimensional pain complaints, based on information from history-taking and physical examination. This setting is appealing because prognostic models appear to generally refine clinicians' predictions of recovery in this domain [1,26,31] and thus hold promise to make a large impact on clinical decision-making and outcomes.

Synthetic data can be generated through a variety of mechanisms. However, only few algorithms have served as a benchmark for a broad range of feature selection and classification methods [24,25] or have been recently used in the context of best subset selection with mixed-integer programming [46]. Therefore, we consider the algorithm that was designed to generate the MADELON dataset of the NIPS 2003 feature selection challenge to be a solid baseline for data synthesis [22]. In MADELON, training examples are grouped in multivariate normal clusters around vertices of a hypercube and randomly labeled. Informative features form the clusters; uninformative features add Gaussian noise. The training examples and labels can also be conceptualized to be sampled from specific instantiations of structural causal models (SCMs). Prior causal knowledge in the form of synthetic data sampled from SCMs has been recently used to meta-train complex deep learning architectures, *i.e.*, transformers, for tabular classification problems, allowing them to very efficiently learn from only a limited amount of real-world (for instance clinical) data [30].

To derive sensible parameter settings for the MADELON algorithm, we scanned the prognostic model literature for multidimensional pain complaints, based on information from history-taking and physical examination, and chose the number of training examples and features according to the minimum and maximum values that we encountered. We thus varied the number of training examples between 82 [20] and 1090 [17] and the number of features between 14 [3,21] and 53 [20]. We also varied the label noise level between 0% and a relatively small value of 5% [45] to simulate the effect of faulty missing label imputations, by randomly flipping the defined fraction of labels. This yielded 8 diverse synthetic datasets in total. We then made the classification on these datasets more challenging by two means. First, we added one "redundant" feature as a linear combination of informative features, which can arise in modeling when both questionnaire subscores and the summary score are included, for example [20]. Second, it is quite common that classes are imbalanced in prognostic model research, which is why we added a class imbalance of 23% / 77% [20]. The number of labels in the smaller class per feature, or outcome events per variable (EPV), thus roughly formed a geometric sequence from 0.36 to 17.91. We did not aim to investigate how to effectively engineer nonlinear features for our generalized linear logistic regression model, therefore we limited the number of clusters per class to one. Finally, as a preprocessing step, we "min-max" scaled each feature.

Each of these 8 datasets was used to assess our cardinality- and budget-constrained feature selection procedures. For our cardinality-constrained feature selection, we performed 4 runs with the number of informative features set to 2, 3, 4, or 5. For our budget-constrained feature selection, we performed 5 runs with the number of informative features set to 5 and each feature being assigned an integer cost between 1 and 10 sampled from a uniform distribution, *i.e.*, $c_n \sim \mathcal{U}(1, 10)$ [36]. For illustrative purposes, we assume that c_n corresponds to the time in minutes that it takes to collect the feature in practice.

5 Experiments

In this section, we describe the experimental setup to evaluate our cardinality- and budget-constrained feature selection procedures, including the mixed-integer conic optimization formulations (Sect. 3), on our 8 synthetic datasets (Sect. 4), and our cardinality-constrained best subset selection on real-world clinical data. We used MOSEK 10 [41] from Julia 1.8.3 via the JuMP 1.4.0 interface [18] for all experiments, with the default optimality gaps and no runtime limit. Finally, we present and discuss the results. In summary, interpretable results are only achieved at $EPV = 17.91$, in line with other EPV recommendations of at least 10 [29] to 15 [27], and without label noise for our optimal selection strategies.

5.1 How Does Our Approach Perform Compared to Other Models?

Experimental Setup. Prior to our extensive evaluation on the 8 synthetic datasets, we conducted a baseline test for our mixed-integer conic optimization formulation on the Oxford Parkinson's disease detection dataset [37,46]. The authors used a nonlinear support vector machine to classify subjects into healthy controls or people with Parkinson's disease based on voice measures. After correlation-based filtering and an exhaustive search, they achieved their top accuracy of 0.91 (95% confidence interval [0.87, 0.96]) with only 4 out of 17 features. For our experiments, we therefore used our cardinality-constrained best subset selection procedure with $k = 4$ and a classification threshold at 0.5. We generated 4 cubic splines per feature, with 3 knots distributed along the quartiles of each feature [45], and applied "min-max" scaling. In order to assess the predictive performance, we used a nested, stratified, 3-fold cross-validation in combination with nonparametric bootstrapping. Nested cross-validation consists of an inner loop for hyperparameter tuning, and an outer loop for evaluation. Nonparametric bootstrapping was used to obtain robust 95% confidence intervals (CIs) that provide a very intuitive interpretation, *i.e.*, that the value of interest will lie in the specified range with a probability of approximately 95% [28]. In the inner loop, we set our regularization hyperparameter λ to one value of the geometric sequence [0.1, 0.02, 0.004], optimized our objective function on 2 of the 3 folds, and used the third fold for validation with the area under the receiver operating characteristic curve (AUC). We repeated this procedure such that each fold was used for validation once, and averaged the validation fold performance over the 3 folds. We repeated these steps for each hyperparameter setting, and then used the best setting to train our predictive model on all 3 folds. In the outer loop, we ran our inner loop on 2 of the 3 folds, and used the third fold for testing with the classification accuracy. We repeated this procedure such that each fold is used for testing once, and averaged the test fold performance over the 3 folds. To obtain robust 95% CIs, we repeated these steps on 399 nonparametric bootstrap samples, *i.e.*, 399 random samples from the whole data with replacement, and calculated the mean test accuracy 2.5% and 97.5% quantiles [13,35]. Additionally, we registered the selected features.

Results and Discussion. We achieved a mean test accuracy of 0.87 (95% CI [0.84, 0.90]), similar to [37]. In fact, the 95% CIs between the original authors' and our feature selection approach overlap by a large amount, meaning that there is no statistically significant difference between the approaches ($p > 0.05$). While the selected subsets are different, the newly introduced and indeed best performing voice measure in [37], the pitch period entropy, was selected in both approaches. Additionally, our subset included the 5-point and 11-point amplitude perturbation quotients as well as the correlation dimension. In contrast to [37], our best subset selection approach uses a single-step selection procedure and allows for probabilistic risk estimates, which supports transparency.

5.2 Comparing Optimal and Greedy Cardinality-Constrained Selection

Experimental Setup. For the evaluation on the synthetic data, we solved our mixed-integer conic optimization problem (8) for each run with the cardinality constraint matching the number of informative features on each of the 8 synthetic datasets. We compared the performance of our best subset selection strategy to a greedy feature selection strategy because it is commonly used in prognostic model research [20,38,45] and recommended as a baseline [23]. As a heuristic, the greedy strategy selected the most valuable features, again matching the number of informative features, based on their univariable association with the outcome (*i.e.*, their parameter magnitude in unregularized univariable logistic regression). This often yields a good, but not necessarily the best subset. The final prognostic model was then built with the filtered features solving eq. (8) with $\boldsymbol{z} = \mathbf{1}$, but without eq. (7). To assess the mean test AUC of both strategies, we performed a nested, stratified, 3-fold cross-validation in combination with nonparametric bootstrapping. Via bootstrapping, we also evaluated the relative number of informative, redundant, and uninformative features selected[1], as well as how the selected predictors changed with slight perturbations in the data during bootstrapping, *i.e.*, the stability of the selection [10,45].

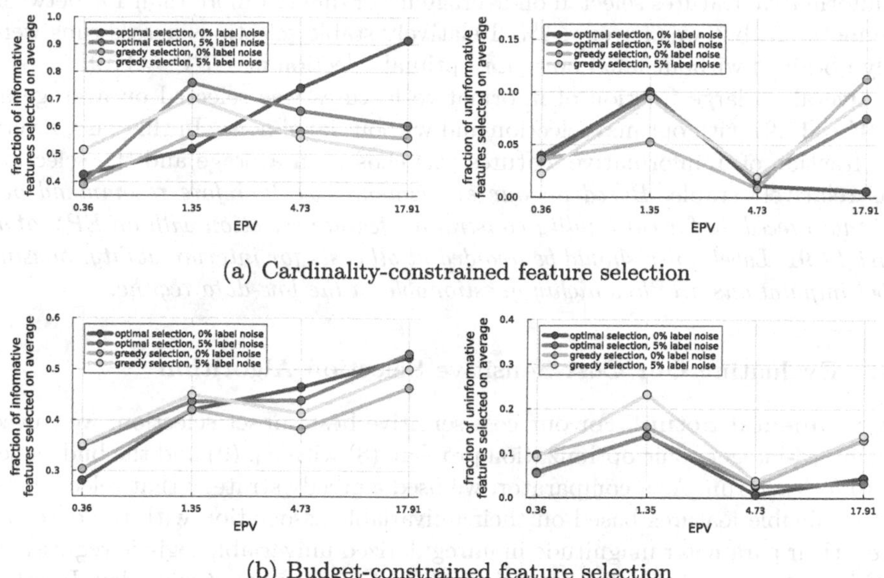

(a) Cardinality-constrained feature selection

(b) Budget-constrained feature selection

Fig. 1. Feature selection performance with increasing levels of EPVs

Results and Discussion. Mean test AUCs ranged from excellent (0.80, 95% CI [0.79, 0.98]) to outstanding (1.0, 95% CI [1.0, 1.0]) for optimal selection and from poor (0.65, 95% CI [0.62, 0.99]) to outstanding (1.0, 95% CI [1.0, 1.0]) for greedy selection. 95% CIs almost always overlapped to a large extent, meaning that there were generally no statistically significant differences in terms of discrimination between the approaches ($p > 0.05$). Without label noise, the fraction of informative features selected on average with optimal selection progressively increased as the EPV increased, from 42% at $EPV = 0.36$ to 91% at $EPV = 17.91$ (see Fig. 1a). The largest fraction of informative features selected on average with greedy selection was 58% at $EPV = 4.73$. While the fraction of redundant features was always less for optimal than for greedy selection, the fraction of uninformative features selected was mostly comparable - the largest

[1] We note that selecting some of the "uninformative" features could in principle still reduce noise and improve prognostic performance [23]. However, it would still be preferable for interpretability if indeed only the informative feature were recovered.

difference was observed at $EPV = 17.91$, with $< 1\%$ on average for optimal selection and 11% on average for greedy selection. Adding 5% label noise drastically decreased the largest fraction of informative features selected on average to 76% for optimal selection at $EPV = 1.35$ and increased it to 70% for greedy selection at $EPV = 1.35$. The fraction of redundant features was almost always less for optimal selection than for greedy selection in this case; the fraction of uninformative features selected on average never differed more than 1% between the methods, but was never $< 2\%$. Relatively stable selections across runs were only obtained without label noise for optimal selection at $EPV = 17.91$.

Overall, a large fraction of informative features was selected on average at $EPV = 17.91$ with optimal selection and without label noise. In this case, a very low fraction of uninformative features was chosen on average and the selection was relatively stable. *Based on our experiments, we therefore recommend our optimal procedure for cardinality-constrained feature selection with an EPV of at least 17.91. Label noise should be avoided at all costs for interpretability, missing label imputations are thus highly questionable in the low-data regime.*

5.3 Evaluating Our Cost-Sensitive Selection Algorithm

Experimental Setup. For our cost-sensitive best subset selection, we solved the mixed-integer conic optimization problem (8) with eq. (9) and the budget set to 10 for each run. As a comparator, we used a greedy strategy that selected the most valuable features based on their univariable association with the outcome (*i.e.*, their parameter magnitude in unregularized univariable logistic regression) divided by their individual cost such that the budget was not exceeded. In other words, we chose features with the most bang for the buck. The final prognostic model was built with the filtered features solving eq. (8) with $\boldsymbol{z} = \mathbf{1}$, but without eq. (9). Like before, we performed a nested, stratified, 3-fold cross-validation in combination with nonparametric bootstrapping to assess the mean test AUC, the relative number of informative, redundant, and uninformative features selected, as well as the selection stability.

Results and Discussion. Mean test AUCs ranged from acceptable (0.75, 95% CI [0.71, 1.0]) to outstanding (1.0, 95% CI [0.99, 1.0]) for optimal selection and from acceptable (0.76, 95% CI [0.61, 0.94]) to outstanding (1.0, 95% CI [0.98, 1.0]) for greedy selection. 95% CIs almost always overlapped to a large extent, so there were generally no statistically significant differences in terms of discrimination between the approaches ($p > 0.05$). The fraction of informative features selected on average with optimal selection progressively increased as the EPV increased, and was mostly comparable to the fraction selected on average with greedy selection (see Fig. 1b). The fraction of redundant features selected on average was always less for greedy selection than for optimal selection without label noise, with equivocal results when label noise was added. The fraction of uninformative features selected on average was again mostly comparable between the approaches, but in favor of optimal selection especially at $EPV = 17.91$.

Relatively stable selections across runs were only observed without label noise for optimal selection at $EPV = 4.73$ and $EPV = 17.91$.

Overall, relatively stable selections were only achieved at $EPV = 4.73$ and $EPV = 17.91$ with optimal selection and without label noise. A larger fraction of informative features was chosen at $EPV = 17.91$, though. In this case, the fraction of uninformative features selected was very low in most runs. *We therefore recommend our optimal cost-sensitive feature selection strategy with an EPV of at least 17.91. As in the cardinality-constrained case, label noise harms interpretability and should be avoided. For this reason, we caution against imputing missing labels even when the sample size is low.*

6 Conclusion

Best subset selection is (NP-)hard. Recent advances in mixed-integer programming [8,14,15,46] have made it possible to select the best feature subsets with certificates of optimality, though, even when (sub)set sizes are moderate. We have presented how both cardinality- and budget-constrained best subset selection problems for logistic regression can be formulated from a mixed-integer conic optimization perspective, and how the more general budget constraints are particularly appealing in situations when different costs can be assigned to individual features, for instance when some take longer to measure than others.

Additionally, we have designed a synthetic data generator for clinical prognostic model research. In particular, we have derived sensible parameter settings from the relevant literature for multidimensional pain complaints with information from history-taking and physical examination. We have pointed out conceptual connections of our data synthesis approach to sampling from structural causal models and mentioned how such a prior causal knowledge has been recently used for transformer-based meta-learning [30]. More broadly, we see that prior knowledge has a huge potential to improve predictive performance when data is limited, as is often the case in the clinical setting. In our view, deep learning models that were pretrained on big data, either real-world or synthetic, and can be adapted to a wide range of downstream tasks, *i.e.*, foundation models, hold a lot of promise in this regard, by capturing effective representations for simpler machine learning models like logistic regression [44,50].

Last but not least, we have compared our optimal feature selection approaches to heuristic approaches on diverse synthetic datasets. In line with the literature, we have observed that interpretable results for logistic regression, *i.e.*, solutions with a large fraction of informative features selected, a small fraction of uninformative features selected, and relatively stable selections across runs, were only achieved at an EPV of at least 10 to 15, in our case 17.91 with an optimal selection strategy. We have also observed that a label noise level of already 5% is detrimental to interpretability, which is why we caution against missing label imputations even when the sample size is low. It is also important to note that different feature sets (for example, with interaction terms or spline transformations), objective functions (for example, with different regularization hyperparameter settings), or machine learning models (for example, nonlinear yet interpretable predictive models like decision trees) would not necessarily yield similar "best" subsets (Sect. 5.1), which makes the decision of whether or not a feature is truly relevant even more challenging. Given a predefined feature set and objective function, though, we have shown that cardinality- and budget-constrained best subset selection is indeed possible for logistic regression using mixed-integer conic programming for a sufficiently large EPV without label noise, providing practitioners with transparency and interpretability during predictive model development and validation as commonly required [16,19]. Our optimization problem could be also modified to additionally select a small representative subset of training examples, *i.e.*, a coreset. In the big-data regime, such a reformulation would be beneficial to circumvent training time and memory constraints. This is work in progress, and indeed prior work shows that coresets for our L2-regularized softplus loss can be constructed with high probability by uniform sampling [12]. We thank the reviewers for this hint.

References

1. Abbott, J.H., Kingan, E.M.: Accuracy of physical therapists' prognosis of low back pain from the clinical examination: a prospective cohort study. J. Manual Manip. Therapy **22**(3), 154–161 (2014)
2. Aytug, H.: Feature selection for support vector machines using generalized benders decomposition. Eur. J. Oper. Res. **244**(1), 210–218 (2015)
3. Bakker, E.W., Verhagen, A.P., Lucas, C., Koning, H.J., Koes, B.W.: Spinal mechanical load: a predictor of persistent low back pain? A prospective cohort study. Eur. Spine J. **16**, 933–941 (2007)
4. Ben-Tal, A., Nemirovski, A.: Lectures on modern convex optimization: analysis, algorithms, and engineering applications. In: SIAM (2001)
5. Bertsimas, D., Copenhaver, M.S.: Characterization of the equivalence of robustification and regularization in linear and matrix regression. Eur. J. Oper. Res. **270**(3), 931–942 (2018)
6. Bertsimas, D., Dunn, J.: Machine Learning Under a Modern Optimization Lens. Dynamic Ideas, LLC, Charlestown (2019)
7. Bertsimas, D., Dunn, J., Pawlowski, C., Zhuo, Y.D.: Robust classification. INFORMS J. Optim. **1**(1), 2–34 (2019)
8. Bertsimas, D., Pauphilet, J., Van Parys, B.: Sparse classification: a scalable discrete optimization perspective. Mach. Learn. **110**, 3177–3209 (2021)
9. Boyd, S., Boyd, S.P., Vandenberghe, L.: Convex Optimization. Cambridge University Press, Cambridge (2004)
10. Breiman, L.: Statistical modeling: the two cultures (with comments and a rejoinder by the author). Stat. Sci. **16**(3), 199–231 (2001)
11. Christodoulou, E., Ma, J., Collins, G.S., Steyerberg, E.W., Verbakel, J.Y., Van Calster, B.: A systematic review shows no performance benefit of machine learning over logistic regression for clinical prediction models. J. Clin. Epidemiol. **110**, 12–22 (2019)
12. Curtin, R.R., Im, S., Moseley, B., Pruhs, K., Samadian, A.: On coresets for regularized loss minimization. arXiv preprint arXiv:1905.10845 (2019)

13. Davidson, R., MacKinnon, J.G.: Bootstrap tests: how many bootstraps? Economet. Rev. **19**(1), 55–68 (2000)
14. Dedieu, A., Hazimeh, H., Mazumder, R.: Learning sparse classifiers: continuous and mixed integer optimization perspectives. J. Mach. Learn. Res. **22**(1), 6008–6054 (2021)
15. Deza, A., Atamturk, A.: Safe screening for logistic regression with l0–l2 regularization. arXiv preprint arXiv:2202.00467 (2022)
16. DIN, DKE: Deutsche Normungsroadmap Künstliche Intelligenz (Ausgabe 2) (2022). https://www.din.de/go/normungsroadmapki/
17. Dionne, C.E., Le Sage, N., Franche, R.L., Dorval, M., Bombardier, C., Deyo, R.A.: Five questions predicted long-term, severe, back-related functional limitations: evidence from three large prospective studies. J. Clin. Epidemiol. **64**(1), 54–66 (2011)
18. Dunning, I., Huchette, J., Lubin, M.: JuMP: a modeling language for mathematical optimization. SIAM Rev. **59**(2), 295–320 (2017)
19. European Commission: Proposal for a Regulation Of The European Parliament and of the Council Laying Down Harmonised Rules on Artificial Intelligence (Artificial Intelligence Act) and Amending Certain Union Legislative Acts (2021). https://artificialintelligenceact.eu/the-act/
20. Evans, D.W., et al.: Estimating risk of chronic pain and disability following musculoskeletal trauma in the united kingdom. JAMA Netw. Open **5**(8), e2228870–e2228870 (2022)
21. van der Gaag, W.H., et al.: Developing clinical prediction models for nonrecovery in older patients seeking care for back pain: the back complaints in the elders prospective cohort study. Pain **162**(6), 1632 (2021)
22. Guyon, I.: Design of experiments of the nips 2003 variable selection benchmark. In: NIPS 2003 Workshop on Feature Extraction and Feature Selection, vol. 253, p. 40 (2003)
23. Guyon, I., Elisseeff, A.: An introduction to variable and feature selection. J. Mach. Learn. Res. **3**(Mar), 1157–1182 (2003)
24. Guyon, I., Gunn, S., Ben-Hur, A., Dror, G.: Result analysis of the NIPS 2003 feature selection challenge. In: Advances in Neural Information Processing Systems, vol. 17 (2004)
25. Guyon, I., Li, J., Mader, T., Pletscher, P.A., Schneider, G., Uhr, M.: Competitive baseline methods set new standards for the NIPS 2003 feature selection benchmark. Pattern Recogn. Lett. **28**(12), 1438–1444 (2007)
26. Hancock, M.J., Maher, C.G., Latimer, J., Herbert, R.D., McAuley, J.H.: Can rate of recovery be predicted in patients with acute low back pain? Development of a clinical prediction rule. Eur. J. Pain **13**(1), 51–55 (2009)
27. Harrell, F.E.: Regression Modeling Strategies: with Applications to Linear Models, Logistic Regression, and Survival Analysis. Springer, Cham (2015). https://doi.org/10.1007/978-3-319-19425-7
28. Hastie, T., Tibshirani, R., Friedman, J.H., Friedman, J.H.: The Elements of Statistical Learning: Data Mining, Inference, and Prediction. Springer, New York (2009). https://doi.org/10.1007/978-0-387-84858-7
29. Heinze, G., Wallisch, C., Dunkler, D.: Variable selection-a review and recommendations for the practicing statistician. Biom. J. **60**(3), 431–449 (2018)
30. Hollmann, N., Müller, S., Eggensperger, K., Hutter, F.: TabPFN: a transformer that solves small tabular classification problems in a second. arXiv preprint arXiv:2207.01848 (2022)

31. Kennedy, C.A., Haines, T., Beaton, D.E.: Eight predictive factors associated with response patterns during physiotherapy for soft tissue shoulder disorders were identified. J. Clin. Epidemiol. **59**(5), 485–496 (2006)
32. Kuijpers, T., van der Windt, D.A., Boeke, A.J.P., Twisk, J.W., Vergouwe, Y., Bouter, L.M., van der Heijden, G.J.: Clinical prediction rules for the prognosis of shoulder pain in general practice. Pain **120**(3), 276–285 (2006)
33. Kuijpers, T., van der Windt, D.A., van der Heijden, G.J., Twisk, J.W., Vergouwe, Y., Bouter, L.M.: A prediction rule for shoulder pain related sick leave: a prospective cohort study. BMC Musculoskelet. Disord. **7**, 1–11 (2006)
34. Labbé, M., Martínez-Merino, L.I., Rodríguez-Chía, A.M.: Mixed integer linear programming for feature selection in support vector machine. Discret. Appl. Math. **261**, 276–304 (2019)
35. LeDell, E., Petersen, M., van der Laan, M.: Computationally efficient confidence intervals for cross-validated area under the roc curve estimates. Elect. J. Statist. **9**(1), 1583 (2015)
36. Lee, I.G., Zhang, Q., Yoon, S.W., Won, D.: A mixed integer linear programming support vector machine for cost-effective feature selection. Knowl. Based Syst. **203**, 106145 (2020)
37. Little, M., McSharry, P., Hunter, E., Spielman, J., Ramig, L.: Suitability of dysphonia measurements for telemonitoring of Parkinson's disease. In: Nature Precedings, p. 1 (2008)
38. Moons, K.G., et al.: Transparent reporting of a multivariable prediction model for individual prognosis or diagnosis (tripod): explanation and elaboration. Ann. Intern. Med. **162**(1), W1–W73 (2015)
39. Moons, K.G., et al.: Probast: a tool to assess risk of bias and applicability of prediction model studies: explanation and elaboration. Ann. Intern. Med. **170**(1), W1–W33 (2019)
40. MOSEK ApS: MOSEK modeling cookbook (2022)
41. MOSEK ApS: MOSEK optimizer API for Python (2023)
42. Natarajan, B.K.: Sparse approximate solutions to linear systems. SIAM J. Comput. **24**(2), 227–234 (1995)
43. Scheele, J., et al.: Course and prognosis of older back pain patients in general practice: a prospective cohort study. PAIN® **154**(6), 951–957 (2013)
44. Steinberg, E., Jung, K., Fries, J.A., Corbin, C.K., Pfohl, S.R., Shah, N.H.: Language models are an effective representation learning technique for electronic health record data. J. Biomed. Inform. **113**, 103637 (2021)
45. Steyerberg, E.W.: Clinical Prediction Models: A Practical Approach to Development, Validation, and Updating. Springer, Cham (2019). https://doi.org/10.1007/978-3-030-16399-0
46. Tamura, R., Takano, Y., Miyashiro, R.: Feature subset selection for kernel SVM classification via mixed-integer optimization. arXiv preprint arXiv:2205.14325 (2022)
47. Tibshirani, R.: Regression shrinkage and selection via the LASSO. J. Roy. Stat. Soc.: Ser. B (Methodol.) **58**(1), 267–288 (1996)
48. Tillmann, A.M., Bienstock, D., Lodi, A., Schwartz, A.: Cardinality minimization, constraints, and regularization: a survey. arXiv preprint arXiv:2106.09606 (2021)
49. Wippert, P.M., et al.: Development of a risk stratification and prevention index for stratified care in chronic low back pain. Focus yellow flags (MiSpEx network). Pain Rep. **2**(6), e623 (2017)

50. Wornow, M., et al.: The shaky foundations of clinical foundation models: a survey of large language models and foundation models for EMRs. arXiv preprint arXiv:2303.12961 (2023)

Detecting Floors in Residential Buildings

Aruscha Kramm(✉), Julia Friske, and Eric Peukert

Center for Scalable Data Analytics and Artificial Intelligence (ScaDS.AI) Dresden/Leipzig, University of Leipzig, Leipzig, Germany
{kramm,friske,peukert}@informatik.uni-leipzig.de

Abstract. Knowing the number of floors of all buildings in a city is vital in many areas of urban planning such as energy demand prediction, estimation of inhabitant numbers of specific buildings or the calculation of population densities. Also, novel augmented reality use cases strongly rely on exact numbers and positions of floors. However, in many cases floor numbers are unknown, its collection is mostly a manual process or existing data is not up-to-date. A major difficulty in automating floor counting lies in the architectural variety of buildings from different decades. So far approaches are only rough geometric approximations. More recently approaches apply neural networks to achieve more precise results. But, these neural network approaches rely on various sources of input that are not available to every municipality. They also tend to fail on building types they have not been trained on and existing approaches are completely black-box so that it is difficult to determine when and why the prediction is wrong.

In this paper we propose a grey-box approach. In a stepwise process we can predict floor counts with high quality and remain explainable and parametrizable. By using data that is easy to obtain, namely the image of a building, we introduce two configurable methods to derive the number of floors. We demonstrate that the correct prediction quality can be significantly improved. In a thorough evaluation we analyze the quality depending on a number of factors such as image quality or building types.

Keywords: floor detection · buildings · street view · urban digital twin

1 Introduction

Knowing the number of floors of a building makes it possible to derive further information about it to be used in city contexts such as urban planning and reconstruction, building topology analysis or modelling building energy consumption in Digital Twins or the generation of 3D-models [3,9]. Combining the number of floors and ground-plan area allows the estimation of living area and the average number of inhabitants, and further the modelling of energy consumption, a critical aspect for optimizing energy efficiency and reaching carbon neutrality. The number of floors affects energy consumption of a building via factors such as heating, cooling, and ventilation needs [7]. The density of the

D. Seipel and A. Steen (Eds.): KI 2023, LNAI 14236, pp. 130–143, 2023.
https://doi.org/10.1007/978-3-031-42608-7_11

urban area is a crucial factor in urban planning and affects the availability of resources such as water, energy, and transportation. By estimating the number of inhabitants in an area using the building's number of floors, urban planners can design more sustainable and efficient urban areas [4]. As cities face increasing demand for living space and must reduce carbon emissions [12], building vertically is becoming necessary [5]. To determine eligible buildings for vertical expansion and monitor changes, knowing the number of floors before and after such a change is crucial. Many of the use cases mentioned are carried out in the smart city project *Connected Urban Twins* (CUT)[1], which we take part in cooperation with the city of Leipzig and which motivated this paper.

Retrieving the number of floors for each building manually is cost-expensive and laborious. While the task is intuitive for humans, it is very challenging for computers [13]. Even humans sometimes cannot correctly assign the number of floors. Estimating the number of floors by dividing a building's height by an average ceiling height contains sources of error. For instance, new buildings have lower ceiling heights than old buildings, resulting in more floors for the same height, which is presenting only one source of error for this approach. Existing machine learning approaches require more data than an image as input such as the work of Roy et al. [10] or have difficulty with data on which they have not been trained on. Floor Level Net (FLN) by Wu et al. [13], which has an image as input, similar to our proposed methods, shows significant problems in recognizing the floor lines for our dataset (s. Fig. 1). It misdetects floors or returns misdirected floor-level lines. The floor detection process in FLN occurs entirely within a neural network, making it challenging to identify when and why errors occur. We present two approaches to infer the number of floors from images of buildings. Both approaches treat floor level detection as a combination of three tasks: classification (detecting facade elements using a neural network), floor-level line detection (separate adjacent floors on a building facade) and the calculation of the number of floors. We argue that by performing part of the detection outside the neural network, we are able to parameterize and adapt our methods to different scenarios. We are further able to detect attic floors and add them to the number of floors with an addition "+ attic". We derive the number

Fig. 1. Testing Floor Level Net on our dataset. For the image on the left, FLN correctly recognized the floors. It cannot detect the attic. The rest of the images are incorrectly detected which shows either by a wrong number of floors or a misdirected floor line.

[1] www.connectedurbantwins.de/en/.

of floors by estimating a line through the points belonging to one floor from the bounding box points of windows detected by the neural network in the first step. By deriving the number of floors from detected windows, our approach can be more easily applied to similar buildings without retraining the neural network. By splitting the task into several steps, we can perform a thorough error analysis. The code to our work is publicly available[2].

The main contributions of this work include:

1. We propose a workflow where we generate floor level lines and floor counts for a given input image.
2. We introduce two methods for line detection to separate adjacent floors on a building facade.
3. We compare our proposed methods to *Floor level net* on various datasets.
4. We conduct a comprehensive evaluation to identify the factors on which a correct prediction depends, including building type and image quality.

The remaining paper is structured as follows: Sect. 2 analyzes and illustrates the problems faced when detecting floors for different buildings. Thereafter the proposed methods are presented in Sect. 4 and evaluated in Sect. 5.

2 Problem Definition

Detecting the number of floors in city buildings poses challenges due to varying architectural styles and features such as decorations or ceiling heights that are reflected in the facade (e.g. through window heights). These architectural styles make it difficult to find a universal approach that works for all building types. Considering for example old buildings built between 1890 and 1920 we often find decorations such balconies or bow-fronts, making the facade deviate from a flat surface. Looking at an image of such a bow-front facade, the deviation from the flat surface manifests in the windows not being on the same level even though they belong to the same floor (see Fig. 2(a)).

As cities grow in population, buildings become taller to accommodate the demand for living space [12]. However, tall buildings pose a challenge for detecting the number of floors and windows since higher buildings tend to appear smaller in the image, making it more difficult to detect windows. In addition to these aspects, occlusions such as greenery or cars can conceal parts of a building. The bigger the concealed part, the higher the risk of a false prediction.

Our dataset consists of street view imagery of Leipzig which is accessible via API with different image rendering methods. Using two different rendering methods resulted in either a perspective image of the building or a rectified surface image (see Fig. 2(b)). The images differ in quality as the perspective image (Fig. 2(b) left) is missing information because the top of the building is cut off.

[2] https://github.com/aruscha-k/detecting-floors-in-residential-buildings.

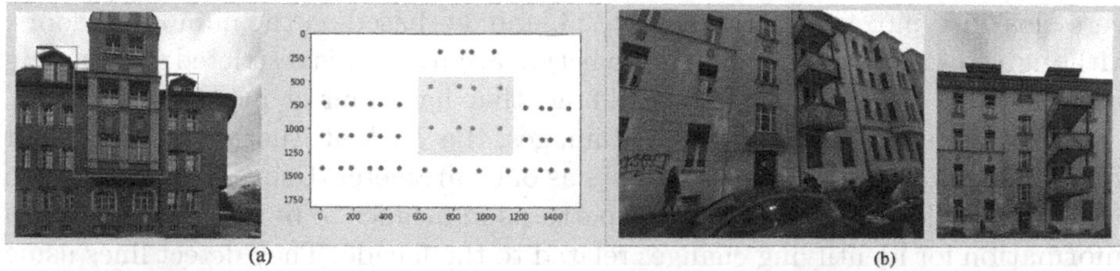

Fig. 2. (a) Building with bow-front and its upper two window points for each detected window. Windows points of the bow-front (red square) are not in line with the others although belonging to the same floor. (b) Two images of the same building rendered based on GPS coordinates (left) and based on facade coordinates (right). (Color figure online)

Although machine learning approaches outperform geometric approximation in recognizing the number of floors in buildings, we observed poor results when applying FLN to our dataset. Since this method is entirely black-box, it is challenging to explain the poor performance on our data, which makes it difficult to adjust parameters accordingly.

3 Related Work

As best as we know, the topic of floor recognition is a relatively new field of research in which there are not yet many existing publications. In contrast, extracting information from street view imagery in the context of GIS applications and urban analytics has increased enormously over the last few years. Due to the advancements in computer vision and availability of computing power, street view imagery has been used for geospatial data collection ranging from analyzing vegetation to transportation to health and social-economic studies [3].

The authors of Floor Level Net introduce a supervised deep learning approach that has been developed for recognizing floor-level lines in street view images [13]. They compile a synthetic dataset and a data augmentation scheme to integrate semantics of the building. The process is completely implemented using machine learning. FLN can only detect floors for buildings with a minimum number of floors of two and is not able to detect attic floors. The output of the method is an image with recognized floor-level lines, not an actual number.

Roy et al. [10] aim to improve the prediction of the number of floors in residential buildings in the Netherlands using various datasets such as footprints, 3D models, cadastral and census data. The authors achieve a good detection rate using regression algorithms, yet their approach requires many different data sources that may not be available to every city. Since the prediction is done by machine learning models, the whole process is hidden inside a black-box.

Ianelli et al. [6] proposed a method to automatically infer the number of floors in buildings using street-level images. To train their deep learning model, they created a labelled dataset for residential houses in the USA: The images were classified into five categories 0,1,2,3, and 4+ based on the number of floors. Although the approach has a high prediction accuracy, it is restricted to assigning only one class (4+) for high-rise buildings that have four or more floors.

Although using different types of images, the work of Auer et al. [2] aims at finding structures in facades using lines as one parameter. Using High-Resolution SAR images the authors aim to find patterns in facades to be used as a source of information for identifying changes related to the facade. They detect lines using Hough-Transform and setting custom constraints such as limiting the number of possible angles for a line. On evaluation, they face the problem of having to select certain parameters manually.

Loosely related is the issue of detecting lines or curves, a topic Schrödl et al. [11] addressed in the context of refining the shape of roads compared to their shape in online maps. The authors refine the course of roads by using routes driven by test vehicles. These routes, represented by GPS points, are used to calculate a median line from several routes using a weighted least-square spline.

4 Methods

The presented approaches in this work are based on detecting facade elements in images such as windows, dormers or skylights. We trained a neural network that outputs bounding boxes for each detected facade object. Assuming that windows of one floor are on one level the aim is to find floor-level lines by fitting a line through all window bounding boxes points that belong to one floor. The floor-level lines are then used to calculate the number of floors. The existence of attics or dormers in the image is used to add the specification "+ attic".

4.1 Detection of Façade Elements

We trained a Detectron2 net [1] to detect the facade elements *windows, front doors, shops, dormers and skylights.* We manually labelled about 400 images to train the network. To lower the error rate within this step, we define a detection certainty threshold of 85%. Detecting facade elements represents the first step in our pipeline to detect the number of floors. The input is a single RGB image $I \in W \times H \times 3$. The outputs are the detected classes and corresponding bounding boxes, that locate the detected objects in the image. The points of the window class boxes are then further used in two ways: the nearest neighbours approach uses the centroid of each box whereas the Ransac method uses the two upper left and right corner vertices of each box.

4.2 Nearest Neighbours (NN) Approach

The algorithm's objective is to horizontally group the center points of bounding boxes (centroids) of windows so that the number of groups represents the number of floors. Initially, the centroid of each bounding box is calculated using the x, y coordinate pairs of its lower left and upper right corners (s. Fig. 3(a)). The grouping of centroids in a horizontal manner involves the following steps:

1. For each centroid the Euclidean distances to all other centroids are calculated. Given a list of n centroid points $P = \{P_1, ..., P_n\}$ in a two-dimensional space, where $Pi = (x_i, y_i)$, the Euclidean distance matrix of dimension $n \times n$ can be calculated (s. Fig. 3(b)):

$$D_{i,j} = d(P_i, P_j) = \sqrt{(x_j - x_i)^2 + (y_j - y_i)^2} \quad (1)$$

2. Based on the obtained distances, a fixed number of the nearest neighbours (e.g., 20) is selected.
3. The slopes for all potential lines between two points are calculated in pairs. The slope is calculated using the standard line gradient formula:

$$S_{i,j} = s(P_i, P_j) = \frac{y_j - y_i}{x_j - x_i} \quad (2)$$

   Only the neighboring points that are considered to be "horizontal" are chosen using a threshold value for the slope of the line (*slope_threshold_range*), resulting in a list of horizontal neighboring point indices for each centroid (s. Fig. 3(c)).
4. These lists are considered duplicates if they contain the same centroids in a different order. Therefore lists that share common indices are merged. For example [[0, 2, 3, 1], [1, 2, 3], [2, 3, 1, 0] is merged to [0, 1, 2, 3]. The output is a list of lines, which represents the number of floors (s. Fig. 3(d)).

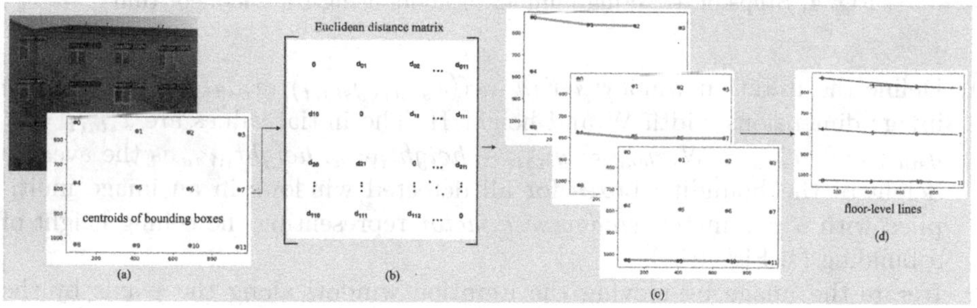

Fig. 3. Steps of detecting number of floors using NN-algorithm

Depending on the position and height of the windows, the bounding boxes' centroids of one floor may not be aligned. This inconsistency in alignment is the reason why the parameter *slope_threshold_range* is needed in step 3.

4.3 Ransac Approach

Ransac (RANdom SAmple Consensus) is an algorithm proposed in 1981 for robust estimation of the model parameters in a presence of outliers, that is data points which are noisy and wrong [8]. We apply this method by putting in the upper left and right points of each predicted window bounding box and generate a set of lines, denoted by $L = L_1, ..., L_n$, where each line represents a floor-level line. Typically, Ransac fits a single line through all data points, resulting in one line per image. To obtain a line for each floor, a horizontal constraint is added to enforce horizontal alignment. This is achieved by iterating the image in horizontal "windows" (see Fig. 4), and running the Ransac algorithm only on the window box points within each iteration window. The approach is divided into these steps:

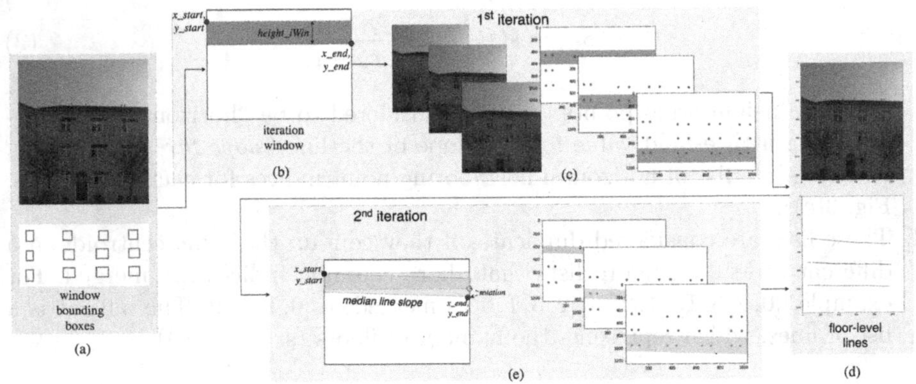

Fig. 4. Steps of detecting number of floors using Ransac-algorithm

1. Define the iteration window $iWin = ((x_{start}, y_{start}), (x_{end}, y_{end}))$ using the image dimensions width W and height H. The initial values are $x_{start} = 0$, $y_{start} = 0$, $x_{end} = \text{W}$, $y_{end} = y_{start} + height_{iWin}$. $height_{iWin}$ is the average height of the bounding boxes for all detected windows in an image multiplied with a parameter *window_size_factor* representing the ceiling height of a building (s. Fig. 4(b)).
2. Iterate the image by moving the iteration window along the y-axis by the value of $height_{iWin}$. In each iteration apply the Ransac algorithm returning a list of points of the fitted line and a list of points of outliers (s. Fig. 4(c)).
3. Two separate lists containing the candidate lines for floor-levels (s. Fig. 4(d)) and outlier points are returned.

If the edges of a building are not parallel to the edges of the image (and therefore the edges of the iteration window), then on the first iteration the iteration window may contain window points from multiple floors. This can cause issues with the resulting fitted line, which may incorporate window points from two or more floors, leading to incorrect results. Therefore the candidate lines are processed further:

4. The slope of each candidate line and a median of all slopes are calculated. Candidate lines are rejected whose slope is deviating from the median slope by a threshold parameter *thresh_slope_deviation*.
5. Iterate all resulting candidate lines pairwise and use the Ransac algorithm to determine if they belong to the same floor. If a line is fitted, merge the two lines; otherwise, leave them as they were.

After completing the steps, there are floor-level lines and outliers. The median slope of all lines is calculated and if it overshoots a threshold parameter *repeat_slope*, steps 1–5 are repeated with an adjusted iteration window. The new window is a parallelogram based on the median slope of the floor-level lines (s. Fig. 4(e)). Considering that the building is photographed from below and not centered, the upper part always appears smaller than the lower part. Therefore the height of the adjusted iteration window will grow by a factor *window_size_grow* on each iteration. The final result is a set of lines where each line represents one floor and a set of points that have been marked as outliers.

5 Evaluation

To test our methods against existing ones we looked at several other publications. Roy et al. [10] used the footprint and 3D model as point cloud of the building along with census data to determine the number of floors. As we lack the necessary information to use their approach on the buildings in our dataset, we cannot compare the results of our methods to the work of Roy et al. The authors of Floor Level Net [13] indeed do not mention the recognition of the correct number of floors as a direct goal, yet for their listed target use cases, all possible floors should be recognized as well. The etrims-dataset they used for training FLN consists of buildings of European cities, which should have similar architectural styles as the buildings in Leipzig. Therefore we compare our methods with FLN. For the comparison we used the model linked in the publication without retraining.

5.1 Dataset

The city of Leipzig collects street view imagery which is accessible via API with various image rendering methods. We tested two rendering methods resulting in different views of a building. The first method renders an image based on a GPS-coordinate which is challenging since differing street width and building height have to be balanced out for each building by using specific API parameters such as camera angle and horizontal field of view. The resulting image is more likely to show a perspective image or to be cut off at the top (see Fig. 2(b)). Furthermore, it can show several buildings making it hard to identify the building to detect the floors for. The images obtained by this method can be compared to images retrieved via Google street view. The second method renders the image based on facade GPS coordinates ($lowerLeft, lowerRight, upperRight, upperLeft$) resulting in a rectified facade image. The output of this method displays only one building, making it easier to attribute the number of floors to the correct building. (see Fig. 2(b)). Moreover, the building is always fully visible, since the view is automatically zoomed out to project the whole building.

To acquire rectified facade images, we were granted access to the API and to additional information on floor plans and building heights. This allowed us to extract 124 images each for evaluating our floor detection methods on perspective facade images and rectified facade surface images. To test our methods on unseen data we randomly chose images from the openly available etrims-dataset FLN was trained and tested on.

Fig. 5. The different building types from left to right: old building, single-family home, industrialized building, modern building and occluded building.

Our testset reflects various difficulties in recognizing floors. First, we divide buildings into informal building types of *old building* (OLD), *single-family home* (SFH), *industrialized building* (IB) and *modern building* (MOD) (s. Fig. 5). In doing so we can relate the prediction quality with the building type. Furthermore, we evaluate the floor detection for *occluded buildings* (OCC).

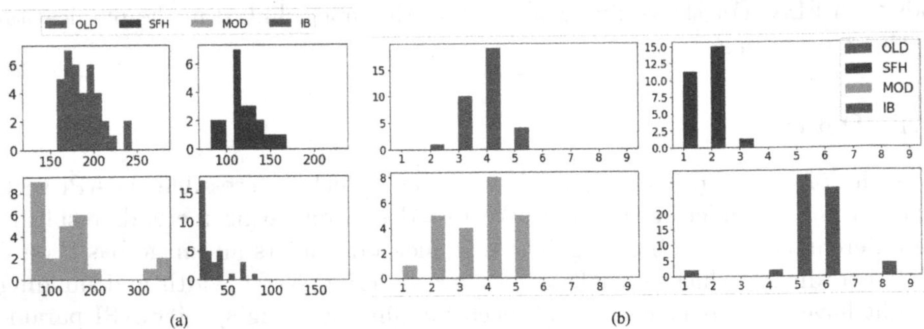

Fig. 6. (a) Different window heights (in px) evaluated from rectified surface facade images. (b) Distribution of number of floors for each building type.

The mentioned building types differ in the arrangement and height of windows. Industrial buildings have lower window heights than old buildings or single-family houses, while modern buildings have the highest variation in window heights (see Fig. 6(a)). Industrial buildings have more floors than the other building types, which means they will be photographed from a greater distance to capture the entire building (see Fig. 6(b)) which may have an impact on the detection of the windows.

For each of the mentioned challenges, we compare the results for rectified facade images and perspective images. We manually labelled the correct number of floors for each building. Images, where we could not specify the number of floors or agree on the same number, were excluded. In this work, attic floors are defined by the visibility of dormers or skylights. Since FLN does not explicitly detect attics, we split the results into buildings with and without attics. Our testset respectively contains 34 images of type old building, 22 images of type single-family home, 34 images of type industrialized building, 23 images of type modern buildings and 11 images of buildings with occlusions of each rectified facade and perspective images. In total, our testset contains 248 images. We randomly chose another 29 images from the etrims-dataset.

5.2 Results

When comparing the prediction results it must be noted that FLN was trained on buildings with a minimal number of two floors. The FLN results for buildings of the category SFH can therefore only be considered to a limited extent since we often find one floor in single-family houses (see Fig. 6(b)). Looking at the prediction quality with respect to the rendering methods, rectified surface images yield higher correct prediction rates for our proposed methods NN and Ransac. As expected FLN has a higher correct prediction rate for perspective images in comparison to surface images since it was trained on perspective images. On perspective images FLN performs better than Ransac only for *modern buildings*, due to the similarity between these buildings and those on which FLN was trained. For all other building types, our methods work equally well or better. For some building types, Ransac outperforms NN. All three methods have higher accuracy for buildings without attics, indicating the challenge in accurately

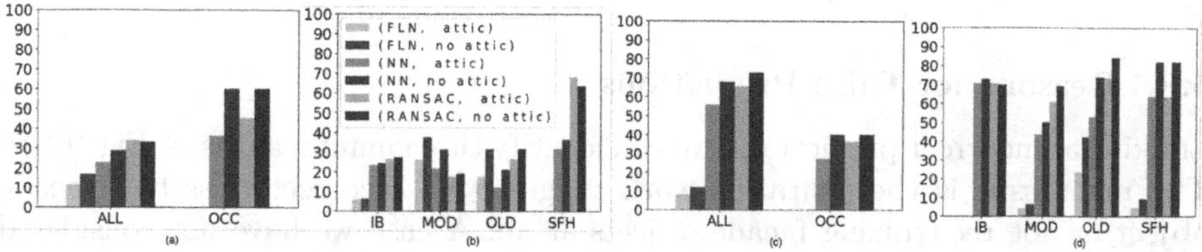

Fig. 7. Correct prediction results (in %) for perspective images (a & b) and rectified surface images (c & d) for all methods and building types. "ALL" forms the average over OCC, IB, MOD, OLD and SFH respectively.

identifying attics. Surprisingly correct prediction results for occluded buildings are higher than the average ("ALL") in perspective images, which could be because the occlusion does not cover the entire building due to the perspective (see Fig. 7(a)).

Using the knowledge of varying window arrangements in different building types the methods' parameters can be adjusted to the building type. Since FLN is a black-box method, we cannot adjust any parameters here. Also *occluded buildings* are excluded from this step since the buildings in this category can be of any type. The NN method's parameter *slope_threshold_range* as well as Ransac's parameters *window_size_factor*, *thresh_slope_deviation* and *window_size_grow* can be adjusted to the building type. Our specifically used values were obtained by testing different values.

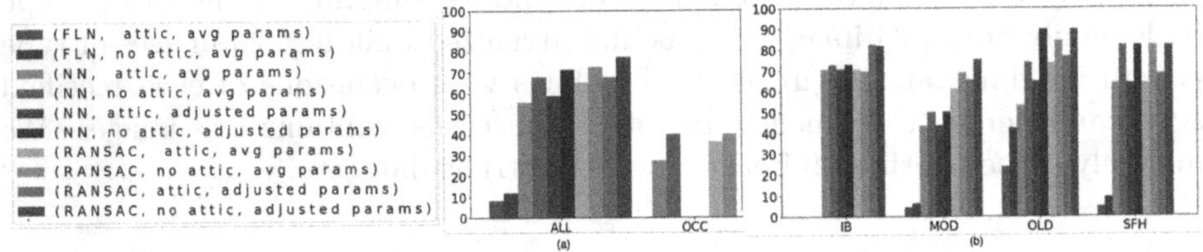

Fig. 8. Detection rates with and without adjusted parameters for each building type for methods Ransac and Nearest neighbour on rectified facade images.

With adjusted parameters, the correct detection rates for the Ransac method can be further increased (s. Fig. 8). NN method's results can only improve for *old buildings*. For old buildings and single-family homes, Ransac and NN achieve equal results of 89.5% and 81.8%, respectively.

The detection rate for industrial and modern buildings is lower than for other types, indicating that predicting their floors is more challenging. Possible reasons include double-story lobbies, stairwell windows, or irregularly placed windows. We also get a poor detection rate for occluded buildings.

The evaluation of the etrims dataset shows similar results. Although trained on this dataset, FLN performs the worst (see Fig. 9(a)). Floor detection using the Ransac method obtains the best results. Also, again a clear difference can be seen between buildings with and without attics.

5.3 Reasons for False Predictions

Looking at incorrect predictions, we can identify three main reasons (s. Fig. 9(b)). The main issue is the neural network assigning an incorrect class to a facade object or not recognizing facade objects at all. A case we have not considered further is the recognition of basement windows, which occur in some buildings. Since the methods take into account all detected windows, basement windows result in an incorrect number of floors. Also, some buildings have stairwell windows that differ in shape and arrangement from the rest of the windows that

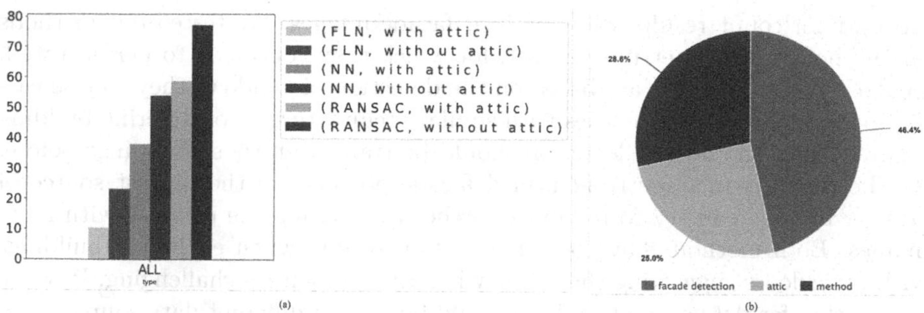

Fig. 9. (a) Detection rates for buildings from etrims dataset. (b) Reasons for false predictions divided into three main categories.

currently are not differentiated between. Further the detection of attic floors is connected with several difficulties. First of all, to the best of our knowledge it is not clearly defined when an attic is considered a full floor. Therefore, we define an attic by the existence of dormers or skylights and append "+ attic" to the detected number of floors. Nevertheless, even this is not straightforward: on some buildings there are only dormers, on others there is a row of windows next to dormers. The third issue is that both methods require manual parameter setting, which significantly affects the prediction results. For the Ransac method, the iteration window size is determined by multiplying the average height of the window bounding boxes with a manually selected factor to represent the ceiling height, which varies between buildings from different time periods. Similarly, the window height affects the NN method. Particularly for modern buildings, window heights vary significantly within a building, resulting in misaligned centroids for the same floor. The NN method defines a threshold value based on the centroid (and therefore the window height) to determine the range of slopes considered "horizontal". However, with prior knowledge of the building type, these parameters can be set automatically.

6 Discussion, Conclusion and Future Work

As street view imagery collection becomes more frequent in the future, cities will have access to regularly updated insights into urban development. With increased data acquisition frequency, there will be a growing demand for automated evaluation methods. We introduced two methods to detect floor numbers in images of buildings, and compared image qualities and predictions for different building types. We further demonstrated difficulties inherent to the task that even humans cannot always solve correctly. We tested our methods against Floor Level Net, the only method we know of that has a similar data input. In using a grey-box approach, our floor prediction can be adjusted to different building types, while staying explainable when errors occur. Further, our methods require only a single data input, namely the image of a facade. We achieved strong floor count results with rectified facade images. We tested our methods on the before unseen etrims-dataset, for which they continued to perform well. Because our approaches are based on the detection of windows they can be easily applied to (European) cities with similar architectures. For differing building architectures, the object detection should be trained on those buildings before. At the moment incorrectly identified facade objects are the biggest source of error, which we can try to minimize further by training the network with more images. Both methods have better detection results when excluding buildings with attic floors suggesting that clearly identifying attics is challenging. Possible approaches for detecting attic floors could be adding different data sources (e.g. aerial data). At the present state our methods contain parameters that have to be set manually, which have the potential to be selected automatically if there is prior knowledge of the building type or year of construction. We were able to demonstrate better detection results with rectified surface images we were able to obtain using additional information of the floor plan and building height. Further work can address the question of whether the required images can be extracted from Google Street View via affine transformation. Thus, the applicability of the methods can be extended as proprietary image data is no longer required. In addition, we are currently investigating to what extent a classification of the building type is possible using a neural network. The classification could then be used prior to detecting floor levels and possibly improve the result. Beyond the scope of this paper, it is worth noting that considering the use case a distinction should be made between residential and office buildings.

Acknowledgement. This use case was worked on in cooperation with the city of Leipzig within the *CUT*-Project. The project was funded under the name "Modellprojekte Smart Cities 436" (10237739). The data used within this work was provided by departments AGB (GeodatenService) and RDS (Digital City Unit).

References

1. Detectron documentation (2023). https://detectron2.readthedocs.io/en/latest/. Accessed 16 May 2023
2. Auer, S., Gisinger, C., Tao, J.: Characterization of facade regularities in high-resolution SAR images. IEEE Trans. Geosci. Remote Sens. **53**(5), 2727–2737 (2015)
3. Biljecki, F., Ito, K.: Street view imagery in urban analytics and GIS: a review. Landsc. Urban Plan. **215**, 104217 (2021)
4. Cheshmehzangi, A., Butters, C.: Sustainable living and urban density: the choices are wide open. Energy Procedia **88**, 63–70 (2016)
5. Gillott, C., Davison, J.B., Densley Tingley, D.: The potential of vertical extension at the city scale. IOP Conf. Ser. Earth Environ. Sci. **1078**(1), 012079 (2022)
6. Iannelli, G., Dell'Acqua, F.: Extensive exposure mapping in urban areas through deep analysis of street-level pictures for floor count determination. Urban Sci. **1**(2), 16 (2017)
7. Kontokosta, C.E., Tull, C.: A data-driven predictive model of city-scale energy use in buildings. Appl. Energy **197**, 303–317 (2017)

8. Mishkin, D.: OpenCV RANSAC is now USAC (2023). https://opencv.org/evaluating-opencvs-new-ransacs/
9. Resch, E., Bohne, R.A., Kvamsdal, T., Lohne, J.: Impact of urban density and building height on energy use in cities. Energy Procedia **96**, 800–814 (2016)
10. Roy, E., Pronk, M., Agugiaro, G., Ledoux, H.: Inferring the number of floors for residential buildings. Int. J. Geogr. Inf. Sci. **37**(4), 938–962 (2023)
11. Schroedl, S., Wagstaff, K., Rogers, S., Langley, P., Wilson, C.: Mining GPS traces for map refinement. Data Mining Knowl. Discov. **9**(1), 59–87 (2004)
12. Szolomicki, J., Golasz-Szolomicka, H.: Technological advances and trends in modern high-rise buildings. Buildings **9**(9), 193 (2019)
13. Wu, M., Zeng, W., Fu, C.W.: FloorLevel-Net: recognizing floor-level lines with height-attention-guided multi-task learning. IEEE Trans. Image Process. **30**, 6686–6699 (2021)

A Comparative Study of Video-Based Analysis Using Machine Learning for Polyp Classification

Adrian Krenzer(✉) and Frank Puppe

Julius-Maximilians-Universität Würzburg, 97070 Würzburg, Germany
adrian.krenzer@uni-wuerzburg.de

Abstract. Colorectal carcinoma is a leading cause of mortality worldwide and predominantly originates from colon polyps. Not all polyps metamorphose into carcinomas, therefore polyps are categorized via various classification systems. The advent of deep learning and the proliferation of video data have given rise to a plethora of model architectures for automated video classification of polyps. However, the selection of an appropriate model for specific tasks requires careful consideration of various factors, including performance metrics and computational efficiency. In this paper, we present a comparative study of six state-of-the-art model architectures. Capitalizing on the strengths of several state-of-the-art models, a newly developed voting system enhances classification accuracy while maintaining computational efficiency, demonstrated across multiple distinct datasets. The paper explores the integration of such a voting system within the broader framework of video-based polyp identification and provides an empirical evaluation of its performance when juxtaposed with contemporary models. The findings underscore the potential of the proposed system in advancing colorectal polyp classification methodologies, aiming to contribute to early and accurate polyp classification, which is vital in preventing colorectal cancer.

Keywords: Machine learning · Deep learning · Endoscopy · Automation · Video Classification

1 Introduction

Colorectal cancer (CRC) represents the second most prevalent etiology of oncology-associated fatalities globally [2]. The genesis of this malignancy is typically rooted in colon-based lesions referred to as polyps. It's noteworthy, however, that not all polyps metamorphose into carcinomas, hence necessitating their categorization via various classification systems. The categorization, in turn, informs subsequent therapeutic interventions and procedures. Given the nascent

Supported by Bayerisches Staatsministeriums für Wirtschaft, Landesentwicklung und Energie.

D. Seipel and A. Steen (Eds.): KI 2023, LNAI 14236, pp. 144–156, 2023.
https://doi.org/10.1007/978-3-031-42608-7_12

clinical experience of young physicians, the development of computer-assisted modalities capable of aiding polyp classification has been accelerated [16].

Pertinently, the advancement of automated gastroenterology-assistance systems has witnessed substantial research into polyp detection via deep learning methodologies. Polyps, which are mucosal protrusions found in organs such as the intestine and stomach, sometimes undergo anomalous transformations that may escalate into carcinomas [11]. Deep learning-based object recognition techniques, such as Convolutional Neural Networks (CNNs), have been designed to identify and classify these polyps autonomously during diagnostic evaluations, thereby supporting endoscopists [1,19,23]. This technological progression could enhance the precision of automated polyp detection in the future, facilitating or corroborating prognostic evaluations for appropriate polyp management.

The categorization of polyps plays an important role, as it guides endoscopists toward suitable therapeutic strategies. Various approaches are employed for polyp classification, including the shape-based Paris system [14] or the surface structure-based NICE system [9]. These taxonomies can provide preliminary indications of polyp malignancy potential and pertinent therapeutic options [14]. Moreover, van Doorn et al. have reported moderate interobserver agreement among Western international experts regarding the Paris classification system, underscoring the potential of automated classification systems in enhancing interobserver consensus [28].

In this study, we present a comprehensive investigation of a selection of colorectal polyp classification models, with particular attention paid to the inclusion of a novel voting mechanism specifically designed for the video-based identification and classification of polyps. In the rapidly advancing field of medical imaging, the ability to accurately classify colorectal polyps through video analysis stands as a crucial step in early disease detection and timely treatment. The complexity of video data and the intrinsic variability of polyps pose significant challenges, necessitating the development of highly effective computational models. The state-of-the-art video classification models include Image CNN, ViST, 3D-ResNet18, R(2+1)D, S2D, and MVitv2, which were tested under different configurations and on various datasets. The code for this paper is available online[1].

2 Related Work

The objectives of computational polyp research extend beyond localization to include classification based on specific characteristics. For instance, Ribeiro et al. leveraged the feature extraction capabilities of Convolutional Neural Networks (CNNs) to dichotomize polyps into "normal" (average) and "abnormal" (adenoma) categories, utilizing Kudo's pit-pattern classification, a surface-structure based categorization of polyp types [13]. The authors reported an accuracy of 90.96% using the CNN-based classification [25].

[1] https://github.com/Adrian398/A-Comparative-Study-of-Video-Based-Analysis-using-Machine-Learning-for-Polyp-Classification.

In another study [27], a deep learning model was proposed for polyp classification into "Benign", "Malignant", and "Nonmalignant" using pit-pattern classification. The model was trained using a proprietary dataset, achieving a reliability of 84%. Similarly, a popular CNN-based polyp classification method was presented in [30], where the authors employed the Narrow-Band Imaging International Colorectal Endoscopic (NICE) classification [9]. Comparable to pit-pattern classification, the NICE system uses surface features for classification. However, this system additionally integrates color and vascular structure characteristics to categorize polyps into types 1 or 2. Consequently, an initial prognosis regarding whether the polyp is hyperplastic or an adenoma can be made. The authors utilized a CNN combined with a Support Vector Machine (SVM), pre-training the CNN on a non-medical dataset due to a lack of polyp data. Their proposed model achieved an accuracy of approximately 86% [30].

In a different study, Bryne et al. employed the NICE classification to characterize polyps, categorizing them as hyperplastic or adenoma polyps. The authors devised a CNN model for real-time application, trained and validated solely on narrow-band imaging (NBI) video frames. They reported an accurate prediction of 94% [3] on a sample of 125 testing polyps. Komeda et al. and Lui et al. presented specific CNN models to classify polyps into "adenoma" and "non-adenoma" polyps and curable versus noncurable lesions, respectively, using NBI and white-lighted images [12,19]. Ozawa et al. developed a CNN model based on a single-shot multibox detector for polyp detection and classification, achieving a true-positive rate of 92% in detection and an accuracy of 83% in polyp characterization [23]. In 2021, Hsu et al. proposed a custom classification network embedded in a detection and classification pipeline using grayscale images to classify polyp pathology, achieving accuracies of 82.8% and 72.2% using NBI and white light, respectively [10]. Chung Ming et al. discriminated between hyperplastic and adenomatous polyps using different deep learning models and classic feature extraction algorithms, with their best model, AlexNet, achieving an accuracy of 96.4% [18]. A recent work of Sierra-Jerez et al. [26] in 2022 introduced a robust frame-level strategy that achieved a full characterization of polyp patterns to differentiate among sessile serrated lesions (ssls), ads (adenomas), and hps (hyperplastic polyps).

The studies mentioned encompass different yet complementary approaches to computational polyp research, with each leveraging unique models, features, and classifications, creating a diverse landscape of methodologies. Although they all strive towards improved polyp classification, their distinct approaches in terms of feature extraction, categorization techniques, and training datasets suggest a degree of orthogonality, enhancing the comprehensive understanding of the field.

3 Methods

The models were developed and executed on an Nvidia A100 GPU with 40GB of RAM. The model implementations are grounded in the PyTorch framework [24], alongside PyTorch Lightning. Figure 1 shows an overview of all the models used for the comparative study in this paper.

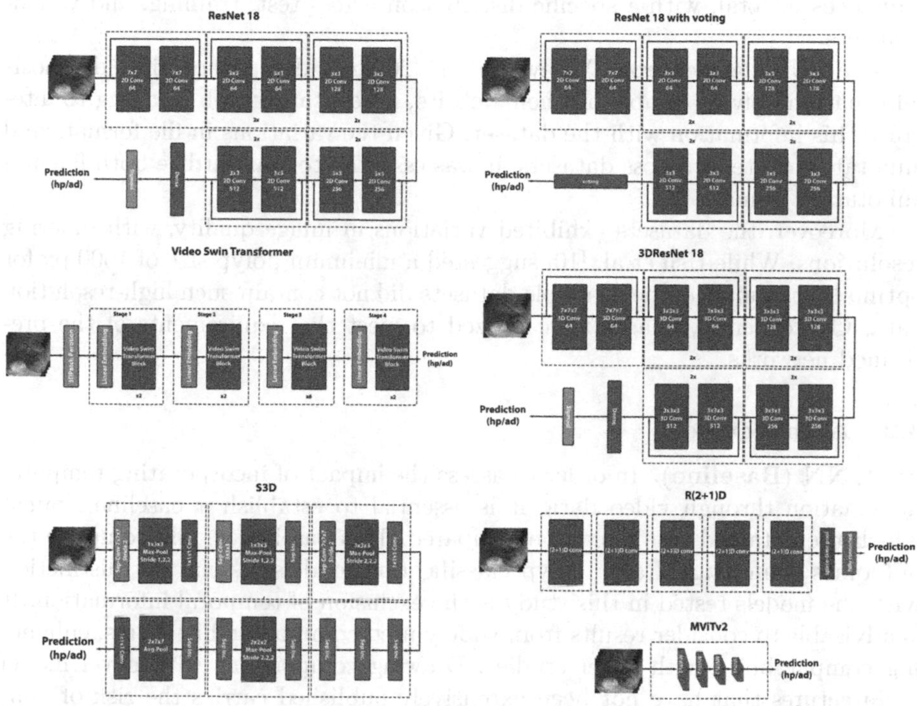

Fig. 1. Overview of the models used in this paper. ResNet18 [8], ResNet18 with voting [8], 3DResNet18 [6], Video Swin Transformer [17], R(2+1)D [6], S3D [29], MVitv2 [15]

3.1 Datasets

In the literature, there are two video datasets containing classification information available online: the KUMC and SUN datasets. To obtain realistic results, the experiments were initially trained and tested on distinct datasets, specifically employing the KUMC dataset for training and the SUN dataset for validation and testing. The SUN dataset was randomly split into a validation and test dataset, maintaining class balance at a 10% split. This division resulted in the selection of specific cases for the hp and ad classes. An additional 20% split was applied to create mixed training, validation, and test datasets, extracting the test set of SUN, leading to the identification of various cases.

The KUMC dataset splits from Li et al. [15] were employed, as they were already generated according to class weights, facilitating comparability. The KUMC dataset contains 155 video sequences, but due to the authors' focus on frame-based models, the training data sequences were combined into one folder and shuffled. However, the annotation files contained adequate information regarding sequence affiliation and position to reverse this process, enabling the use of training images as sequence input. Notably, the data comprises 153 sequences in total, with a specific distribution across test, training, and validation sets.

The SUN dataset from Misawa et al. [20] provides pathological diagnosis information only as a table on their website, necessitating web scraping to integrate this information with the dataset. Given the variations in file formats and annotation systems across datasets, it was essential to standardize both file and annotation formats.

Moreover, the datasets exhibited variations in image quality, with differing resolutions. While Hsu et al. [10] suggested a minimum polyp size of 1600 px for optimal performance, the available datasets did not contain such high-resolution data. Consequently, images were resized to meet the requirements of the pre-trained networks.

3.2 Architectures

2D CNN (Baseline). In order to assess the impact of incorporating temporal information through video data, it is essential to establish a baseline against which the obtained results can be compared. This baseline should comprise the outcomes from single-frame polyp classification models, as the key distinction with the models tested in this study is the exclusion of temporal information. It is advisable to consider results from widely used common architectures, enhancing comparability with other studies. Drawing comparisons with novel model architectures that have not been extensively published carries the risk of non-replicable results.

In terms of establishing a straightforward baseline, a ResNet18 model based on the study by He et al. [8] was implemented and evaluated using random single images. The model, initially pre-trained on the ImageNet dataset [4], was adapted to the new classification task by removing the last layer and retaining the remainder of the model with unmodified parameters to serve as a backbone for feature extraction. A simple linear classification layer with the appropriate number of classes was then trained on the output of the feature extractor.

3D-ResNet18. An additional elementary method to incorporate multi-image information into a model involves expanding existing operations from a 2D paradigm (single image) to a 3D paradigm (multiple images). This could be achieved for commonplace operations such as 2D convolution or 2D max-pooling.

It is established that 3D operations, possessing multiple parameters of their 2D counterparts, consequently demand more training examples and time to attain equivalent performance [21].

Given its popularity and performance in prior research studies, a 3D-ResNet architecture, based on the implementation from Tran et al. [6], was implemented. The *r3d* model is a 3D extension of the renowned ResNet18 architecture, celebrated for its capability to manage deep networks without succumbing to the vanishing gradient problem.

Video Swin Transformer (ViST). Liu et al. [17] published their innovative work, aimed at elucidating video comprehension and discerning actions within videos through a pioneering methodology named *Video Swin Transformer (ViST)*. Their realization builds upon the mmaction2 framework [22] and is publicly accessible. ViST employs the Swin Transformer [5] architecture, originally conceptualized for image recognition tasks, and enhances it for video comprehension by integrating temporal information.

ViST fragments the input video frames into non-overlapping segments and subjects them to the Swin Transformer, thereby capturing spatial and temporal features concurrently. The stratified structure of the Swin Transformer enables the model to effectively apprehend long-range dependencies. The inclusion of temporal data is executed by performing a temporal shift operation on the features extracted by the spatial Swin Transformer layers. This procedure facilitates the model to efficiently consolidate information from a multitude of frames, culminating in superior video comprehension.

ViST was contrasted with other prevailing video understanding methodologies, encompassing 3D CNNs and other transformer-oriented models. The outcomes indicated that ViST outperformed its counterparts, underscoring the benefits of amalgamating the Swin Transformer architecture with temporal information for video understanding tasks.

R(2+1)D. Tran et al. [6] published their research in 2018, emphasizing action recognition in videos. They utilized 3D convolutions as the foundational approach and advanced the field with the development of a novel method referred to as *R(2+1)D convolutions*. They extended the widely adopted ResNet architecture [8] to address video understanding tasks by integrating spatiotemporal convolutions.

The authors advocated a systematic methodology to construct spatiotemporal convolutional networks, investigating varying combinations of models incorporating both mixed 2D & 3D convolutions and complete 3D convolutions within the ResNet framework.

They presented the notion of factorized spatiotemporal convolutions that decompose 3D convolutions into independent spatial and temporal convolutions, thereby diminishing the computational complexity, culminating in the R(2+1)D convolutions.

The R(2+1)D convolutions approach was compared with other prevalent video understanding strategies, such as 2D CNNs, 3D CNNs, and two-stream CNNs. The experimental results demonstrated that the R(2+1)D convolution method exhibited superior performance, accentuating the advantages of employing spatiotemporal convolutions within a residual network for action recognition tasks.

S3D. Xie et al. [29] presented their approach to video classification in their publication in 2018. The research delved into the balancing act between speed and accuracy in video classification. This innovative work tackled the challenges of video classification utilizing convolutional neural networks (CNNs) and put forth the S3D model. The paper primarily dealt with three critical impediments: the representation of spatial information, the representation of temporal information, and the trade-offs between model complexity and speed during both the training and testing phases.

To address these issues, the authors embarked on an extensive investigation of various 3D CNNs and proposed a novel model known as S3D, an acronym for "separable 3D CNN". The S3D model substitutes conventional 3D convolutions with spatially and temporally separable 3D convolutions, bearing similarities with the R(2+1)D model. This led to a model with fewer parameters, higher computational efficiency, and unexpectedly superior accuracy compared to the original I3D model. They further proposed a top-heavy model design, in which 3D temporal convolutions are preserved in higher layers and 2D convolutions are employed in lower layers. This configuration resulted in models that were both faster and more accurate.

The S3D model, when integrated with a spatiotemporal gating mechanism, led to the formation of a new model architecture known as S3D-G. This architecture significantly outperformed baseline methods across various challenging video classification datasets. Furthermore, the model exhibited excellent performance in other video recognition tasks, such as action localization.

MVit2. Li et al. [15] introduced their refined model architecture, MVitv2, in their paper towards the end of 2021. The MViTv2 paradigm represents an enhancement of the MViTv2 model by Fan et al. [7], which is essentially an expansion of the ViT [5] framework, characterized by a feature hierarchy ranging from high to low resolution.

MViTv2 stands as a unified architecture for both image and video classification, along with object detection. The major challenges addressed in this paper include the intricacies of designing architectures for a myriad of visual recognition tasks and the considerable computational and memory demands associated with Vision Transformers (ViT) in the context of high-resolution object detection and space-time video understanding.

Two primary enhancements underpin MViTv2: decomposed relative positional embeddings and residual pooling connections. These modifications have driven a marked improvement in both accuracy and computational efficiency.

To enhance the information flow and streamline the training of pooling attention blocks in MViTv2, with a minimal impact on computational complexity, the authors integrated a residual pooling connection.

3.3 Majority Voting with 2D CNN

In an effort to enhance the performance of the baseline model and emulate the inclusion of temporal information, a majority voting mechanism is employed. This strategy evaluates the baseline model based on not only the current frame but also on the preceding n frames, thereby enabling the model to consider a series of frames as opposed to a single frame. The predicted class for each frame is collated, and the final output is determined by the majority vote among these predictions.

To realize this approach, the aforementioned ResNet18 baseline model was utilized. A sliding window of size n was applied to the video frames, and the model's predictions were documented for each frame within this window. The class that garnered the highest number of votes was then designated as the final prediction for the current frame. This process was reiterated for each frame in the video, with the overall performance of the majority voting mechanism subsequently evaluated.

The majority voting technique confers several advantages. Firstly, it offers a straightforward and efficient method to integrate temporal information into the baseline model, circumventing the need for complex alterations or retraining. Secondly, it augments the model's robustness to noise and outliers, as majority voting can assist in filtering out sporadic incorrect predictions. Lastly, the majority voting approach can be readily adapted to varying window sizes, thereby providing an opportunity to examine the effect of different degrees of temporal information on model performance. The application of this technique can incorporate an equal or unequal number of frames within the sliding window. The sliding window method retains its efficiency and can be scaled up to accommodate a greater number of potential categories.

Nonetheless, this method also presents certain limitations. The majority voting mechanism presumes that the relationships between successive frames remain consistent, a condition that may not always be fulfilled. Moreover, the selection of window size n can significantly influence performance, and determining the optimal value may necessitate trial-and-error and cross-validation.

By juxtaposing the results derived using the majority voting method with those obtained using other methods, it becomes feasible to garner a more profound understanding of the significance of temporal information for polyp detection and diagnosis.

4 Results and Discussion

This comparative analysis facilitates an assessment of the models' ability to generalize across diverse data sources and conditions.

Table 1. Evaluation on the seperated datasets. Inference time (Inf. Time) is givin in ms while all other values (F1, Accuracy (Acc.), Precision (Pre.), Recall (Rec.)) are given in %.

Model	F1	Acc.	Prec.	Rec.	Inf. Time
Image CNN	95.3	93.2	94.2	**96.4**	**3**
Image CNN with voting	**95.9**	93.9	**96.7**	95.1	15
ViST	93.3	93.4	93.3	93.4	68
3D-ResNet18	93.7	**94.1**	93.3	94.1	27
R(2+1)D	92.6	91.3	94.0	91.3	36
S2D	93.9	93.8	94.0	93.8	122
MViTv2	92.4	90.7	94.1	90.7	81

Table 2. Evaluation on the mixed datasets for generalization. Inference time (Inf. Time) is givin in ms while all other values (F1, Accuracy (Acc.), Precision (Pre.), Recall (Rec.)) are given in %.

Model	F1	Acc.	Prec.	Rec.	Inf. Time
Image CNN	89.6	88.2	91.0	88.2	**3**
Image CNN with voting	**91.3**	89.4	**97.0**	86.2	15
ViST	89.6	85.3	94.3	85.3	68
D-ResNet18	87.2	85.2	93.5	85.2	27
R(2+1)D	88.9	86.3	91.7	86.3	36
S2D	91.1	87.9	94.6	87.9	122
MViTv2	91.9	**89.7**	94.3	**89.7**	81

The impetus for contrasting separate and mixed datasets resides in the assessment of the models' capability to generalize across disparate data sources, a pivotal factor in real-world applications. In such practical scenarios, training data often originates from distinct sources compared to the test data, potentially leading to varied characteristics.

It is postulated that models trained on mixed datasets might exhibit superior generalization abilities given their exposure to a broader spectrum of examples during training. Conversely, models trained on separate datasets could potentially demonstrate enhanced performance if evaluated on their respective data sources, as they may acquire more dataset-specific features and dependencies.

To evaluate the performance of models trained on separate and mixed datasets, mixed training, validation, and test datasets were constructed. These were assembled with a 10%/20%/80% allocation for validation/test/training respectively, and the partitioning was carried out randomly, maintaining class ratios as similar to the original dataset as feasible. Retaining identical hyperparameters and training configurations across the two sets of splits might not invariably lead to optimal performance for each approach, as the ideal settings could fluctuate depending on the dataset composition.

The outcomes of the experiment contrasting the performance of models trained on separate and mixed datasets are tabulated in Table 1 and 2.

The performance of various models was assessed on two distinct dataset categories: separated and mixed datasets. The performance metrics evaluated include F1 score, accuracy (Acc), precision (Prec), recall (Rec), and inference time (Inf Time). The models under consideration are Image CNN, ViST, 3D-ResNet18, R(2+1)D, S2D, and MViTv2.

The results presented in Table 1 show that, for the separated datasets, the Image CNN with voting model demonstrated the highest F1 score at 95.9%. In terms of accuracy, the 3D-ResNet18 model was the standout performer with an accuracy of 94.1%. However, it's worth noting that inference time varied significantly among the models, with Image CNN offering the fastest time of 3 ms, while S2D had the slowest at 122 ms.

Turning our attention to the mixed datasets (Table 2) the image CNN with voting and the MViTv2 model outperformed the others in terms of F1 score (91.3%), precision (97.0%), and recall (89.7%). However, its accuracy score (89.7%) was marginally higher than that of the S2D model (87.9%). As with the separated datasets, the inference time for the mixed datasets was fastest for the Baseline model (3 ms) and slowest for the S2D model (122 ms).

Upon examining the overall results, it's evident that while Image CNN with voting and MViTv2 delivered the highest F1 scores for the separated and mixed datasets respectively, their computational speed, as measured by inference time, was not the fastest. This highlights a critical trade-off between model performance and computational efficiency. Nevertheless, in border cases, the majority voting mechanism may face challenges due to insufficient frames at the start or end of a video sequence. To address this, padding strategies or dynamic window resizing could be employed, each with its unique implications on the model's performance. Additionally, situations where there's an equal vote among classes, particularly with even-sized windows, necessitate a strategy for breaking ties, possibly using prediction confidence levels. Therefore, despite its advantages, the majority voting mechanism requires further exploration to optimize its performance in border cases and other complex scenarios.

It is also apparent that 3D-ResNet18 offers an appealing balance between high accuracy and reasonable inference time. For situations where time is a critical factor, the Baseline model or Image CNN might be more suitable due to their faster inference times, albeit at the cost of some model performance.

The findings suggest that the choice of model should be guided by the specific requirements of the task at hand, including the type of dataset and the relative importance of performance metrics and inference time. Future research could explore ways to optimize these models further or investigate other model architectures that may offer superior performance or computational efficiency.

5 Conclusion

In conclusion, this study has yielded insights into the field of colorectal polyp classification through a comprehensive examination of various state-of-the-art models. Central to our findings is the novel voting system introduced specifically for video-based polyp classification, which demonstrated a noteworthy improvement in classification performance across numerous models. The system's effectiveness was empirically validated across diverse datasets, exhibiting enhanced performance in both accuracy and computational efficiency.

The results underscore the potential of integrating robust ensemble methods, such as the proposed voting system, within video classification tasks in medical imaging. The system has been shown to offer a means of leveraging the collective strengths of different models to enhance the overall predictive performance, consequently augmenting the diagnostic accuracy of colorectal polyp detection.

References

1. Bour, A., Castillo-Olea, C., Garcia-Zapirain, B., Zahia, S.: Automatic colon polyp classification using convolutional neural network: a case study at basque country. In: 2019 IEEE International Symposium on Signal Processing and Information Technology (ISSPIT), pp. 1–5 (2019). https://doi.org/10.1109/ISSPIT47144.2019.9001816
2. Bray, F., Ferlay, J., Soerjomataram, I., Siegel, R.L., Torre, L.A., Jemal, A.: Global cancer statistics 2018: GLOBOCAN estimates of incidence and mortality worldwide for 36 cancers in 185 countries. CA Cancer J. Clin. **68**(6), 394–424 (2018). https://doi.org/10.3322/caac.21492
3. Byrne, M., et al.: Real-time differentiation of adenomatous and hyperplastic diminutive colorectal polyps during analysis of unaltered videos of standard colonoscopy using a deep learning model. Gut **68**, gutjnl-2017 (2017). https://doi.org/10.1136/gutjnl-2017-314547
4. Deng, J., Dong, W., Socher, R., Li, L.J., Li, K., Fei-Fei, L.: Imagenet: a large-scale hierarchical image database. In: Staff, I. (ed.) 2009 IEEE Conference on Computer Vision and Pattern Recognition, pp. 248–255. IEEE (2009). https://doi.org/10.1109/CVPR.2009.5206848
5. Dosovitskiy, A., et al.: An image is worth 16x16 words: transformers for image recognition at scale. https://arxiv.org/pdf/2010.11929
6. Tran, D., Wang, H., Torresani, L., Ray, J., LeCun, Y., Paluri, M.: A closer look at spatiotemporal convolutions for action recognition. https://arxiv.org/pdf/1711.11248
7. Fan, H., et al.: Multiscale vision transformers. https://arxiv.org/pdf/2104.11227

8. He, K., Zhang, X., Ren, S., Sun, J.: Deep residual learning for image recognition. Tech. https://arxiv.org/pdf/1512.03385
9. Hewett, D.G., et al.: Validation of a simple classification system for endoscopic diagnosis of small colorectal polyps using narrow-band imaging. Gastroenterology **143**(3), 599–607 (2012)
10. Hsu, C.M., Hsu, C.C., Hsu, Z.M., Shih, F.Y., Chang, M.L., Chen, T.H.: Colorectal polyp image detection and classification through grayscale images and deep learning. Sensors **21**(18), 5995 (2021). https://doi.org/10.3390/s21185995
11. Khan, M.A., et al.: Gastrointestinal diseases segmentation and classification based on duo-deep architectures. Pattern Recognit. Lett. **131**, 193–204 (2020). https://doi.org/10.1016/j.patrec.2019.12.024. https://www.sciencedirect.com/science/article/pii/S016786551930399X
12. Komeda, Y., et al.: Computer-aided diagnosis based on convolutional neural network system for colorectal polyp classification: preliminary experience. Oncology **93**, 30–34 (2017). https://doi.org/10.1159/000481227
13. Kudo, S., et al.: Colorectal tumours and pit pattern. J. Clin. Pathol. **47**(10), 880–885 (1994). https://doi.org/10.1136/jcp.47.10.880. https://jcp.bmj.com/content/47/10/880
14. Lambert, R.F.: Endoscopic classification review group. Update on the Paris classification of superficial neoplastic lesions in the digestive tract. Endoscopy **37**(6), 570–578 (2005)
15. Li, K., et al.: Colonoscopy polyp detection and classification: Dataset creation and comparative evaluations. PLoS ONE **16**(8), e0255809 (2021). https://doi.org/10.1371/journal.pone.0255809
16. Liaqat, A., Khan, M.A., Shah, J.H., Sharif, M.Y., Fernandes, S.L.: Automated ulcer and bleeding classification from WCE images using multiple features fusion and selection. J. Mech. Med. Biol. **18**, 1850038 (2018)
17. Liu, Z., Ning, J., Cao, Y., Wei, Y., Zhang, Z., Lin, S., Hu, H.: Video swin transformer. https://arxiv.org/pdf/2106.13230
18. Lo, C.M., Yeh, Y.H., Tang, J.H., Chang, C.C., Yeh, H.J.: Rapid polyp classification in colonoscopy using textural and convolutional features. Healthcare **10**(8) (2022). https://doi.org/10.3390/healthcare10081494. https://www.mdpi.com/2227-9032/10/8/1494
19. Lui, T., Wong, K., Mak, L., Ko, M., Tsao, S., Leung, W.: Endoscopic prediction of deeply submucosal invasive carcinoma with use of artificial intelligence. Endosc. Int. Open **07**, E514–E520 (2019). https://doi.org/10.1055/a-0849-9548
20. Misawa, M., et al.: Development of a computer-aided detection system for colonoscopy and a publicly accessible large colonoscopy video database (with video). Gastrointest. Endosc. **93**(4), 960-967.e3 (2021). https://doi.org/10.1016/j.gie.2020.07.060
21. Mittal, S.: Vibhu: A survey of accelerator architectures for 3d convolution neural networks. J. Syst. Architect. **115**, 102041 (2021). https://doi.org/10.1016/j.sysarc.2021.102041
22. MMAction2 Contributors: Openmmlab's next generation video understanding toolbox and benchmark (2020)
23. Ozawa, T., Ishihara, S., Fujishiro, M., Kumagai, Y., Shichijo, S., Tada, T.: Automated endoscopic detection and classification of colorectal polyps using convolutional neural networks. Therap. Adv. Gastroenterol. **13**, 175628482091065 (2020). https://doi.org/10.1177/1756284820910659
24. Paszke, A., et al.: Pytorch: an imperative style, high-performance deep learning library. https://arxiv.org/pdf/1912.01703

25. Ribeiro, E., Uhl, A., Häfner, M.: Colonic polyp classification with convolutional neural networks. In: 2016 IEEE 29th International Symposium on Computer-Based Medical Systems (CBMS), pp. 253–258 (2016). https://doi.org/10.1109/CBMS.2016.39
26. Sierra-Jerez, F., Martínez, F.: A deep representation to fully characterize hyperplastic, adenoma, and serrated polyps on narrow band imaging sequences. Heal. Technol. (2022). https://doi.org/10.1007/s12553-021-00633-8
27. Tanwar, S., Goel, P., Johri, P., Diván, M.: Classification of benign and malignant colorectal polyps using pit pattern classification. SSRN Electron. J. (2020). https://doi.org/10.2139/ssrn.3558374
28. Van Doorn, S.C., et al.: Polyp morphology: an interobserver evaluation for the Paris classification among international experts. Official J. Am. College Gastroenterol. ACG **110**(1), 180–187 (2015)
29. Xie, S., Sun, C., Huang, J., Tu, Z., Murphy, K.: Rethinking spatiotemporal feature learning: speed-accuracy trade-offs in video classification. https://arxiv.org/pdf/1712.04851
30. Zhang, R., Zheng, Y., Mak, W., Yu, R., Wong, S., Poon, C.: Automatic detection and classification of colorectal polyps by transferring low-level CNN features from nonmedical domain. IEEE J. Biomed. Health Inform. **21**(1), 41-47 (2016). https://doi.org/10.1109/JBHI.2016.2635662

Object Anchoring for Autonomous Robots Using the Spatio-Temporal-Semantic Environment Representation SEEREP

Mark Niemeyer[1(✉)], Marian Renz[1], and Joachim Hertzberg[1,2]

[1] Plan-Based Robot Control Group, German Research Center for Artificial Intelligence, Osnabrück, Germany
{mark.niemeyer,marian.renz}@dfki.de

[2] Knowledge -Based Systems Group, Osnabrück University, Osnabrück, Germany
joachim.hertzberg@uni-osnabrueck.de

Abstract. For single-plant specific weed regulation, robotic systems and agricultural machinery in general have to collect a large amount of temporal and spatial high-resolution sensor data. *SEEREP*, the Spatio-Temporal-Semantic Environment Representation, can be used to structure and manage such data more efficiently. *SEEREP* deals with the spatial, temporal and semantic modalities of data simultaneously and provides an efficient query interface for all three modalities that can be combined for high-level analyses. It supports popular robotic sensor data such as images and point clouds, as well as sensor and robot coordinate frames changing over time. This query interface enables high-level reasoning systems as well as other data analysis methods to handle partially unstructured environments that change over time, as for example agricultural environments. But the current methodology of *SEEREP* cannot store the result of the analysis methods regarding specific objects instances in the world. Especially the results of the anchoring problem which searches for a connection between symbolic and sub-symbolic data cannot be represented nor queried. Thus, we propose a further development of the *SEEREP* methodology in this paper: For a given object, we link the existing semantic labels in different datasets to a unique common instance, thereby enabling queries for datasets showing this object instance and with this enabling the efficient provision of datasets for object-centric analysis algorithms. Additionally, the results of those algorithms can be stored linked to the instance either by adding facts in a triple-store like manner or by adding further data linked to the instance, like a point, representing the position of the instance. We show the benefits of our anchoring approach in an agricultural setting with the use-case of single-plant specific weed regulation.

Keywords: anchoring · environment representation · selective weeding

The DFKI Niedersachsen (DFKI NI) is sponsored by the Ministry of Science and Culture of Lower Saxony and the VolkswagenStiftung. This work is supported by the Federal Ministry for the Environment, Nature Conservation, Nuclear Safety and Consumer Protection (BMUV) within the CognitiveWeeding project (grant number: 67KI21001B).

D. Seipel and A. Steen (Eds.): KI 2023, LNAI 14236, pp. 157–169, 2023.
https://doi.org/10.1007/978-3-031-42608-7_13

1 Introduction

Fig. 1. RGB image (left) and point cloud (right) showing the same section of simulated maize plants and weeds. The individual plants in both data types can be linked and accessed by assigning instances in *SEEREP*. This enables the retrieval of all available data of a specific plant at once for analysis purposes.

Mobile robots and agricultural machinery, which want to perform plant-specific weed regulation, need a lot of information about their environment. For that, they have to acquire spatial and temporal high-resolution data. Cameras and laser scanners, but also others, are popular sensor types to sense the environment. If an object or specific regions of the environment are perceived by multiple sensors, maybe even in a multi-robot situation, at multiple points in time those sensor readings from various viewpoints can be combined. Depending on the different analysis algorithms and their requirements, the various information types can be used jointly.

For the analysis it is useful to filter the data based on the spatial, temporal, and semantic information. Data in these three modalities can be stored and queried using the semantic environment representation *SEEREP*[1] [10]. When data is added to *SEEREP*, indices are created based on the position and spatial extent, the timestamp and semantic information of the data. Based on those indices, queries for robotic 3D sensor data can be performed efficiently, in a specific region, within a given time frame and with defined semantic annotations.

Currently, the analysis of the sensor data by an external program is not performed in one step but rather by multiple consecutive analysis steps. The addition of semantic labels to the raw sensor data, for example, is one of the first steps. *SEEREP* already supports semantic labels, but other analysis results like the recognition of an object instance in different datasets or attributes of those instances are not yet supported. Thus, this paper presents further development of the methodology of *SEEREP* which allows the linkage of all datasets

[1] https://github.com/agri-gaia/seerep.

containing an object instance. Furthermore, attributes (analysis results regarding an instance generated by an external program) can be stored in a triple-store like fashion attached to instances. Additionally, the datatype *point* is added to the existing datatypes of *SEEREP*. The *point* datatype can be used to store a single position like the position of an instance which was calculated based on some sensor data.

With the linkage of the data to an instance, the results of the anchoring problem can be stored. The anchoring problem tries to find the connection between symbolic (e.g., the semantic description of a specific object) and sub-symbolic data (e.g., sensor data showing the object). For the implementation, it is assumed that subsets of the sub-symbolic data are already labelled with semantic annotations and that with this the connection to the symbolic representation already exists. Based on this, the generic semantic annotation has to be linked to a specific instance. Another result of the anchoring problem is commonly the position of the object. With the storage of *point* data, this position can be saved and linked to the object instance.

As a whole, the sensor data and the analysis results are stored in one location and the sensor data as well as the analysis results can be queried by further analysis algorithms. For example, if an agent intends to analyze the movement of chairs in an office during a specific day, it could just query for data showing the chairs (semantic query) in the given room (spatial query) during the given time interval (temporal query) and analyze it. The anchoring result could afterwards be stored in *SEEREP* using the instances and the datatype *point*. Another algorithm, which wants to calculate the distance traveled by the chair, could then query the positions of the chair over time easily.

Besides the indoor scenario, outdoor applications can also benefit from the unified storage of sensor data and analysis results. Especially in agriculture, there is a high demand for analysis of the cropping system [17].

In smart agriculture with plant-specific treatment, it is relevant to be able to query this single-plant specific data. This enables, for example, growth analysis of plants, because the data of a plant at different points in time can be queried easily. Also, different sensor readings from different sensor types of the same plant instance are easily accessible by queries (see Fig. 1).

Additionally, the distinction of plants into the categories crop, harmful and not harmful weeds can be done with this, because the position and other attributes of the plant instances can be easily queried. This distinction of the plants is used as an application example of the presented further development of *SEEREP* in Sect. 4. The application is motivated by the research project *Cognitive Weeding*, which aims to categorize weeds into harmful and not-harmful weeds so that only the harmful weeds are removed. Plant specific analysis becomes crucial in the mentioned weed categorization, since single plants are targeted in this example. It benefits from *SEEREP*'s instances, since it simplifies the retrieval of plant-specific information for spatial analysis. Additionally, the example requires plants to be counted, which can only be achieved with fused plant instances.

The anchoring problem is simplified in this use-case, because the plants do not move. They only change their position when they are planted or removed/harvested. Also, they change their shape only through growth gradually or when they are partially harvested.

The main contribution of our paper is a practical approach to the anchoring problem in robotics, based on the further development of the methodology of the environment representation *SEEREP*. To demonstrate the advantage of storing the anchoring results in a spatio-temporal-semantic environment representation, a simulated dataset as well as the anchoring result, known from the simulation, are stored in *SEEREP*. Based on this, queries for data and instances are used to enable the use-case. All needed data is provisioned to the use-case in a unified manner, in contrast to existing solutions.

Section 2 discusses related work, and our further development of the *SEEREP* architecture is presented in Sect. 3. An example usage of the further development is demonstrated in Sect. 4 on the basis of a simulated dataset. Finally, in Sect. 5 the conclusions are presented and future work is discussed.

2 Related Work

Anchoring is a symbol grounding problem [6], which arises when attempting to assign meanings to symbols. A well-known special case is the tracking problem [14] which consists of finding an object in different sensor data. Several approaches for highly dynamic indoor environments exist. Elfring [3] introduced a semantic world model with probabilistic multiple hypothesis anchoring, and Persson et al. [13] also proposes a probabilistic semantic world model for object tracking. For this, explicit object detection is used to assign labels to the objects in the data. Günther et al. [5] also uses a probabilistic multiple hypothesis anchoring and extends it by semantic context-awareness. Another approach [8] creates an object-based semantic world model without explicit labels. They assume that objects are always on planes (like the ground and tables). Thus, they detect objects on planes and describe them by their 2D convex hull. All these approaches focus on the current state of the world and especially on the current state and position of the recognized objects, while our approach focuses on connecting the object instances with the corresponding data at all timestamps.

In another use-case, anchoring is needed to get all the data regarding an object so that the data can be used to analyze the object. An example is plant phenotyping [2,7]. For phenotyping based on robotic sensor data, it is necessary to match the data within one data acquisition run as well as the data from multiple runs from different points in time. Most analysis methods focus on the analysis itself and do not address the input data storage and provision, and neither do they address the storage of the results for easy usage by following algorithms.

For the storage of the data, at least spatio-semantic environment representations are needed to store the numerical sensor data as well as the semantic labels and anchoring results. To address the different requirements of the numerical data and the semantic information, Oliviera et al. [11] proposed a dual memory approach. They store the numerical data in a NoSQL database, where the data is stored in a key-value store. The symbolic data is stored in subject-predicate-object triples in a triple store. This approach has the advantage that both numerical and symbolic data can be stored appropriately. Though, the storage of the numerical data in the NoSQL database has the disadvantage that no spatial indices are created and it is not possible to query the data based on its spatial position. Deeken et al. [1] also uses a dual memory approach, but instead of the NoSQL database, they store the numerical data in a GIS, enabling 2D spatial queries. In contrast to those, our approach is not just spatio-semantic but spatio-temporal-semantic. It can also handle 3D instead of just 2D data and with the temporal modality the data from different points in time can be used in the anchoring result.

The spatio-temporal-semantic environment representation *SEEREP* [10] does not only enable 3D spatial queries instead of just 2D, but it also stores and creates indices for the temporal component (the timestamp) as well as for the semantic annotations of the robotic sensor data. Through the combined information in spatio-temporal-semantic queries, data can be easily provisioned to anchoring algorithms.

It is worth mentioning that the current methodology of *SEEREP* cannot link data to specific object instances and with this it cannot store the result of the anchoring problem. Thus, it is not possible to store the sensor data used for the anchoring problem and the anchoring results in a unified manner. The next Sect. 3 presents our further development of the *SEEREP* architecture. This will address the missing capability of *SEEREP* to store the anchoring result and to provide object-centric queries, thus, enabling other analysis algorithms to get the needed data for the analysis of an object easily and to also store the analysis result afterwards attached to the object instance.

3 SEEREP Architecture Extension

SEEREP is designed to run on a server with more storage and computing power than the average robotic system, because the *SEEREP* server has to store all the data a robotic system creates over time and it has to create indices to enable efficient queries. These indices are created for the spatial, the temporal and the semantic modality of the data. Thus, the server can handle all three modalities at once and it enables combined queries in all three modalities. The architecture of *SEEREP* is sketched in Fig. 2. The existing parts of the architecture are referenced by the letters A-I in the figure and the contribution in this paper regarding the instances is referenced by the letters J-K. The additional contribution of the datatype *point* does not extend the architecture, because it is already included in the generic description of data types [C-D]. For further information regarding the general architecture, take a look at [10]. The remainder of this section will elaborate the instance overview and the instances in detail. Furthermore, the new data type *point* is presented.

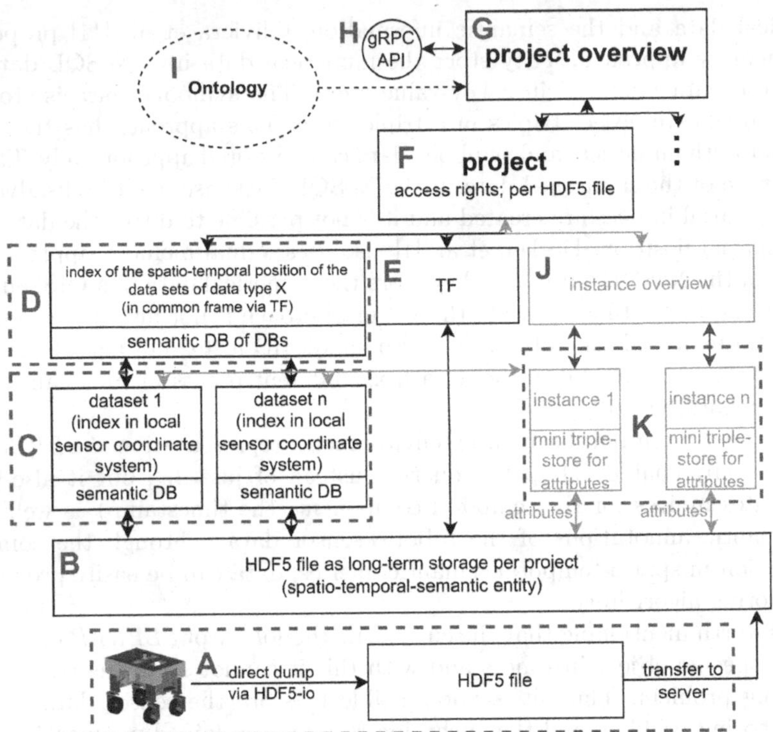

Fig. 2. ***SEEREP* architecture overview.** In this diagram, the sensor data is dumped directly into HDF5 files on the robot and the later transfer to the server [A] and the structure of the *SEEREP* server itself [B-K]. The server consists of an HDF5 file [B] per project [F], indices [C] per dataset and indices [D] of the bounding boxes of the datasets. Transformations (TF) [E] are used to support multiple coordinate frames. A project overview [G] combines multiple projects, serves an API [H] and has a connection to an ontology [I]. The existing architecture is extended by the instance overview [J] and the individual instances [K] with a mini-triple store for attributes and their references to the datasets.

3.1 Instances

The current approach of the semantic labels in the datasets only assigns a concept of an ontology to a dataset or parts of a dataset. With this a link between the sub-symbolic, numerical data and the symbolic ontology is already created, but the linkage of the data to a specific object in the real world is not possible. Therefore, it is also not feasible to store results of the anchoring problem. To address this, we introduce instances [K]. Each instance represents a specific object in the world on a symbolic level. To implement this in the existing architecture, the labels of the semantic annotations in the datasets have to be linked to the instance. For this, the labels are extended by a universally unique identifier (UUID). This UUID identifies a specific object instance and is irrevocably connected to the ontology concept of the object. For example, a region in an image was labeled as *a tree* beforehand. With this UUID, the label of the region changes from *a tree* to *tree with a given UUID*.

When this label with UUID is assigned to multiple datasets, all of them are linked to a common instance. Thus, the instance knows all the datasets containing labels with its instance UUID. This link between the data [C] and the instance [K] is visualized in the architecture in Fig. 2. To avoid redundant information in the HDF5 file, the link between the datasets and the instances is only stored as the instance UUID attached to the labels of the datasets. The mapping the other way around from the instances to all the datasets containing the instance is created when the HDF5 file is loaded.

The linkage of all the data regarding an instance enables algorithms to get all the data regarding a specific instance and to use this data to perform analysis. With this, different information types like color information from images and spatial information from point clouds are easily accessible. Possible analysis results in the agricultural domain are, for example, the biomass, the growth rate or the position of a plant. To store those results, each instance has a mini triple-store, in which the subject is always the instance. The predicate and the object can be used to store the analysis result as an attribute to the instance. For persistence, the triple-store is stored in the HDF5 file.

3.2 Data Type Point

The position of an instance calculated by an analysis algorithm can be stored in the triple-store of an instance. Though, no spatial index is created for the positions which are stored in the triple-store. Therefore, the data type *point* is added besides the already existing data types *images* and *point clouds*. This means for the architecture in Fig. 2 that another vertical stack for a data type at [C-D] is added.

The spatio-temporal-semantic index of the data type *point* allows efficient queries in the three domains for this data type, thus, enabling efficient spatial queries of instance positions in particular. The semantic labels of the *points* are extended by the UUID of the corresponding instance, like it was already introduced for all data types in general. Hence, it can be closed from the result of the spatial query of the *points* to the instances. With these, spatial queries regarding the instances are possible indirectly.

4 Application Example

We test *SEEREP* for data management in a selective weeding example using a synthetic data set of maize plants and weeds on a field. The advantages of the synthetic data are that perfect labels and the solution to the object anchoring problem are available directly. Thus, this application example does not rely on an object anchoring algorithm and the error-prone data preparation steps for real robotic sensor data can be omitted. This includes that the synthetic data

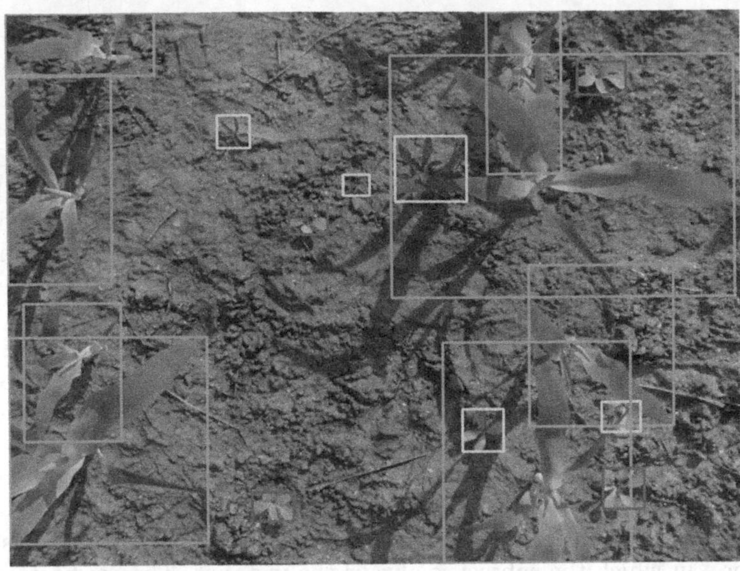

Fig. 3. Result of the rule inference for the selective weeding example. Detected plants are indicated by bounding boxes. Different colors indicate the inferred category. Green marks a plant as crop, yellow as harmless and red as harmful weed. The visualization is created using *SEEREP*'s query interface, enabling efficient retrieval of image, bounding boxes and categories via instances. (Color figure online)

can be uploaded to the *SEEREP* server directly and that the data capturing on the robot, the data preparation, as well as the transfer from the robot to the server is not needed.

Based on the data, the selective weeding example categorizes, by using a rule-based system, each individual plant as one of four categories: harmful weed, non-harmful weed, crop and endangered species. For this categorization, the rule-based system requires plant specific information to reason about whether a specific plant is considered harmful or harmless. This information could be derived by any arbitrary data processing pipeline, but *SEEREP*'s query interface simplifies the access to the sensor data for each individual plant by linking the respective data to the occurring plant instances. Therefore, anchoring can be done beforehand, the resulting plant instances can be stored in *SEEREP* and instance specific information can be retrieved easily.

The rule-based system differentiates between different weed species and identifies the harmful individuals among them based on their number per square meter and their distance to the next crop. Contrary to other precision weeding applications like [12], the actual weed removal can therefore not be done as soon as the weeds are detected but requires an additional processing step. This additional step allows keeping selected weeds which is thought to benefit biodiversity [16] as well as soil quality [9].

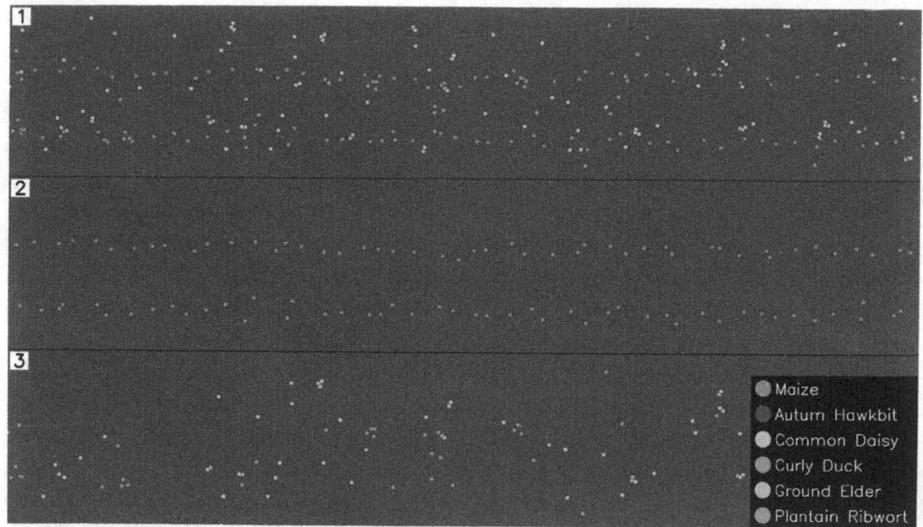

Fig. 4. Returned points for different queries displayed as shapefiles. Different colors display different plant species. (1) show all available plants, (2) shows plants returned by a query for the semantic label "Maize" and (3) shows plants returned by a query for the semantic label "Harmful Weed".

The synthetic data set is created using Blender [4]. A photorealistic 3D model of a field scene with maize plants and weeds is created. The maize models are placed in rows, the weeds are distributed randomly over the field. A virtual camera (1280×960 px) is placed facing down at 2 m above the field. From the simulation, RGB images are rendered. Laser scans and labeled bounding boxes with plant names are generated for each frame and used as ground truth. For each frame, the camera is moved vertically along the crop line for 0.5 m, resulting in 40 adjacent frames covering 20 m. Additionally, the simulation provides the global position of each plant with a unique identifier, also linking each plant to its appearance in the rendered frames. To summarize, the simulation generates 40 RGB images and point cloud frames with labeled bounding boxes and unique identifiers for each plant. Note that multiple bounding boxes can be linked to the same plant.

The given global position of each plant simplifies the anchoring process, since each bounding box can easily be linked to its instance. Nevertheless, in the real world scenario of this example the anchoring process still remains a bigger challenge, especially when monitoring plants over a period of time. Plants change due to growth, new plants may appear or be removed. The obvious way to properly link a plant's appearance to a semantically annotated instance would be the constant position, anchoring plants over a period of time regardless of the change of size and shape. *SEEREP* can easily be used for such kind of anchoring by utilizing queries, returning the plants at a specific position over a given time frame and linking occurrences of the same species at the same position to a single instance.

While this could be an approach for real world data, it is not necessary for the synthetic data set. The ground truth quality labeling and the given global position provides a shortcut for the anchoring problem, while the resulting plant instances can still be stored as instances in *SEEREP*.

The rule-based system for the selective weeding example is set up as described in [15]. It requires each plant's distance to the closest crop and the count of plants per species and square meter. To calculate these values, the position and the species name of each plant instance are needed, which makes *SEEREP*'s spatial queries highly valuable. Specifically, the position of each plant and the species name are necessary for the rule-based systems reasoning. Additionally, the areal extent of each plant is needed as information for the selective removal, so that a selective sprayer or a selective how can act precisely on the right area.

The individual plant instances are retrieved using a query for points (Fig. 4 (1)). Each point represents a plant instance, containing the instance UUID, the species name and the position in the camera coordinate frame. By not specifying spatial and temporal constraints, all available plant instances are returned. A second query is used to obtain the rigid transformation from the respective camera frame into the global coordinate frame. The transformation is necessary since each point is represented in the camera's coordinate frame and needs to be placed into a grid map, which is represented in the global coordinate frame. The map representation is used to calculate necessary values for the rule-based system, which then infers the category for each plant instance. The resulting category can be written back to *SEEREP* by adding another semantic label to the respective instance. Additionally, the position of the weeds categorized as harmful, and the maize plants are stored as points in the shapefiles format to be used as an application map for a selective sprayer or hoe. This requires the areal extent of each plant as an attribute for the point in the shapefiles. The areal extent is defined as the circular area around the plant's position that includes the whole plant as visible from above. To calculate the areal extent, the bounding box of each plant is used. Bounding boxes are obtained by *SEEREP*'s query interface using an image query. For efficiency, a flag is set in the query to retrieve the bounding boxes without the actual image data. Each bounding box is assigned to its respective plant instance using the instance UUID and each bounding box is to calculate the diameter in *cm* using the camera's intrinsic projection matrix. If a plant instance is represented by multiple bounding boxes and therefore has multiple areal extents, only the biggest areal extent is assigned as attribute.

Obviously, the creation of the shapefiles requires only points belonging to "Maize" and "Harmful Weed". Again, *SEEREP*'s query interface can be used for efficient retrieval, using queries for points, that contain the respective label (Fig. 4 (2) & (3)).

Furthermore, the query interface is used to visualize the effect of each rule of the rule-based system (Fig. 3). Again, an image query is used to retrieve all images with bounding boxes and plant instance UUID attached to each bounding box. Each instance UUID is used to fetch the plant instance's stored category with an instance query from *SEEREP*. Depending on the category, the bounding box is colored in the image. *SEEREP* allows plant instance specific analysis, giving easy and efficient access to all data.

Using *SEEREP* makes the design of the overall weeding system more flexible, since new data streams can easily be integrated. An algorithm for estimating the plant's height for example could access the laser scan data using the link made by the plant instance, without having to match sub parts of the point cloud with the individual plants.

5 Conclusion

This paper presented the further development of the methodology of *SEEREP*, a spatio-temporal-semantic environment representation for autonomous mobile robots, to allow the storage and efficient query of the results of the object anchoring problem. Thus, we added instances and the datatype point to the methodology. With this extension of the methodology, it is possible to store the results of the object-anchoring problem and to store other object-related analysis results attached to the instance as a triple. Using the instances, it is possible to link the annotations of several datasets showing the very same real-world object. With this linkage, object-centric queries are enabled and analysis algorithms in need of data regarding a specific objects can retrieve the data easily.

Those analysis algorithms calculate results regarding the specific object about which the algorithm queried the data beforehand. The results can be stored in the mini triple-store that is attached to each instance. With this, arbitrary results can be stored if it can be represented in the triple format with the instance as the subject.

For the specific result of the anchoring problem, the data type *point* was also presented in this paper. The position of the object can be stored as a 3D point and this point can be linked to the object instance. Using the temporal modality of *SEEREP* it is also possible to calculate the position of an object at multiple points in time and store those positions with their timestamp linked to the object instance. With this, the object anchoring result as well as the data used to generate the solution are linked to the instance.

The development of *SEEREP* is an ongoing project and future work includes tasks such as: (1) Adding further data types to *SEEREP* to enable the representation of further modalities of the environment. Of special interest are meshes, because they can be used for surface analysis of objects and for 3D navigation. (2) Supporting the workflow of linking the data showing the same object to an instance. The refinement of the query interface to enable queries like "Give me all images showing the region of this pointcloud" is needed for this. (3) For the previous task, an extension of the already implemented data type image may be helpful. In this extension, the frustum of the field of view of the camera taking the image may be calculated with a use-case dependent viewing distance. Using this frustum, the question if a region may be visible in general in an image can be answered easily.

References

1. Deeken, H., Wiemann, T., Hertzberg, J.: Grounding semantic maps in spatial databases. Robot. Auton. Syst. **105**, 146–165 (2018). https://doi.org/10.1016/j.robot.2018.03.011
2. Dong, J., Burnham, J.G., Boots, B., Rains, G., Dellaert, F.: 4D crop monitoring: Spatio-temporal reconstruction for agriculture. In: 2017 IEEE ICRA, pp. 3878–3885. Singapore (2017). https://doi.org/10.1109/ICRA.2017.7989447
3. Elfring, J., van den Dries, S., van de Molengraft, M.J.G., Steinbuch, M.: Semantic world modeling using probabilistic multiple hypothesis anchoring. Robot. Auton. Syst. **61**(2), 95–105 (2013). https://doi.org/10.1016/j.robot.2012.11.005
4. Blender Foundation: blender.org - home of the blender project - free and open 3D creation software. https://www.blender.org/
5. Günther, M., Ruiz-Sarmiento, J.R., Galindo, C., González-Jiménez, J., Hertzberg, J.: Context-aware 3D object anchoring for mobile robots. Robot. Auton. Syst. **110**, 12–32 (2018). https://doi.org/10.1016/j.robot.2018.08.016
6. Harnad, S.: The symbol grounding problem. Phys. D **42**, 335–346 (1990)
7. Magistri, F., Chebrolu, N., Stachniss, C.: Segmentation-based 4D registration of plants point clouds for phenotyping. In: 2020 IEEE/RSJ IROS, pp. 2433–2439. IEEE (2020). https://doi.org/10.1109/IROS45743.2020.9340918
8. Mason, J., Marthi, B.: An object-based semantic world model for long-term change detection and semantic querying. In: 2012 IEEE/RSJ IROS, pp. 3851–3858 (2012). https://doi.org/10.1109/IROS.2012.6385729
9. Moreau, D., Pointurier, O., Nicolardot, B., Villerd, J., Colbach, N.: In which cropping systems can residual weeds reduce nitrate leaching and soil erosion? Eur. J. Agron. **119**, 126015 (2020). https://doi.org/10.1016/j.eja.2020.126015
10. Niemeyer, M., Pütz, S., Hertzberg, J.: A spatio-temporal-semantic environment representation for autonomous mobile robots equipped with various sensor systems. In: 2022 IEEE International Conference on Multisensor Fusion and Integration for Intelligent Systems (MFI-2022) (2022). https://doi.org/10.1109/MFI55806.2022.9913873
11. Oliveira, M., Lim, G.H., Seabra Lopes, L., Kasaei, S.H., Tomé, A., Chauhan, A.: A perceptual memory system for grounding semantic representations in intelligent service robots. In: Proceedings of the IEEE/RSJ IROS, IEEE (2014). https://doi.org/10.1109/IROS.2014.6942861
12. Partel, V., Charan Kakarla, S., Ampatzidis, Y.: Development and evaluation of a low-cost and smart technology for precision weed management utilizing artificial intelligence. Comput. Electron. Agric. **157**, 339–350 (2019). https://doi.org/10.1016/J.COMPAG.2018.12.048
13. Persson, A., Martires, P.Z.D., Loutfi, A., De Raedt, L.: Semantic relational object tracking. IEEE Trans. Cogn. Dev. Syst. **12**(1), 84–97 (2020). https://doi.org/10.1109/TCDS.2019.2915763, arXiv:1902.09937 [cs]
14. Reid, D.: An algorithm for tracking multiple targets. IEEE Trans. Automat. Contr. **24**(6), 843–854 (1979). https://doi.org/10.1109/TAC.1979.1102177
15. Renz, M., Niemeyer, M., Hertzberg, J.: Towards model-based automation of plant-specific weed regulation. 43. GIL-Jahrestagung, Resiliente Agri-Food-Systeme (2023)

16. Storkey, J., Westbury, D.B.: Managing arable weeds for biodiversity. Pest Manage. Sci. **63**(6), 517–523 (2007). https://doi.org/10.1002/PS.1375
17. Yang, X., et al.: A survey on smart agriculture: development modes, technologies, and security and privacy challenges. IEEE/CAA J. Autom. Sinica **8**(2), 273–302 (2021). https://doi.org/10.1109/JAS.2020.1003536

Associative Reasoning for Commonsense Knowledge

Claudia Schon(✉)

Institut für Informatik, Hochschule Trier, Trier, Germany
C.Schon@hochschule-trier.de

Abstract. Associative reasoning refers to the human ability to focus on knowledge that is relevant to a particular problem. In this process, the meaning of symbol names plays an important role: when humans focus on relevant knowledge about the symbol *ice*, similar symbols like *snow* also come into focus. In this paper, we model this associative reasoning by introducing a selection strategy that extracts relevant parts from large commonsense knowledge sources. This selection strategy is based on word similarities from word embeddings and is therefore able to take the meaning of symbol names into account. We demonstrate the usefulness of this selection strategy with a case study from creativity testing.

1 Introduction

According to Kahneman [7], humans rely on two different systems for reasoning. System 1 is fast, emotional and less accurate, system 2 is slow, more deliberate and logical. Reasoning with system 2 is much more difficult and exhausting for humans. Therefore, system 1 is usually used first to solve a task and system 2 is only used for demanding tasks. Examples of tasks that system 1 does are solving simple math problems like 3 + 3, driving a car in an empty street, or associating a certain profession with the description like *a quiet, shy person who prefers to deal with numbers rather than people*. Especially, the associative linking of information with each other falls within the scope of system 1. In contrast, we typically use system 2 for tasks that require our full concentration such as driving in a crowded downtown area or following complex logical reasoning.

Humans have vast amounts of background knowledge that they skillfully use in reasoning. In doing so, they are able to focus on knowledge that is relevant for a specific problem. Associative thinking and priming play an important role in this process. These are things handled by system 1. The human ability of focusing on relevant knowledge is strongly dependent on the meaning of symbol names. When people focus on relevant background knowledge for a statement like *The pond froze over for the winter*, semantic similarities of words play an important role. For this statement, a human will certainly not only focus on background knowledge that relates exactly to the terms *pond*, *froze*, and *winter*, but also knowledge about similar term such as *ice* and *snow*. We refer to the process of focusing on relevant knowledge as *associative reasoning*.

D. Seipel and A. Steen (Eds.): KI 2023, LNAI 14236, pp. 170–183, 2023.
https://doi.org/10.1007/978-3-031-42608-7_14

If we want to model the versatility of human reasoning, it is necessary to model not only different types of reasoning such as deductive, abductive, and inductive reasoning, but also the ability to focus on relevant background knowledge using associative reasoning.

This aspect of human reasoning, focusing on knowledge relevant to a problem or situation, is what we model in this paper. For this purpose, we develop a selection strategy that extracts relevant parts from a large knowledge base containing background knowledge. We use background knowledge formalized in knowledge graphs like ConceptNet [20], ontologies like Adimen SUMO [1], and Cyc [8], or knowledge bases. A nice property of commonsense knowledge sources is that the used symbol names are often based on natural language words. For example in Adimen SUMO you find symbol names like *c__SecondarySchool*. To model the associative nature of human focusing, we exploit this nice property and use word similarities from word embeddings.

In word embeddings, large amounts of text are used to learn a vector representation of words. These so called word vectors have the nice property that similar words are represented by similar vectors. We propose a representation of background knowledge in terms of vectors such that similar statements in the background knowledge are represented by similar vectors.

Based on this vector representation of background knowledge, we present a new selection strategy, the *vector-based selection* that pays attention to the meaning of symbol names and thus models associative reasoning as it is done by humans. The main contributions of this paper are:

- The introduction of the vector-based selection strategy, a statistical selection technique for commonsense knowledge which is based on word embeddings.
- A case study using benchmarks for creativity testing in humans which demonstrates that the vector-based selection allows us to model associative reasoning and selects commonsense knowledge in a very focused way.

The paper is structured as follows: after discussing related work and preliminaries, we briefly revise SInE, a syntax-based selection strategy for first-order logic reasoning with large theories. Next, we turn to the integration of statistical information into selection strategies where, after revising distributional semantics, we introduce the vector-based selection strategy. Then we present experimental results. Finally, we discuss future work.

2 Related Work

Selecting knowledge that is relevant to a specific problem is also an important task in automated theorem proving. In this area, often a large set of axioms called a knowledge base is given as background knowledge, together with a much smaller set of axioms $F_1, \ldots, F_n$ and a query Q. The reasoning task of interest is to show that the knowledge base together with the axioms $F_1, \ldots, F_n$ implies the query Q. This corresponds to showing that $F_1 \wedge \ldots \wedge F_n \rightarrow Q$ is entailed by the knowledge base. $F_1 \wedge \ldots \wedge F_n \rightarrow Q$ is usually referred to as *goal*. As soon as the size of the knowledge base forbids to use the entire knowledge base to show that Q follows from the knowledge base using an automated theorem prover, it is necessary to select the axioms from the knowledge base that are necessary for this reasoning task. However, identifying these axioms is not trivial, so common selection strategies are based on heuristics and are usually incomplete. This means that it is not always possible to solve the reasoning task with the selected axioms: If too few axioms have been selected, the prover cannot find a proof. If too many have been selected, the reasoner may be overwhelmed with the set of axioms and run into a timeout.

Most strategies for axiom selection are purely syntactic like the SInE selection [6], lightweight relevance filtering [12] and axiom relevance ordering [17]. A semantic strategy for axiom selection is SRASS [21] which is a model-based approach. This strategy is based on the computation of models for subsets of the axioms and consecutively extends these sets. Another interesting direction of research is the development of metrics for the evaluation of selection techniques [9] which allow to measure the quality of selection strategies without having to actually run the automated theorem prover on the selected axioms and the conjecture at hand. Another approach to axiom selection is the use of formula metrics [10] which measure the dissimilarity of different formulae and lead to selection strategies which allow to select the k axioms from a knowledge base most similar to a given problem. None of the selection methods mentioned so far in this section take the meaning of symbol names into account.

An area where the meaningfulness of symbol names was evaluated is the semantic web [18]. The authors come to the conclusion that the semantics encoded in the names of IRIs (Internationalized Resource Identifiers) carry a kind of social semantics which coincides with the formal meaning of the denoted resource. Furthermore, [5] aims at simulating human associations using linked data. For given pairs of sources and targets, evolutionary algorithms are used to learn a model able to predict a target entity in the considered linked data suitable for a new given source entity.

Similarity SInE [4] is an extension of SInE selection which uses a word embedding to take similarity of symbols into account. Similarity SInE works in such a way that it always selects all formulae that SInE would select and, in addition, selects formulae for symbols that are similar to the symbols occurring in the goal. By this mixture of syntactic and statistical methods, Similarity SInE represents a hybrid selection approach.

In contrast, the vector-based selection presented in this paper is a purely statistical approach.

3 Preliminaries and Task Description

Numerous sets of benchmarks exist for the area of commonsense reasoning. Typically these problems are multiple choice questions about everyday situations which are given in natural language. Figure 1 shows a commonsense reasoning problem from the choice of plausible alternative challenge (COPA) [11]. Usually, for these commonsense reasoning problems it is not the case that one of

The pond froze over for the winter. What happened as a result?

1. **People brought boats to the pond.**
2. **People skated on the pond.**

Fig. 1. Example from the *Choice of Plausible Alternative Challenge* (COPA) [11].

the answer alternatives can actually be logically inferred. Often only one of the answer alternatives is more plausible than the others. To solve these problems, a broad background knowledge is necessary. For the example given in Fig. 1, knowledge about winter, ice, frozen surfaces and boats is necessary. In humans, system 1 with associative reasoning is responsible to focus on relevant background knowledge for a specific problem.

In this paper, we aim at modeling the human ability to focus on background knowledge relevant for a specific task. We introduce a selection strategy based on word embeddings to achieve this. For this, we assume that the background knowledge is given in first-order logic. One reason for this assumption is that this allows us to use already existing automated theorem provers for modeling human reasoning in further steps. Furthermore, this allows us to easily compare our approach to selection strategies for first-order logic theorem proving like SInE. Moreover, this assumption is not a limitation, since knowledge given in other forms like for example in the form of a knowledge graph can be easily transformed into first-order logic [19].

We furthermore assume the description of the commonsense reasoning problem to be given as a first-order logic formula. Again, this is not a limitation since, for example, the KnEWS [2] system can convert natural language into first-order logic formulas. Following terminology from first-order logic reasoning, we refer to the formula for the commonsense reasoning problem as *goal*. Referring to the example from Fig. 1, we would denote the first-order logic formula for the statement *The pond froze over for the winter.* as F, the formula for *People brought boats to the pond.* as Q_1, and the formula for *People skated on the pond.* as Q_2. This leads to the two goals $F \rightarrow Q_1$ and $F \rightarrow Q_2$ for the commonsense reasoning problem from Fig. 1. For these goals, we could now select from knowledge bases with background knowledge using first-order logic axiom selection techniques.

Axiom selection for a given goal in first-order logic as described at the beginning of the related work section is very similar to the problem of selecting background knowledge relevant for a specific problem in commonsense reasoning. Both problems have in common that large amounts of background knowledge are given that is too large to be considered completely. The main difference is the fact that in commonsense reasoning we cannot necessarily assume that a proof for a certain goal can be found. Therefore, drawn inferences are also interesting in this domain. In both cases, the task is to select knowledge that is relevant for the given goal formula.

In the case study in our evaluation, we compare the vector-based selection strategy introduced in this paper with the syntax-based SInE selection strategy which is broadly used in first-order logic theorem proving. Therefore, we briefly introduce the SInE selection in the next section.

We denote the set of all predicate and function symbols occurring in a formula F by $sym(F)$. We slightly exploit notation and use $sym(KB)$ for the set of all predicate and function symbols occurring in a knowledge base KB.

4 SInE: A Syntax-Based Selection Strategy

In [6] the SInE selection strategy is introduced which is successfully used by many automated theorem provers. Since this selection strategy does not consider the meaning of symbol names, we classify this strategy as a *syntax-based* selection. The basic idea of SInE is to determine a set of symbols for each axiom in the knowledge base which is allowed to *trigger* the selection of this axiom. For this a *trigger* relation is defined as follows:

Definition 1 (Trigger relation for the SInE selection [6]**).** *Let KB be a knowledge base, A be an axiom in KB and $s \in sym(A)$ be a symbol. Let furthermore $occ(s, KB)$ denote the number of axioms in which s occurs in KB and $t \in \mathbb{R}$, $t \geq 1$. Then the triggers relation is defined as*

$$\begin{aligned} triggers(s, A) \text{ iff for all symbols } s' \text{ occurring in } A \text{ we} \\ \text{have } occ(s, KB) \leq t \cdot occ(s', KB) \end{aligned}$$

Note that an axiom can only be triggered by symbols occurring in the axiom. Parameter t specifies how strict we are in selecting the symbols that are allowed to trigger an axiom. For $t = 1$ (the default setting of SInE), a symbol s may only trigger an axiom A if there is no symbol s' in A that occurs less frequently in the knowledge base than s. This prevents frequently occurring symbols such as *subClass* from being allowed to trigger all axioms they occur in.

The *triggers* relation is then used to select axioms for a given goal. The basic idea is that starting from the symbols occurring in the goal, the symbols occurring in the goal are considered to be relevant and an axiom A is selected if A is triggered by some symbol occurring in the set of relevant symbols. The symbols occurring in the selected axioms are added to the set of relevant symbols and if desired, the selection can be repeated.

Definition 2 (Trigger-based selection [6]**).** *Let KB be a knowledge base, A be an axiom in KB, $s \in sym(KB)$ and G be a goal to be proven from KB.*

1. *If s is a symbol occurring in the goal G, then s is 0-step triggered.*
2. *If s is n-step triggered and s triggers A ($triggers(s, A)$), then A is $n+1$-step triggered.*
3. *If A is n-step triggered and s occurs in A, then s is n-step triggered, too.*

An axiom or a symbol is called triggered if it is n-step triggered for some $n \geq 0$.

For a given knowledge base, goal G and some $n \in \mathbb{N}$ SInE selects all axioms which are n-step triggered. In the following the SInE selection selecting all n-step triggered axioms is called SInE with recursion depth n.

SInE selection can also be used in commonsense reasoning to select background knowledge relevant to a statement: to do this, we just need to convert this statement into a first-order logic formula, and use the formula as a goal and select with SInE for it. Further details on SInE and some examples can be found in [6].

5 Use of Statistical Information for the Selection of Axioms

SInE selection completely ignores the meaning of symbol names. For SInE it makes no difference whether a predicate is called p or *dog*. If we consider knowledge bases with commonsense knowledge, the meaning of symbol names provides information that can be exploited by a selection strategy. For example, the symbol *dog* is more similar to the symbol *puppy* than to the symbol *car*. If a goal containing the symbol *dog* is given, it is more reasonable to select axioms containing the symbol *puppy* than axioms containing the symbol *car*. This corresponds to human associative reasoning, which also takes into account the meaning of symbol names and similarities.

5.1 Distributional Semantics

To determine the semantic similarity of symbol names, we rely on distributional semantics of natural language, which is used in natural language processing. The basis of distributional semantics is the distributional hypothesis [15], according to which words with similar distributional properties on large texts also have similar meaning. In other words: Words that occur in a similar context are similar.

An approach used in many domains which is based on the distributional hypothesis are word embeddings [13,14]. Word embeddings map the words of a vocabulary to vectors in $\mathbb{R}^n$. Typically, word embeddings are learned using neural networks on very large text sets. Since we use existing word embeddings in the following, we do not go into the details of creating word embeddings. An interesting property of word embeddings is that semantic similarity of words corresponds to the relative similarities of the vector representations of those words. To determine the similarity of two vector representations the cosine similarity is usually used.

Definition 3 (Cosine similarity of two vectors). *Let $u, v \in \mathbb{R}^n$, both non-zero. The* cosine similarity *of u and v is defined as:*

$$cos\_sim(u,v) = \frac{u \cdot v}{||u|| \; ||v||}$$

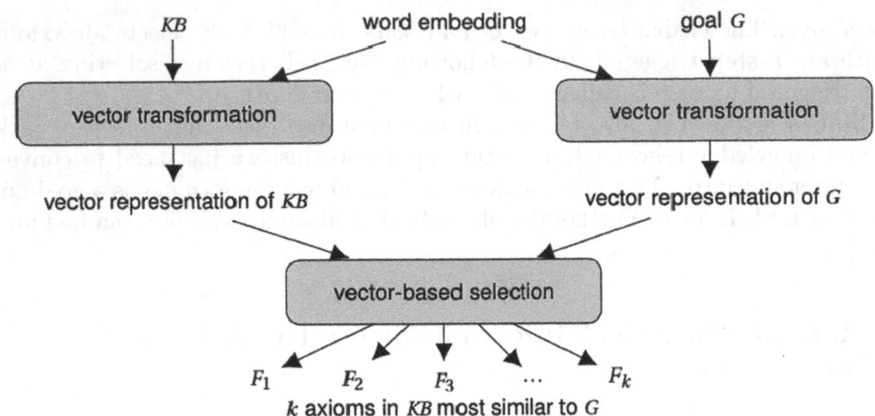

Fig. 2. Overview of the vector-based selection strategy. The vector transformation of the knowledge base KB and the vector transformation of the goal use the same word embedding.

The more similar two vectors are, the greater is their cosine similarity. For example in the ConceptNet Numberbatch word embedding [20], the cosine similarity of *dog* and *puppy* is 0.84140545 which is much larger than the cosine similarity of *dog* and *car* that is 0.13056317. Based on these similarities, word embeddings can furthermore be used to determine the k words in the vocabulary most similar to a given word for some $k \in \mathbb{N}$.

5.2 Vector-Based Selection: A Statistical Selection Strategy

Word embeddings represent words as vectors in such a way that words that are frequently used in a similar context are mapped to similar vectors. Vector-based selection aims to represent the axioms of a knowledge base as vectors in such a way that similar axioms are mapped to similar vectors. Where we consider two axioms of a knowledge base to be similar if they represent similar knowledge.

Figure 2 gives an overview of the vector-based selection strategy. In a preprocessing step, vector representations are computed for all axioms of the knowledge base KB using an existing word embedding. This preprocessing step has to be performed only once. pro

Given a goal G for which we want to check if it is entailed by the knowledge base, we transform G into a vector representation using the same word embedding as for the vector transformation of the knowledge base. Next, vector-based selection determines the k vectors in the vector representation of the knowledge base most similar to the vector representation of G. The corresponding k axioms form the result of the selection. Various metrics can be used for determining the k vectors that are most similar to the vector representation of G. We use cosine similarity, which is widely used in word embeddings.

One way to represent an axiom as a vector is to look up the vectors of all the symbols occurring in the axiom in the word embedding and represent the

$$\forall X, Y, Z\Big((instance(X, carnivore) \wedge instance(Y, eating) \wedge agent(Y, X) \wedge patient(Y, Z)) \rightarrow instance(Z, animal)\Big)$$

Symbol Name:	Frequency:
instance	4237
agent	140
patient	183
carnivore	5
eating	6
animal	63

Fig. 3. Example axiom from Adimen SUMO together with frequencies of the symbols of the axiom in Adimen SUMO. To increase readability, we omitted prefixes of symbols.

axiom by the average of these vectors. However this treats all symbols occurring in an axiom equally. This is not always useful, as the axiom in Fig. 3 from Adimen SUMO illustrates for which it seems desirable that the symbols *instance*, *agent* and *patient* contribute less to the computation of the vector representation than the symbols *carnivore*, *eating* and *animal*. The reason for this lies in the frequency of the symbols in the knowledge base which are given in the Table in Fig. 3. Symbols *carnivore*, *eating* and *animal* occur much less frequently in Adimen SUMO than *instance*, *agent* and *patient*. This suggests that *carnivore*, *eating* and *animal* are more important for the statement of the axiom. This is similar to the idea in SInE that only the least common symbol in an axiom is allowed to trigger the axiom.

We implement this idea in the computation of the vector representation of axioms by weighting the influence of a symbol using inverse document frequency (idf). In the area of information retrieval, for the task of rating the importance of word w to a document d in a set of documents D, idf is often used to diminish the weight of a word that occurs very frequently in the set of documents. Assuming that there is at least one document in D, in which w occurs, $idf(w, D)$ is defined as:

$$idf(w, D) = \log \frac{|D|}{|\{d \in D \mid w \text{ occurs in } d\}|}$$

If w occurs in all documents in D, the fraction is equal to 1 and $idf(w, D) = 0$. If w occurs in only one of the documents in D, the fraction is equal to $|D|$ and $idf(w, D) > 0$. The higher the proportion of documents in which w occurs, the lower $idf(w, D)$.

We transfer this idea to knowledge bases by interpreting a knowledge base as a set of documents and each axiom in this knowledge base as a document. The resulting computation of idf for a symbol in a knowledge base is given in Definition 4.

For the often used tf-idf (term frequency - inverse document frequency) the idf value is multiplied by the term frequency of the term in a certain document. Where the term frequency of a term t in a document d corresponds to the relative frequency of t in d. However the number of occurrences of a symbol in a single axiom does not necessarily correspond to its importance to the axiom (as illustrated by the axiom given in Fig. 3). Multiplying the idf value of a symbol with its tf value in a formula could even increase the influence of frequent symbols like *instance*, since they often appear more than once in a formula. This is why we omit this multiplication and use idf for the weighting instead.

For simplicity, we assume that $sym(F)$ is a subset of the vocabulary of the word embedding in the following definition.

Definition 4 (idf-based vector representation of an axiom, a knowledge base). *Let* $KB = \{F_1, \ldots, F_n\}$, $n \in \mathbb{N}$ *be a knowledge base,* $F \in KB$ *be an axiom.* V *be a vocabulary and* $f : V \rightarrow \mathbb{R}^n$ *a word embedding. Let furthermore* $sym(F) \subseteq V$. *The* idf *value for a symbol* $s \in sym(F)$ *w.r.t.* KB *is defined as*

$$idf(s, KB) = \log \frac{|KB|}{|\{F' \in KB \mid s \in sym(F')\}|}$$

The idf-based vector representation *of* F *is defined as*

$$v_{idf}(F) = \frac{\sum_{s \in sym(F)}(idf(s, KB) \cdot f(s))}{\sum_{s \in sym(F)} idf(s, KB)}$$

Furthermore, $V_{idf}(KB) = \{v_{idf}(F_1), \ldots, v_{idf}(F_n)\}$ *denotes the idf-based vector representation of* KB.

Please note that $v_{idf}(F)$ is a vector and denotes the vector representation of formula F. Given a goal G and a knowledge base, we can use the vector representations of the knowledge base and G to select the k axioms from the knowledge base most similar to the vector representation of G for some $k \in \mathbb{N}$ (see Fig. 2).

Definition 5 (Vector-based selection). *Let* KB *be a knowledge base,* G *be a goal with* $sym(G) \subseteq sym(KB)$ *and* $f : V \rightarrow \mathbb{R}^n$ *a word embedding. Let furthermore* V_{KB} *be a vector representation of* KB *and* v_G *a vector representation for* G *both constructed using* f. *For* $k \in \mathbb{N}$, $k \leq |KB|$ *the* k *axioms in* KB *most similar to* G *denoted as* $mostsimilar(KB, G, k)$ *are defined as the following set:*

$$\{F_1, \ldots, F_k \mid \{F_1, \ldots, F_k\} \subseteq KB \text{ and } \forall F' \in KB \setminus \{F_1, \ldots, F_k\} \\ cos\_sim(v_{F'}, v_G) \leq \min_{i=1,\ldots,k} cos\_sim(v_{F_i}, v_G)\}.$$

For KB, G *and* $k \in \mathbb{N}$ *given as above described, vector-based selection selects* $mostsimilar(KB, G, k)$.

Definition 5 is intentionally very general and allows other vector representations besides idf-based vector representation. In addition to that, the similarity measure cos_sim can be easily replaced by some other measure like euclidean distance.

In the previous section we assumed the set of symbols in a knowledge base to be a subset of the vocabulary of the used word embedding. However in practice this is not always the case and in many cases it might be necessary to construct a mapping for this. Each combination of knowledge base and word embedding requires a specific mapping. For the case study we present in the next section such a mapping is not necessary which is why we refrain from presenting it.

Table 1. Examples from the fRAT dataset. Given the three query words, the task is to determine the target word which establishes a functional connection.

Query Words w_1, w_2 and w_3	Target Word w_t
tulip, daisy, vase	flower
sensitive, sob, weep	cry
algebra, calculus, trigonometry	math
duck, sardine, sinker	swim
finger, glove, palm	hand

6 Evaluation: A Case Study on Commonsense Knowledge

In areas where commonsense knowledge is used as background knowledge, automated theorem provers can be used not only for finding proofs, but also as inference engines. One reason for this is that even if there are large ontologies and knowledge bases with commonsense knowledge, this knowledge is still incomplete. Therefore, it is likely that not all the information needed for a proof is represented. Nevertheless, automated theorem provers can be very helpful on commonsense knowledge, because the inferences that a prover can draw from a problem description and selected background knowledge provide valuable information. How well these inferences fit the problem description depends strongly on the selected background knowledge. Here it is very important that the selected background knowledge is broad enough but still focused.

6.1 Functional Remote Association Tasks

The benchmark problems we use to evaluate the vector-based selection introduced in this paper are the functional Remote Association Tasks (fRAT) [3] which were developed to measure human creativity. In fRAT, three words like *tulip*, *daisy* and *vase* are given and the task is to find a fourth connecting word, called target word (here *flower*). The words are chosen in such a way that a functional connection must be found between the three words and the target word. To solve these problems, a broad background knowledge is necessary. The solution of the above fRAT task requires the background knowledge that tulips and daisies are flowers and that a vase is a container in which flowers are kept.

The dataset [16] used for this evaluation consists of 49 fRAT tasks. Table 1 gives some examples for tasks in the dataset.

6.2 Experimental Results

For an fRAT task consisting of the words w_1, w_2, w_3 and the target word w_t, we first generate a simple goal $w_1(w_1) \land w_2(w_2) \land w_3(w_3)$ using the query words of the tasks as predicate and constant symbols. For example the goal for the query words $w_1 = duck$, $w_2 = sardine$ and $w_3 = sinker$ would be the formula

Table 2. Results of selecting with vector-based selection for the 49 fRAT tasks: percentage of tasks where the target word w_t occurs in the axioms selected by vector-based selection. Parameter k corresponds to the number of selected axioms.

Vector-based selection on fRAT		
k	% of tasks with in selection w_t	avg. pos. of target word
5	55.10%	2.22
10	69.34%	3.38
25	75.51%	4.46
50	87.75%	8.50
100	89.80%	9.98
150	95.92%	16.82
350	98.00%	24.63
495	100%	33.67

Table 3. Results of selecting with SInE for the 49 fRAT tasks: percentage of tasks where the target word w_t occurs in the axioms selected by SInE.

SInE on fRAT		
rec. depth	% of tasks with w_t in selection	avg. number of selected axioms
1	18.75%	8.76
2	22.45%	50.67
3	28.57%	243.08
4	32.65%	750.92
5	34.69%	1444.45
6	36.74%	2044.16

$duck(duck) \wedge sardine(sardine) \wedge sinker(sinker)$. For the goal formulae created like this we select background knowledge using different selection strategies. Then we check whether the word w_t occurs in the selected axioms. Since we only want to evaluate selection strategies on commonsense knowledge, we do not use a reasoner in the following experiments and leave that to future work. As background knowledge we use ConceptNet [20] which is a knowledge graph containing broad commonsense knowledge in the form of triples. For this evaluation, we use a first-order logic translation [19] of around 125,000 of the English triples of ConceptNet as knowledge base.

For the translation of the knowledge base into vectors we use the pretrained word embedding ConceptNet Numberbatch [20] which combines data from ConceptNet, word2vec, GloVe and OpenSubtitles 2016.

We use both vector-based selection and SInE to select axioms for a goal created for an fRAT task and then check if the target word w_t occurs in the selected axioms. Table 2 shows the results for vector-based selection, Table 3 for SInE. Note that for vector-based selection the k parameter naturally determines the number of axioms contained in the result of the selection. For k values greater than or equal to 495, the target word is always found in the selection. Since the selected axioms are sorted in descending order with respect to the similarity to the goal in vector-based selection, Table 2 furthermore provides the average position of the target word in the selected axioms.

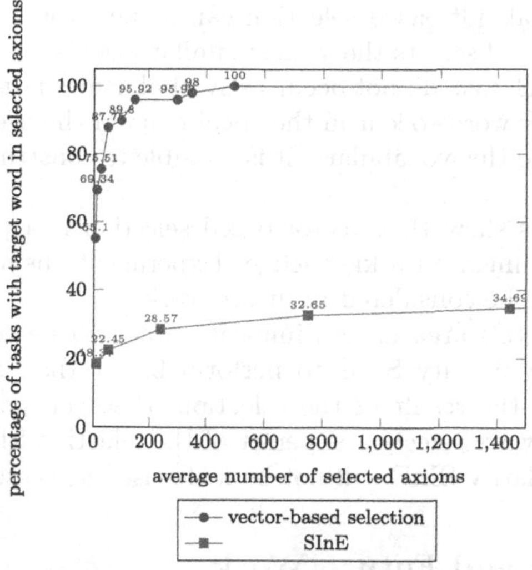

Fig. 4. Percentage of fRat tasks for which the target word occurs in the selected axioms depending on the number of selected axioms. Vector-based selection and SInE were used for the selection.

Please recall that SInE selection selecting all n-step triggered axioms is called SInE with recursion depth n. The results for SInE in Table 3 show that even for recursion depth 6, where SInE selected 2044.16 axioms on average for an fRAT task, in only 36.74% of the tasks the target word occurred in the selection. Compared to that, the result of vector-based selection of only five axioms already contains the target word in 55% of the tasks. As soon as the vector-based selection selects more than 494 axioms, the target word is contained in the selection for all of the tasks. The Fig. 4 illustrates the relationship between the number of axioms selected and the percentage of target words found for the two selection strategies.

Although SInE selects significantly more axioms than vector-based selection, axioms containing the target word are often not selected. In contrast, vector-based selection is much more focused and even small sets of selected axioms contain axioms mentioning the target word.

The experiments revealed another problem specific for the task of selecting background knowledge from commonsense knowledge bases: Since knowledge bases in this area usually are extremely large, it is reasonable to assume that a user looking for background knowledge for a set of keywords is not aware of the exact symbol names used in the knowledge base. For example none of the query words *tulip*, *daisy* and *vase* corresponds to a symbol name in our first-order logic translation of ConceptNet. Therefore a selection using SInE with the goal created from these query words results in an empty selection and in an empty selection result, of course, the target word cannot be found. In contrast to that, vector-based selection constructs a query vector from the symbol names occurring in the goal (idf-based selection can assume the average idf value for unknown symbols) and selects the k most similar axioms even though the query words from the fRAT task do not occur as symbol names in the knowledge base. As long as the query words occur in the vocabulary of the used word embedding or can be mapped to this vocabulary, it is possible to construct the query vector and select axioms.

The experiments show that vector-based selection is a promising approach for selection on commonsense knowledge. Experiments using reasoners on the selected axioms will be considered in future work.

Experiments in this area using Similarity SIne will be considered in future work. We expect Similarity SInE to perform better than SInE w.r.t. finding the target word in the result of the selection. However since Similarity SInE per construction always selects a superset of the selection SInE would perform, selecting with Similarity SInE will not be as focused as vector-based selection.

7 Conclusion and Future Work

Although humans possess large amounts of background knowledge, it is easy for them to focus on the knowledge relevant to a specific problem. Associative reasoning plays an important role in this process. The vector-based selection presented in this paper uses word similarities from word embeddings to model associative reasoning. Our experiments on benchmarks for testing human creativity show that vector-based selection is able to select in a very focused way on commonsense knowledge. In future work, we want to use deductive as well as abductive reasoning and default reasoning on the result of these selections. One problem with using default reasoning will be to determine which rules should be default rules and which should not. Relations in ConceptNet are assigned weights. We will try to exploit this information to decide whether a rule should be a default rule or not.

References

1. Álvez, J., Lucio, P., Rigau, G.: Adimen-SUMO: reengineering an ontology for first-order reasoning. Int. J. Semant. Web Inf. Syst. **8**, 80–116 (2012)
2. Basile, V., Cabrio, E., Schon, C.: KNEWS: using logical and lexical semantics to extract knowledge from natural language. In: Proceedings of the European Conference on Artificial Intelligence (ECAI) 2016 Conference (2016)
3. Blaine, R., Worthen, P.M.C.: Toward an improved measure of remote associational ability. J. Educ. Measure. **8**(2), 113–123 (1971)
4. Furbach, U., Krämer, T., Schon, C.: Names are not just sound and smoke: word embeddings for axiom selection. In: Fontaine, P. (ed.) CADE 2019. LNCS (LNAI), vol. 11716, pp. 250–268. Springer, Cham (2019). https://doi.org/10.1007/978-3-030-29436-6_15
5. Hees, J.: Simulating Human Associations with Linked Data. Ph.D. thesis, Kaiserslautern University of Technology, Germany (2018). https://kluedo.ub.rptu.de/frontdoor/index/index/docId/5430

6. Hoder, K., Voronkov, A.: Sine qua non for large theory reasoning. In: Bjørner, N., Sofronie-Stokkermans, V. (eds.) CADE 2011. LNCS (LNAI), vol. 6803, pp. 299–314. Springer, Heidelberg (2011). https://doi.org/10.1007/978-3-642-22438-6_23
7. Kahneman, D.: Thinking, Fast and Slow. Macmillan, London (2011)
8. Lenat, D.B.: CYC: a large-scale investment in knowledge infrastructure. Commun. ACM **38**(11), 33–38 (1995)
9. Liu, Q., Wu, Z., Wang, Z., Sutcliffe, G.: Evaluation of axiom selection techniques. In: PAAR+SC^2@IJCAR. CEUR Workshop Proceedings, vol. 2752, pp. 63–75. CEUR-WS.org (2020)
10. Liu, Q., Xu, Y.: Axiom selection over large theory based on new first-order formula metrics. Appl. Intell. **52**(2), 1793–1807 (2022). https://doi.org/10.1007/s10489-021-02469-1
11. Maslan, N., Roemmele, M., Gordon, A.S.: One hundred challenge problems for logical formalizations of commonsense psychology. In: Twelfth International Symposium on Logical Formalizations of Commonsense Reasoning, Stanford, CA (2015)
12. Meng, J., Paulson, L.C.: Lightweight relevance filtering for machine-generated resolution problems. J. Appl. Logic **7**(1), 41–57 (2009)
13. Mikolov, T., Chen, K., Corrado, G., Dean, J.: Efficient estimation of word representations in vector space. CoRR abs/1301.3781 (2013). http://arxiv.org/abs/1301.3781
14. Mikolov, T., Sutskever, I., Chen, K., Corrado, G.S., Dean, J.: Distributed representations of words and phrases and their compositionality. In: NIPS, pp. 3111–3119 (2013)
15. Miller, G.A., Charles, W.G.: Contextual correlates of semantic similarity. Lang. Cogn. Process. **6**(1), 1–28 (1991). http://eric.ed.gov/ERICWebPortal/recordDetail?accno=EJ431389
16. Olteteanu, A., Schöttner, M., Schuberth, S.: Computationally resurrecting the functional remote associates test using cognitive word associates and principles from a computational solver. Knowl. Based Syst. **168**, 1–9 (2019). https://doi.org/10.1016/j.knosys.2018.12.023
17. Roederer, A., Puzis, Y., Sutcliffe, G.: Divvy: an ATP meta-system based on axiom relevance ordering. In: Schmidt, R.A. (ed.) CADE 2009. LNCS (LNAI), vol. 5663, pp. 157–162. Springer, Heidelberg (2009). https://doi.org/10.1007/978-3-642-02959-2_13
18. de Rooij, S., Beek, W., Bloem, P., van Harmelen, F., Schlobach, S.: Are names meaningful? Quantifying social meaning on the semantic web. In: Groth, P., et al. (eds.) ISWC 2016. LNCS, vol. 9981, pp. 184–199. Springer, Cham (2016). https://doi.org/10.1007/978-3-319-46523-4_12
19. Schon, C., Siebert, S., Stolzenburg, F.: Using conceptnet to teach common sense to an automated theorem prover. In: ARCADE@CADE. EPTCS, vol. 311, pp. 19–24 (2019)
20. Speer, R., Chin, J., Havasi, C.: Conceptnet 5.5: an open multilingual graph of general knowledge. In: AAAI, pp. 4444–4451. AAAI Press (2017)
21. Hasan, O., Tahar, S.: Formalization of continuous probability distributions. In: Pfenning, F. (ed.) CADE 2007. LNCS (LNAI), vol. 4603, pp. 3–18. Springer, Heidelberg (2007). https://doi.org/10.1007/978-3-540-73595-3_2

Computing Most Likely Scenarios of Qualitative Constraint Networks

Tobias Schwartz[(✉)] and Diedrich Wolter

University of Bamberg, Bamberg, Germany
{tobias.schwartz,diedridch.wolter}@uni-bamberg.de

Abstract. Qualitative constraint networks are widely used to represent knowledge bases in qualitative spatial and temporal reasoning (QSTR). However, inconsistencies may arise in various application contexts, such as merging data from different sources. Here, identifying a consistent constraint network that deviates as little as possible from the over-constrained network is of great interest. This problem of *repairing* an inconsistent network is a challenging optimization problem, as determining network consistency is already NP-hard for most qualitative constraint languages (also called qualitative calculi). But not all repairs are created equal: Unlike previous work, we consider a practical use case in which facts feature different likelihoods of being true. In this paper, we propose an approach to address this problem by extending qualitative constraint networks with a likelihood factor that can, for example, be derived from the credibility of different data sources. Specifically, we present a Partial MaxSAT encoding and a Monte Carlo Tree Search (MCTS) implementation for solving qualitative constraint networks with likelihoods optimally and efficiently. Our experimental evaluation demonstrates the effectiveness of our approach, showing that approximate search techniques can offer a promising trade-off between computational budget and optimality.

Keywords: Qualitative Constraint Networks · Likelihood · MaxSAT · Monte Carlo Tree Search

1 Introduction

The field of Qualitative Spatial and Temporal Reasoning (QSTR) is motivated by a range of research and application contexts. A primary goal of researchers in this field is to develop efficient symbolic representation and reasoning methods for everyday spatial and temporal concepts that align with human conceptualizations of space and time. To this end, a variety of formalisms, known as qualitative calculi, have been proposed and applied in diverse application areas, including robotics, geographic information systems, and language understanding [8,15].

In QSTR, a dominant approach is constraint-based reasoning, and knowledge bases are commonly represented as qualitative constraint networks (QCNs). In this paper, we introduce an extension to QCNs that allows for the representation

D. Seipel and A. Steen (Eds.): KI 2023, LNAI 14236, pp. 184–197, 2023.
https://doi.org/10.1007/978-3-031-42608-7_15

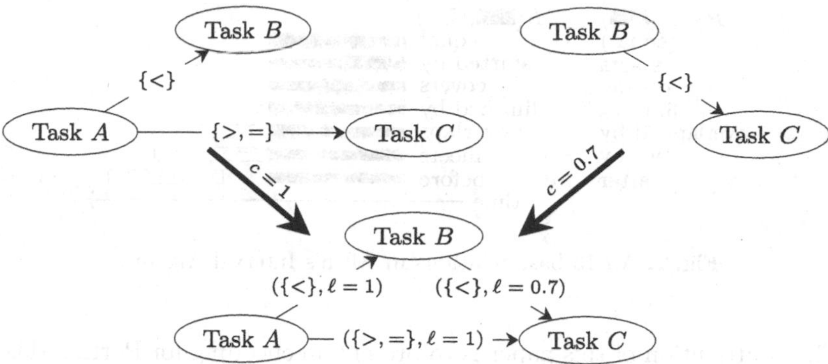

Fig. 1. Two different teams within a project having their own representations of temporal constraints between tasks, represented as QCNs (top). Teams vary in experience and hence are attributed different credibility c. Information is merged into a shared, but inconsistent, network representation with likelihood ℓ, derived from the credibility of their source (bottom).

of a *likelihood* factor, i.e., a notion of how likely a given constraint holds. Consequently, a high likelihood factor indicates that a constraint is likely to hold and should be satisfied by a solution. The likelihood of a constraint can, among other sources, originate from the credibility of the data source. For example, consider a software development project with two independent teams A and B. Team A, comprising experienced developers, and team B, consisting of junior developers, have their own representations of temporal constraints between tasks. Team A's QCN holds higher credibility due to their extensive expertise and successful track record, while team B's QCN carries lower credibility due to their limited experience. When merging the two QCNs, the *likelihood* of certain constraints being true may be a direct consequence of this credibility, resulting from a difference in expertise. See Fig. 1 for an illustration. Note that out notion of likelihood is not to be confused with the probabilistic semantics of likelihood in machine learning where it gives the conditional probability $P(\text{data} \mid \text{model parameters})$. In this paper, likelihood is a scalar measure not subject to the laws of probability.

In previous work, dealing with inconsistent data in a QCN has been addressed by *repairing* the network, i.e., changing the network such that it becomes consistent [5,6,9]. However, not all repairs are created equal: In the above example, we would like to preserve the constraints from team A as much as possible, while we are more willing to change the constraints from team B. Thus, in this paper, we propose an approach to address this problem by extending QCNs with a likelihood factor. We are then interested in computing a most likely solution, called *scenario* in QSTR. The idea of introducing likelihoods is also related to earlier works on merging (conflicting) QCNs. Condotta et al. (2008) [4] introduce a merging operator that can be interpreted as preference ordering for selecting an alternative in the construction of a solution. Dylla and Wallgrün (2007) [9] discuss various measures from an application-oriented point of view.

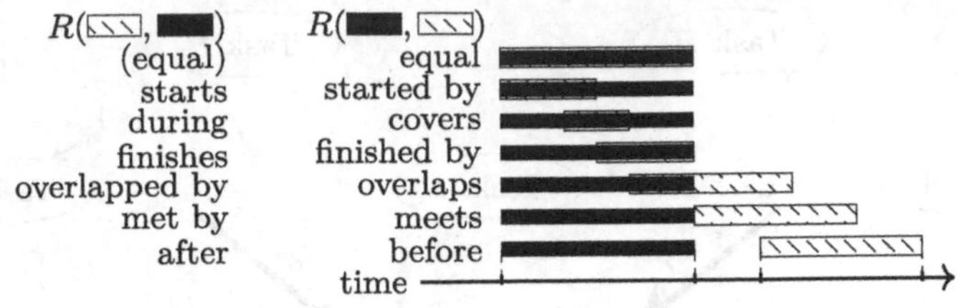

Fig. 2. All 13 base relations in Allen's Interval Algebra

The contribution of this paper is to present an encoding for Partial MaxSAT and a Monte Carlo Tree Search implementation for solving those QCNs with likelihood optimally and efficiently. Our experimental evaluation demonstrates the effectiveness of our approach, showing that approximate search techniques can offer a promising trade-off between computational budget and optimality.

2 Preliminaries and Problem Definition

We first recall some important definitions for qualitative spatial and temporal reasoning and then formulate the problem studied in this paper.

2.1 Qualitative Constraint-Based Reasoning

A *binary* qualitative constraint language is based on a finite set B of *jointly exhaustive and pairwise disjoint* relations, called the set of *base relations* (*atoms*), that is defined over an infinite domain D (e.g., some topological space) [16]. These base relations represent definite knowledge between two entities of D; indefinite knowledge can be specified by a union of possible base relations, and is represented by the set containing them. The set B contains the identity relation Id, and is closed under the *converse* operation ($^{-1}$). The entire set of relations 2^{B} is equipped with the set-theoretic operations of union and intersection, the converse operation, and the *weak composition* operation denoted by $\diamond$ [16]. The weak composition ($\diamond$) of two base relations $b, b' \in \mathsf{B}$ is the smallest relation $r \in 2^{\mathsf{B}}$ that includes $b \circ b'$; formally, $b \diamond b' = \{b'' \in \mathsf{B} : b'' \cap (b \circ b') \neq \emptyset\}$, where $b \circ b' = \{(x, y) \in \mathsf{D} \times \mathsf{D} : \exists z \in \mathsf{D}$ such that $(x, z) \in b \wedge (z, y) \in b'\}$. Finally, for all $r \in 2^{\mathsf{B}}, r^{-1} = \bigcup\{b^{-1} : b \in r\}$, and for all $r, r' \in 2^{\mathsf{B}}, r \diamond r' = \bigcup\{b \diamond b' : b \in r, b' \in r'\}$.

As an illustration, consider Allen's Interval Algebra (IA) [1], which is a calculus for representing and reasoning about temporal information. The domain D of IA consists of time intervals that can be specified by their starting and ending points. A time interval X is represented as a pair (x_s, x_e), where x_s and x_e denote the starting and ending points of X, respectively. Figure 2 displays all 13 base relations of IA that can hold between any two time intervals, such as *before* ($<$), *after* ($>$), or *equal* ($=$). These relations can be used to reason about temporal constraints, such as the temporal ordering of events or tasks (see Fig. 1 for an example).

The problem of representing and reasoning about qualitative spatial or temporal information may be tackled via the use of a *Qualitative Constraint Network*, defined as follows:

Definition 1. (QCN). *A* Qualitative Constraint Network *(*QCN*) is a tuple* (V, C) *where:*

- $V = \{v_1, \ldots, v_n\}$ *is a non-empty finite set of variables, each representing an entity of an infinite domain* D*;*
- *and* C *is a mapping* $C : V \times V \rightarrow 2^{\mathsf{B}}$ *such that* $C(v, v) = \{\mathsf{Id}\}$ *for all* $v \in V$ *and* $C(v, v') = C(v', v)^{-1}$ *for all* $v, v' \in V$*.*

Note that, for clarity, in the QCNs in Fig. 1 neither converse relations nor Id loops are shown, but they are part of any QCN. Further, given a QCN $\mathcal{N} = (V, C)$, we refer to a constraint $C(i, j)$ with $i, j \in V$ also as $c_{i,j}$. The size of $\mathcal{N}$, denoted as $|\mathcal{N}|$, is the number of variables $|V|$. A *scenario* (solution) of $\mathcal{N}$ is a mapping $\sigma : V \rightarrow \mathsf{D}$ such that $\forall v, v' \in V$, $\exists b \in C(v, v')$ such that $(\sigma(v), \sigma(v')) \in b$, and $\mathcal{N}$ is *satisfiable* iff a valid scenario exists [8].

A QCN that admits a valid scenario is also called consistent. For a QCN in which all relations are elements of a so-called tractable subset of IA, consistency can be decided using the *algebraic closure* operation (also called path-consistency) that computes the fixed point of refinements $c_{i,j} \leftarrow c_{i,j} \cap (c_{i,k} \diamond c_{k,j})$ [18,21]. Then, the network is consistent iff algebraic closure does not refine any constraint to the empty relation. Relations not contained in a given tractable subset can be rewritten as disjunctions of relations from the subset. For arbitrary problem instances, algebraic closure is then used for constraint propagation / forward checking (empty relations only occur for inconsistent networks) and combined with a backtracking search that refines disjunctions to individual relations [21].

2.2 Problem Definition

We are interested in finding the most likely scenario of a given QCN $\mathcal{N} = (V, C)$. For this, we introduce a notion of *likelihood* $\ell : C \rightarrow [0, 1]$ that indicates how likely a certain constraint $c \in C$ is true. We note that our notion of likelihood delineates from its usage in probability theory, but instead is similar to the notion of *possibility* introduced in the context of classical constraint satisfaction problems [7,22]. Generally, the possibility $\Pi_\pi(k)$ represents the possibility for the constraint k to be satisfied according to a knowledge base. In our context, we have a possibility distribution π over all base relations $b \in \mathsf{B}$ where $\pi_b \in [0, 1]$ and $\pi_{\mathsf{B}} = \sum_{b \in \mathsf{B}} \pi_b = 1$, since it is guaranteed that at least one of the base relations yields a valid scenario. We then define the *likelihood* ℓ of a given constraint $c_{i,j}$ as the summed possibility of all its base relations $\ell(c_{i,j}) = \sum_{b \in c_{i,j}} \pi_b$. Clearly, this formulation reduces to the standard QCN problem when all likelihoods are set to 1, semantically restricting the set of possibilities to only the constraints present in the original QCN. Hence, the problem of computing the most likely scenario is a straight-forward generalization of the standard task of computing a consistent atomic network.

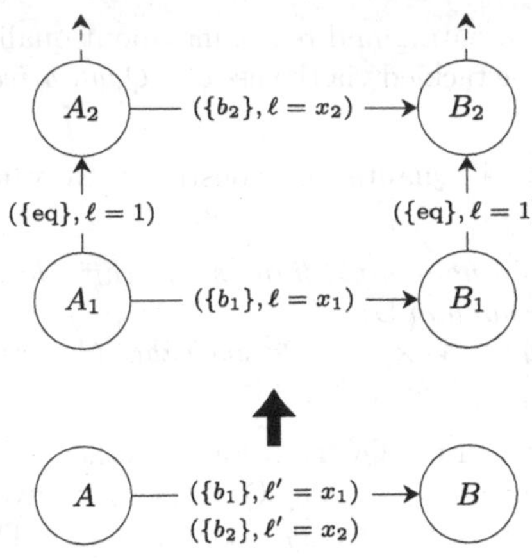

Fig. 3. Illustration for the construction of reducing an individual likelihood most likely QCN solution to a most likely QCN solution.

Definition 2 (most likely scenario with summed possibility). *Given a* QCN $\mathcal{N}$ *with constraints* $c_{i,j}$, $1 \leq i,j \leq |\mathcal{N}|$ *over a calculus with base relations* B *and a likelihood measure* $\ell : c_{i,j} \to [0,1]$, *compute* $\phi : c_{i,j} \to \mathsf{B}$ *such that (1)* ϕ *yields a consistent network and (2)*

$$L(\phi) = \sum_{i,j} \begin{cases} \ell(c_{i,j}) & \text{if } \phi(c_{i,j}) \in c_{i,j} \\ 1 - \ell(c_{i,j}) & \text{otherwise} \end{cases}$$

is maximal. We also call L *the likelihood of an atomic* QCN.

In this problem definition, one simply sums up the likelihood of all constraints satisfied and the complement for those not satisfied by a solution. The maximum likelihood for a given QCN $\mathcal{N} = (V, C)$ is thus limited by the number of its constraints $|C|$. Now assume a more fine-grained model of likelihood is desired, where likelihoods can be defined for base relations individually. As we will see, such problem is equivalent to the one above.

Definition 3 (most likely scenario with individual possibility). *Given a* QCN $\mathcal{N}$ *with constraints* $c_{i,j}$, $1 \leq i,j \leq |\mathcal{N}|$ *over a calculus with base relations* B *and a likelihood measure* $\ell' : [1, 2, \ldots, N]^2 \times B \to [0,1]$, *such that* $\ell'(i,j,b)$ *represents the likelihood that base relation* $b \in B$ *holds in constraint* $c_{i,j}$, *i.e.,* $b \in c_{i,j}$. *Compute* $\phi : c_{i,j} \to \mathsf{B}$ *such that (1)* ϕ *yields a consistent network and (2)*

$$L'(\phi) = \sum_{i,j} \ell(\phi(c_{i,j}))$$

is maximal.

Proposition 1. *Problem definitions (2) and (3) are reducible to one another.*

Proof. (2) → (3): Given an instance of (2), we construct an instance of (3) by keeping the QCN and setting

$$\ell'(i,j,b) = \begin{cases} \ell(c_{i,j}) & \text{if } b \in c_{i,j} \\ 1 - \ell(c_{i,j}) & \text{otherwise} \end{cases}$$

This way we have $L(\phi) = L'(\phi)$ and thus a solution ϕ for that is maximal with respect to L is also minimal with respect to L'.

(3) → (2): Given an instance of (3) $\mathcal{N}$, we construct an instance of (2) by constructing a new QCN. The idea is to introduce a separate constraint per single base relation for any constraint $c_{i,j}$ and is illustrated in Fig. 3. Formally, we construct QCN $\mathcal{N}'$ by introducing variables X_b, $b \in \mathsf{B}$ for all variables X in $\mathcal{N}$ and connect them with equality constraints using the equality relations from the respective calculus, i.e., $c_{X_b,X'_b} = \text{eq}$ and set $\ell(c_{X_b,X'_b}) = 1$. We then introduce constraints $c'_{x_b,y_b} = b$ for all $c_{x,y}$ in $\mathcal{N}$ and set $\ell(c'_{i,j,b}) = \ell'(c_{i,j,b})$. Now, consider solutions for $\mathcal{N}'$ with respect to problem (2). Observe that there exists a solution ϕ' in which all equality constraints introduced by the construction are satisfied and thus exactly one constraint c'_{x_b,y_b}, $b \in \mathsf{B}$ is satisfied. This observation directly follows from the fact that base relations are pairwise disjoint and the likelihood of satisfying equality constraints is not less than satisfying the likelihood of a base relation constraint c_{x_b,y_b}. We can now construct a solution ϕ to the input QCN $\mathcal{N}$ by setting $\phi(c_{x,y}) = b$ whenever c_{x_b,y_b} is satisfied in ϕ'.

2.3 MAX-QCN Problem

For classical QCNs without likelihood, we can only determine whether a QCN is consistent or not. In case of inconsistency, one may be interested in determining the minimal set of changes in constraints, called repairs, that are necessary to make $\mathcal{N}$ consistent. This is also known as the MAX-QCN problem [6]. We recall the definition of the MAX-QCN problem, and reformulate it slightly to better suit our presentation.

Definition 4. **(QCN repair problem).** *Given a* QCN $\mathcal{N}$ *with constraints* $c_{i,j}$, $1 \le i,j \le |\mathcal{N}|$ *over a calculus with base relations* B, *compute a minimal set of* repairs, *also called* relaxations, $c_{i,j} \mapsto \mathsf{B}$ *that, when applied to* $\mathcal{N}$, *make* $\mathcal{N}$ *consistent.*

We note that any repair $c_{i,j} \mapsto \mathsf{B}$ that is necessary corresponds to a single modification $c_{i,j} \mapsto \{b\}$ for some base relation b previously not contained in $c_{i,j}$, since in any solution of a QCN exactly one of the base relations holds. The observant reader will note that our reformulation goes a step further than the definition of the MAX-QCN problem in that we do not only inquire how many repairs are needed, but what these repairs are exactly. Considering the likelihood of all constraints to be set alike, computing the most likely QCN solution is the more general problem.

Corollary 1. *The problem of computing the most likely* QCN *solution also solves the* MAX-QCN *problem.*

The state-of-the-art for solving the MAX-QCN problem is based on an encoding as a *partial MaxSAT* problem [5]. We briefly recall relevant concepts for the partial MaxSAT problem. A literal represents a propositional variable or its negation, and a clause is a disjunction of literals. The MaxSAT problem refers to the task of identifying an assignment that satisfies the maximum number of clauses from a given set of clauses [13]. As such, the QCN repair problem (see Definition 4) can be considered a variation of the MaxSAT problem for QCNs. The partial MaxSAT problem [3,17] extends the MaxSAT problem as follows: A partial MaxSAT instance Ω differentiates between hard and soft clauses, such that a solution ω of Ω is an assignment that satisfies all hard clauses while maximizing the number of satisfied soft clauses.

The general idea of encoding a QCN $\mathcal{N} = (V, C)$ into SAT is to introduce a propositional variable $x^b_{i,j}$ for every constraint $c_{i,j} \in C$ and for all possible base relations $b \in \mathsf{B}$ of the calculus. To ensure that there exists exactly one base relation for every constraint in the solution, at-least-one (ALO) $\bigvee_{b \in \mathsf{B}} x^b_{i,j}$ and at-most-one (AMO) $\forall \{b, b'\} \subseteq \mathsf{B}$ with $b \neq b', \neg x^b_{i,j} \vee \neg x^{b'}_{i,j}$ hard clauses are defined. Further hard clauses are introduced to ensure that the base relations are consistent with the theory of the calculus. Specifically, for every triangle $i, j, k \in V$ in $\mathcal{N}$, we need to ensure that the constraints between them are closed under the weak composition $\diamond$.

Finally, the soft clauses are defined as follows. For every constraint $c_{i,j} \in \mathcal{N}$, all its possible base relations are added as ALO soft clause $\bigvee_{b \in c_{i,j}} x^b_{i,j}$ with weight $w = 1$. Hence, not satisfying any constraint $c \in C$ will admit a cost of 1.

3 Monte Carlo Tree Search for Most Likely Scenarios

In this section, we present a Monte Carlo Tree Search (MCTS) algorithm for computing the most likely scenarios of a given QCN with likelihood. MCTS is a best-first search that estimates the values of nodes in a search tree by conducting a (pseudo-)random exploration of the search space, and therefore does not require a positional evaluation function. Starting with an empty tree, MCTS grows the tree by traversing it from the root node to a leaf node during the *selection step*. The choice of the leaf node determines where the growth will occur, either by exploring alternative options at the same level or by exploiting previous decisions and growing deeper trees. The *expansion step* determines the number of new nodes to be added to the tree, with their values estimated using a simulation during the *simulation step*. The simulation can be as simple as applying available actions randomly until a terminal or known state is reached, or more sophisticated methods based on rules, heuristics or even complete search procedures can be employed. After estimating the values of newly added nodes, the *backpropagation step* updates all nodes along the path from the leaf back up to the root node. These four steps are repeated until a given computational budget is exhausted [2,23].

MCTS has been applied successfully in many domains, such as games, planning, scheduling, and combinatorial optimization [2,23]. Probably most relevant to this work are the application to boolean satisfiability [20] and maximum satisfiability (MaxSAT) [10]. As outlined above, the QCN repair problem can be seen as a version of the MaxSAT problem for QCNs. In the following, we describe our approach and point out similarities and differences with the `UCTMAXSAT` algorithm [10].

We start at the root of the search tree with an empty QCN consisting of the same variables V as the original QCN $\mathcal{N} = (V, C)$. Every node in the search tree then represents a separate QCN $\mathcal{P}' = (V, C')$ that is obtained by adding a single constraint $c_{i,j} \in C$ from $\mathcal{N}$ to the QCN $\mathcal{P} = (V, C'')$ of its parent node p, i.e., QCN $\mathcal{P}' = (V, C'' \cup \{c_{i,j}\})$, where $c_{i,j} \notin C''$. This is similar to the structure of the `UCTMAXSAT` algorithm, where every node represents a partial assignment of the variables in the MaxSAT instance, equally starting with an empty assignment. The remainder of this section describes the four MCTS steps in detail.

Selection. Like `UCTMAXSAT`, we rely on the well-known UCT (Upper Confidence bounds applied to Trees) selection strategy [14]. At any node p, UCT selects the child i according to the formula:

$$\mathsf{argmax}_i \left[v_i + \alpha \cdot \sqrt{\frac{\ln n_p}{n_i}} \right]$$

Here, n_i and n_p denote the number of times node i and its parent node p have been visited, respectively. α is a coefficient that has to be tuned experimentally, with smaller values favoring exploitation and larger ones exploration of the search space. Kocsis and Szepesvári [14] propose $1/\sqrt{2} \approx 0.707$ for values $v \in [0, 1]$. The value v_i of a node estimates the average fraction of constraints that need to be repaired in the subtree rooted at node i; it is computed during simulation. As $v_i \in [0, 1]$, we use $\alpha = 1/\sqrt{2}$ throughout our experiments.

Expansion. `UCTMAXSAT` employs a heuristic to determine the next variable assignment. Once a variable v is selected, the tree gets expanded by two nodes. One sets v to true and the other v to false in the original SAT formula.

Instead, we randomly choose the next constraint $c_{i,j} \in \mathcal{N}$ such that $c_{i,j} \notin$ QCN $\mathcal{P}$ of the current node p, and add it to $\mathcal{P}$. If the resulting network is satisfiable, we add a new child node with QCN $\mathcal{P}' = \mathcal{P} \cup \{c_{i,j}\}$. Additionally, in any case, we add a new child node to represent the case that the constraint $c_{i,j}$ is not added to $\mathcal{P}$, i.e., the constraint is repaired. Essentially, the children of a node represent both possible options (use or repair) of the next constraint $c_{i,j}$, with the exception that we only permit children that represent satisfiable QCNs.

Simulation. To estimate the value v_i of a node i, we repeatedly perform the strategy described in the expansion step. That is, we keep adding constraints from QCN $\mathcal{N}$ to the initially empty QCN $\mathcal{N}'$ of the root. Whenever the addition of a constraint would make $\mathcal{N}'$ unsatisfiable, we do not add the constraint but rather add it to a list of repaired constraints $\mathcal{R}$. The simulation is over once all constraints of the initial QCN $\mathcal{N}$ are either applied to $\mathcal{N}'$ or are repaired.

The likelihood of a node i is then estimated as $\ell_i = L(\phi') + \sum_{c_{i,j} \in \mathcal{R}}(1 - \ell(c_{i,j}))$, where ϕ' is a solution of $\mathcal{N}'$ and $L(\phi')$ its likelihood. For better comparison, we normalize ℓ_i by dividing it by the maximum likelihood possible, i.e., the number of constraints $|C|$ in $\mathcal{N}$. So the value of node i is denoted as $v_i = \ell_i/|C|$ such that $v_i \in [0, 1]$.

Backpropagation. After a successful simulation on node i, the parent node's value is updated using the common method of taking the average of all children's values. This approach is also employed in `UCTMAXSAT` as it has demonstrated superior performance compared to alternative backpropagation strategies.

Solution. Over the course of the search, we keep track of the best solution found so far. Once the computational budget is exhausted, we subsequently return this as the final solution.

4 Experimental Evaluation

We empirically evaluate our approach on IA (cf. Section 2.1). Specifically, we compare our MCTS approach to an adaptation of the partial MaxSAT encoding for the MAX-QCN problem of Condotta et al. (2016) [5]. Our extensions are detailed in Sect. 4.1. The dataset and experimental setup are described in Sect. 4.2 and Sect. 4.3, respectively. We conclude with a discussion of the results in Sect. 4.4.

4.1 Implementation Details

In order to adapt the forbidden covering and triangulation based encoding (FCTE) [5] for solving QCNs with likelihood, we only change the soft clauses. In particular, given a QCN $\mathcal{N} = (V, C)$ with likelihood $\ell(c)$ for every constraint $c \in C$, we adapt the weight w of the soft clause to reflect its likelihood. Since weights are usually represented as integers, we discretize ℓ to the integer range of full percent, i.e., $\ell \in [0, 100]$ by simply multiplying it with 100. So for every $c_{i,j} \in C$ we have the same soft clause as before, but with weight $w = \lfloor \ell(c_{i,j}) * 100 \rceil$. We then add an additional ALO soft clause over all base relations $b \notin c_{i,j}$, specifically $\bigvee_{b \notin c_{i,j}} x^b_{i,j}$ with weight $w = \lfloor (1 - \ell(c_{i,j})) * 100 \rceil$. Generally, this clause is only satisfied if none of the base relations $b \in c_{i,j}$ is selected, i.e., the constraint is repaired. Due to the AMO hard constraint, only one of the soft clauses described above may be satisfied.

The partial MaxSAT solver then tries to minimize the sum of weights of all unsatisfied clauses, which corresponds to minimizing the likelihood of the repaired constraints. Or in other words, it maximizes the likelihood of the constraints that are not repaired. For the case where no likelihood is specified, i.e., $\forall c \in C.\ \ell(c) = 1$, this formulation defaults to the original encoding.

Finally, the likelihood $L(\phi)$ for a found solution ϕ of $\mathcal{N} = (V, C)$ can be derived from the weight W of the partial MaxSAT solver and the number of its constraints $|C|$, such that $L(\phi) = (|C| * 100 - W)/100$.

4.2 Dataset

Since no benchmark for QCNs with likelihood, as studied in this paper, exists, we generate random QCNs using an extension of the well-known $\mathsf{A}(\mathsf{n},\mathsf{d},\mathsf{l})$ model [21]. This model has also been used to generate the datasets for the evaluation by Condotta et al. (2015, 2016) [5,6]. In the $\mathsf{A}(\mathsf{n},\mathsf{d},\mathsf{l})$ model, n denotes the number of variables in the QCN, d the average degree, i.e., constraints incident to a single variable, and l the average label size, i.e., number of base relations per constraint, set to $l = |\mathsf{B}|/2 = 6.5$. Increasing the degree d increases the likelihood of obtaining an inconsistent QCN [21], hence also requiring more repairs on average. The generated QCNs may be seen as the result of merging information present in different data sources and hence feature different degrees of likelihood p that the constraint is actually correct. Likelihood p is sampled from a normal distribution with mean $\mu = 1$ and standard deviation $\sigma = 0.1$, to resemble the fact that most data is likely correct in real-world scenarios.

The whole dataset comprises 1500 QCNs. For each degree $d \in \{8 \dots 15\}$, using a step size of 0.5, we generate 100 QCNs of size $n = 20$. This is similar to the dataset by Condotta et al. [5], but with 100 instead of 10 networks for every degree to account for variations.

4.3 Experimental Setup

Every method has 60 s to compute a solution for each network. All experiments were conducted on virtual machines with 1 CPU core and 4 GB of RAM, running on a server with 2 Intel Xeon Gold 6334 CPUs, each with 8 cores at 3.6GHz. We implemented both the MCTS and the partial MaxSAT encoding in Python and run it using the Just-in-Time (JIT) compiler PyPy. As partial MaxSAT solver we used RC2 [12] as part of the PySAT toolkit [11].

4.4 Experimental Results

When solving the MAX-QCN problem, we see that partial MaxSAT is requiring significantly more run time the higher the degree d of the QCN is. This means that given a time out of 60 s per network, the partial MaxSAT solver is not able to solve all instances with degree $d \geq 9$ and not even one instance for degrees $d \geq 13$ (see Fig. 4). In contrast, our MCTS approach is able to solve all instances within the given time limit, usually returning the first solution after only a fraction of a second. But, while MaxSAT is able to find the optimal solution for smaller degrees comparingly fast and return the result, MCTS always requires the full time given, as the optimal solution is unknown, and the search space is too large to fully explore before reaching the time limit. However, as MCTS is an anytime algorithm, it can be stopped at any time during the search and still return the best solution found so far, which is not possible for MaxSAT. We explore the varying results by comparing three different variants of MCTS, namely MCTS-30, MCTS-60, and MCTS-120, with a computational budget of 30 s, 60 s, and 120 s, respectively. The results are shown in Fig. 5. We see that

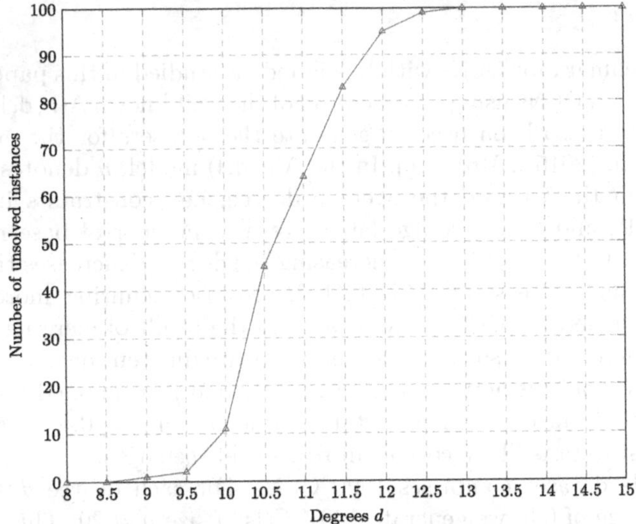

Fig. 4. Number of unsolved instances for Partial MaxSAT by degree d, using a timeout of 60 s per instance.

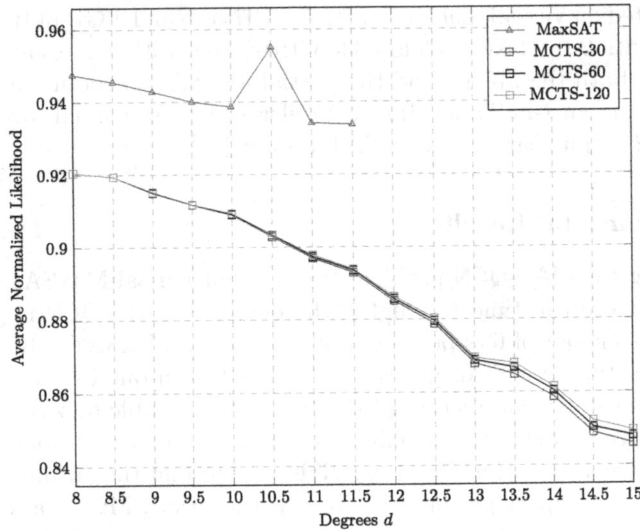

Fig. 5. Qualitative comparison of the MaxSAT encoding using 60 s as time out and three variants of our MCTS algorithm, where MCTS-i has i seconds to compute the most likely scenario.

the average normalized likelihood hardly changes by increasing the run time, indicating that the computational budget does not have a significant impact on the quality of the solution.

For a qualitative comparison, we used the normalized likelihood of the solutions. That is, we divided the likelihood of the solution by the number of constraints in the QCN to normalize it to the range $[0, 1]$. We then compared the average normalized likelihood of the solutions found by MaxSAT and MCTS for each degree d. A higher score is better and 1 the theoretical limit, also used as basis for comparison for instances where MaxSAT fails to compute a solution. The results are shown in Fig. 5. As expected, the average normalized likelihood of the solutions found by MaxSAT is higher than the one found by MCTS. As MaxSAT always computes the optimum, we can also conjecture that for higher degrees d, the likelihood of random scenarios decreases. This is in line with the general observation that the higher the degree d of a QCN is, the more likely it is to be inconsistent [21].

5 Conclusion and Future Work

This paper tackles the problem of computing the most likely scenario of a *Qualitative Constraint Network* (QCN). For this, the notion of *likelihood* is introduced, ascribing a certain possibility that a particular constraint holds. We show that this problem is a generalization of the MAX-QCN problem [5,6], and present a Monte Carlo Tree Search (MCTS) algorithm for solving it. We empirically evaluate our approach on a dataset of generated QCNs. Our results showcase that leveraging approximate search techniques offers a promising trade-off between computational budget and optimality. This finding emphasizes the practical viability of our approach in real-world application contexts, where merging data from different sources often leads to inconsistencies.

Overall, our work contributes to advancing the field of qualitative constraint networks and qualitative calculi by providing a novel perspective on repairing inconsistent networks. By considering likelihood factors, we address an important practical use case and offer a valuable solution for identifying consistent constraint networks while minimizing deviations from the over-constrained network. We believe that our research opens up new avenues for handling inconsistencies in knowledge bases and highlights the significance of integrating likelihood considerations into qualitative reasoning frameworks.

Future work includes the consideration of other qualitative calculi, as well as conducting a more extensive evaluation. For this, we would also like to extend other SAT encodings previously proposed for qualitative temporal reasoning to MaxSAT in order to apply them to our setting. In particular, the ones by Pham et al. (2006) [19] and Westphal et al. (2013) [24] seem promising, as they are specifically developed for the temporal domain of IA.

Acknowledgements. We acknowledge financial support by Technologieallianz Oberfranken (TAO) and thank the anonymous reviewers for their feedback.

References

1. Allen, J.F.: Maintaining knowledge about temporal intervals. Commun. ACM 832–843 (1983). https://doi.org/10.1145/182.358434
2. Browne, C.B., et al.: A survey of Monte Carlo tree search methods. IEEE Trans. Comput. Intell. AI Games **4**(1), 1–43 (2012)
3. Cai, S., Luo, C., Thornton, J., Su, K.: Tailoring local search for partial maxsat. In: Brodley, C.E., Stone, P. (eds.) Proceedings of the Twenty-Eighth AAAI Conference on Artificial Intelligence, pp. 2623–2629. AAAI Press (2014)
4. Condotta, J.F., Kaci, S., Schwind, N.: A framework for merging qualitative constraints networks. In: Proceedings of the Twenty-First International FLAIRS Conference (2008)
5. Condotta, J., Nouaouri, I., Sioutis, M.: A SAT approach for maximizing satisfiability in qualitative spatial and temporal constraint networks. In: Baral, C., Delgrande, J.P., Wolter, F. (eds.) Principles of Knowledge Representation and Reasoning: Proceedings of the Fifteenth International Conference, pp. 432–442. AAAI Press (2016)
6. Condotta, J.F., Mensi, A., Nouaouri, I., Sioutis, M., Saïd, L.B.: A practical approach for maximizing satisfiability in qualitative spatial and temporal constraint networks. In: Proceedings of International Conference on Tools with Artificial Intelligence (ICTAI), pp. 445–452 (2015). https://doi.org/10.1109/ICTAI.2015.73
7. Dubois, D., Fargier, H., Prade, H.: Possibility theory in constraint satisfaction problems: handling priority, preference and uncertainty. Appl. Intell. **6**(4), 287–309 (1996). https://doi.org/10.1007/bf00132735
8. Dylla, F., et al.: A survey of qualitative spatial and temporal calculi: Algebraic and computational properties. ACM Comput. Surv. **50**, 1–39 (2017). article 7
9. Dylla, F., Wallgrün, J.O.: Qualitative spatial reasoning with conceptual neighborhoods for agent control. J. Intell. Robotic Syst. **48**(1), 55–78 (2007). https://doi.org/10.1007/s10846-006-9099-4
10. Goffinet, J., Ramanujan, R.: Monte-carlo tree search for the maximum satisfiability problem. In: Rueher, M. (ed.) CP 2016. LNCS, vol. 9892, pp. 251–267. Springer, Cham (2016). https://doi.org/10.1007/978-3-319-44953-1_17
11. Ignatiev, A., Morgado, A., Marques-Silva, J.: PySAT: a python toolkit for prototyping with SAT oracles. In: SAT, pp. 428–437 (2018). https://doi.org/10.1007/978-3-319-94144-8_26
12. Ignatiev, A., Morgado, A., Marques-Silva, J.: RC2: an efficient maxsat solver. J. Satisf. Boolean Model. Comput. **11**(1), 53–64 (2019). https://doi.org/10.3233/SAT190116
13. Johnson, D.S.: Approximation algorithms for combinatorial problems. J. Comput. Syst. Sci. **9**(3), 256–278 (1974). https://doi.org/10.1016/S0022-0000(74)80044-9
14. Kocsis, L., Szepesvári, C.: Bandit based monte-carlo planning. In: Fürnkranz, J., Scheffer, T., Spiliopoulou, M. (eds.) ECML 2006. LNCS (LNAI), vol. 4212, pp. 282–293. Springer, Heidelberg (2006). https://doi.org/10.1007/11871842_29
15. Ligozat, G.: Qualitative Spatial and Temporal Reasoning. Wiley, Hoboken (2011)
16. Ligozat, G., Renz, J.: What Is a qualitative calculus? A general framework. In: Zhang, C., W. Guesgen, H., Yeap, W.-K. (eds.) PRICAI 2004. LNCS (LNAI), vol. 3157, pp. 53–64. Springer, Heidelberg (2004). https://doi.org/10.1007/978-3-540-28633-2_8

17. Miyazaki, S., Iwama, K., Kambayashi, Y.: Database queries as combinatorial optimization problems. In: Kambayashi, Y., Yokota, K. (eds.) Proceedings of the International Symposium on Cooperative Database Systems for Advanced Applications, pp. 477–483. World Scientific (1996)
18. Nebel, B., Bürckert, H.: Reasoning about temporal relations: a maximal tractable subclass of allen's interval algebra. J. ACM **42**(1), 43–66 (1995). https://doi.org/10.1145/200836.200848
19. Pham, D.N., Thornton, J., Sattar, A.: Towards an efficient SAT encoding for temporal reasoning. In: Benhamou, F. (ed.) CP 2006. LNCS, vol. 4204, pp. 421–436. Springer, Heidelberg (2006). https://doi.org/10.1007/11889205_31
20. Previti, A., Ramanujan, R., Schaerf, M., Selman, B.: Monte-carlo style UCT search for boolean satisfiability. In: Pirrone, R., Sorbello, F. (eds.) AI*IA 2011. LNCS (LNAI), vol. 6934, pp. 177–188. Springer, Heidelberg (2011). https://doi.org/10.1007/978-3-642-23954-0_18
21. Renz, J., Nebel, B.: Efficient methods for qualitative spatial reasoning. J. Artif. Intell. Res. (JAIR) **15**, 289–318 (2001)
22. Schiex, T.: Possibilistic constraint satisfaction problems or "how to handle soft constraints?". In: Dubois, D., Wellman, M.P. (eds.) Proceedings of the Eighth Annual Conference on Uncertainty in Artificial Intelligence (UAI 1992), pp. 268–275. Morgan Kaufmann (1992)
23. Świechowski, M., Godlewski, K., Sawicki, B., Mańdziuk, J.: Monte carlo tree search: a review of recent modifications and applications. Artif. Intell. Rev. (2022). https://doi.org/10.1007/s10462-022-10228-y
24. Westphal, M., Hué, J., Wölfl, S.: On the propagation strength of SAT encodings for qualitative temporal reasoning. In: 25th IEEE International Conference on Tools with Artificial Intelligence, pp. 46–54. IEEE Computer Society (2013). https://doi.org/10.1109/ICTAI.2013.18

PapagAI: Automated Feedback for Reflective Essays

Veronika Solopova[1(✉)], Eiad Rostom[1], Fritz Cremer[1], Adrian Gruszczynski[1], Sascha Witte[1], Chengming Zhang[2], Fernando Ramos López[1], Lea Plößl[2], Florian Hofmann[2], Ralf Romeike[1], Michaela Gläser-Zikuda[2], Christoph Benzmüller[2,3], and Tim Landgraf[1]

[1] Freie Universität Berlin, Berlin, Germany
veronika.solopova@fu-berlin.de

[2] Friedrich-Alexander-Universität Erlangen-Nürnberg, Erlangen, Germany

[3] Otto-Friedrich-Universität Bamberg, Bamberg, Germany

Abstract. Written reflective practice is a regular exercise pre-service teachers perform during their higher education. Usually, their lecturers are expected to provide individual feedback, which can be a challenging task to perform on a regular basis. In this paper, we present the first open-source automated feedback tool based on didactic theory and implemented as a hybrid AI system. We describe the components and discuss the advantages and disadvantages of our system compared to the state-of-art generative large language models. The main objective of our work is to enable better learning outcomes for students and to complement the teaching activities of lecturers.

Keywords: Automated feedback · Dialogue · Hybrid AI · NLP

1 Introduction

Dropout rates as high as 83% among pre-service teachers and associated teacher shortages are challenging the German education system [2,20]. This may be due to learning environments not adequately supporting prospective teachers in their learning process [29]. Written reflective practice may alleviate the problem: By reflecting on what has been learned and what could be done differently in the future, individuals can identify areas for improvement. However, instructors may be overburdened by giving feedback to 200+ students on a weekly basis. With the rise of large language models (LLMs, [30]), automated feedback may provide welcome relief. Students could iteratively improve their reflection based on the assessment of a specialized model and through that, their study performance. Instructors could supervise this process and invest the time saved in improving the curriculum. While current research is seeking solutions to align the responses of LLMs with a given set of rules, it is currently impossible to guarantee an output of a purely learnt model to be correct. Here, we propose "PapagAI", a platform to write reflections and receive feedback from peers,

D. Seipel and A. Steen (Eds.): KI 2023, LNAI 14236, pp. 198–206, 2023.
https://doi.org/10.1007/978-3-031-42608-7_16

instructors and a specialized chatbot. PapagAI uses a combination of ML and symbolic components, an approach known as hybrid AI [10]. Our architecture is based on various natural language understanding modules[1], which serve to create a text and user profile, according to which a rule-based reasoner chooses the appropriate instructions.

2 Related Work

PapagAI employs a number of models for detecting topics contained in -, and assessing the quality and depth of the reflection, as well as for detecting the sentiment and emotions of the author. While extensive previous work was published on each of these tasks, implementations in German are rare. To our knowledge, there is no previous work that combined all in one application. Automated detection of reflective sentences and components in a didactic context has been described previously [12,18,22,24,36,38]. In [18], e.g., the authors analyse the depth of a reflection on the text level according to a three-level scheme (none, shallow, deep). Document-level prediction, however, can only provide coarse-grained feedback. Liu et al. [23], in contrast, also use three levels for predicting reflective depth for each sentence. In emotion detection, all previous works focus on a small set of 4 to 6 basic emotions. In Jena [16], e.g., the author describes detecting students' emotions in a collaborative learning environment. Batbaatar et al. [1] describes an emotion model achieving an F1 score of 0.95 for the six basic emotions scheme proposed by Ekman [9]. Chiorrini et al. [7] use a pre-trained BERT to detect four basic emotions and their intensity from tweets, achieving an F1 score of 0.91. We did not find published work on the German language, except for Cevher et al. [5], who focused on newspaper headlines. With regard to sentiment polarity, several annotated corpora were developed for German [34,37], mainly containing tweets. Guhr et al. [15] use these corpora to fine-tune a BERT model. Shashkov et el. [33] employ sentiment analysis and topic modelling to relate student sentiment to particular topics in English. Identifying topics in reflective student writing is studied by Chen et al. [6] using the MALLET toolkit [28] and by De Lin et al. [8] with Word2Vec + K-Means clustering. The techniques in these studies are less robust than the current state-of-art, such as ParlBERT-Topic-German [19] and Bertopic [14]. Overall, published work on automated feedback to student reflections is scarce, the closest and most accomplished work being AcaWriter [21] and works by Liu and Shum [23]. They use linguistic techniques to identify sentences that communicate a specific rhetorical function. They also implement a 5-level reflection depth scheme and extract parts of text describing the context, challenge and change. The feedback guides the students to the next level of reflective depth with a limited number of questions. In their user study, 85.7% of students perceived the tool positively. However, the impact on the reflection quality over time was not measured and remains unclear.

[1] All ML models are available in our OSF depository (https://osf.io/ytesn/), while linguistic processing code can be shared upon request.

3 Methods, Components and Performances

Data collection. Our data comes from the German Reflective Corpus [35]. The dataset contains reflective essays collected via google-forms from computer science and ethics of AI students in German, as well as e-portfolio diaries describing school placements of teacher trainees from Dundee University. For such tasks as reflective level identification and topic modelling, we enlarged it by computer science education students' essays and pedagogy students' reflections.[2] It consists of reflections written by computer science, computer science education, didactics and ethics of AI students in German and English. Data is highly varied, as didactics students write longer and deeper reflections than e.g. their computer science peers.

Emotions Detection. Setting out from the Plutchik wheel of basic emotions [31], during the annotation process we realised that many of the basic emotions are never used, while other states are relevant to our data and the educational context (e.g. confidence, motivation). We framed it as a multi-label classification problem at the sentence level. We annotated 6543 sentences with 4 annotators. The final number of labels is 17 emotions, with the 18th label being 'no-emotion'. We calculated the loss using binary cross entropy, where each label is treated as a binary classification problem, the loss is calculated for each label independently, which we sum for the total loss. We achieved the best results with a pre-trained RoBERTa [25] , with a micro F1 of 0.70 and a hamming score of 0.67 across all emotion labels. The model achieved the highest scores for "surprise", "approval" and "interest". With a lenient hamming score, accounting for the model choosing similar emotions (e.g. disappointment instead of disapproval) our model achieves up to 0.73.

Gibbs Cycle. [13] illustrates cognitive stages needed for optimal reflective results. It includes 6 phases: *description*, *feelings*, *evaluation*, *analysis*, *conclusion* and *future plans.* We annotated the highest phase present in a sentence and all the phases present. We treated this as a multi-class classification problem and used a pre-trained ELECTRA model. While evaluating, we compared one-hot prediction to the highest phase present and 3 top probability classes with all the phases present. While one-hot matching only managed to score 65% F1 macro, the top 3 predictions achieve up to 98% F1 macro and micro.

Reflective Level Detection. Under the supervision of Didactics specialists two annotators labelled 600 texts according to Fleck & Fitzpatrick's scheme [11], achieving moderate inter-annotators agreement of 0.68. The coding scheme includes 5 levels: *description*, *reflective description*, *dialogical reflection*, *transformative reflection* and *critical reflection*; With 70% of the data used for the training and 30% for evaluation, we used pre-trained BERT large and complete document embeddings for the English and German, resulting in QWK score of 0.71 in cross-validation.

[2] This still non-published data can be obtained upon request.

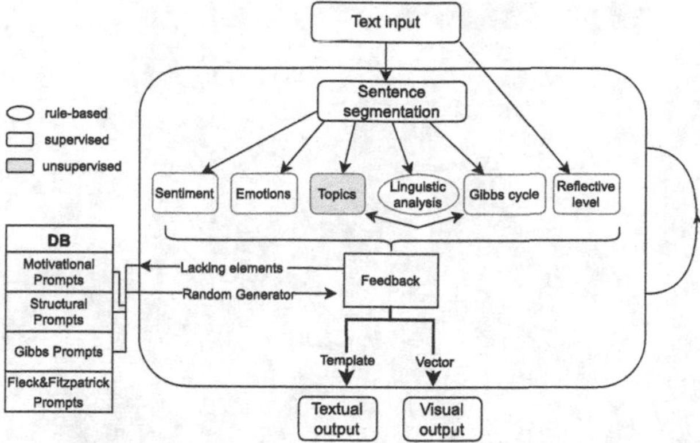

Fig. 1. The diagram of our PapagAI system shows the main productive modules. The legend on the left indicates the nature of the AI modules used.

Topic Modelling. We used BERTopic [14] on the sentence level. First, we tokenized and normalize the input sequence to lowercase and filter out numbers, punctuation, and stop-words using nltk library [3]. Then, we extract embeddings with BERT, reduce dimensionalities with UMAP, cluster reduced embeddings with HDBSCAN, create topic representation with tfidf and fine-tune topic representations with the BERT model. Because we have a lot of data of different origins, we created two clusterings, one more specific to the pedagogy topic and one including various educational topics. You can see our clusters in App.

Linguistic Scoring. Using spacy[3] we tokenized, and lemmatize the sentences, extracted dependencies parcing and part of speech. Additionally, we used RFTagger [32] for parts of speech and types of verbs. We extract sentence length, adverb for verb ratio, adjective for noun ratio, number of simple and complex sentences, types of subordinated clauses and number of discourse connectors[4] used. This information enables us to determine the reflection length, expressivity and variability of the language, as well as surface coherence and structure.

4 System Architecture

In PapagAI (see Fig. 1) the input text of the reflection is received from the AWS server through a WebSocket listener script. To minimize the response time, the models are loaded in the listener script once and then the user request spawn threads with the models already loaded. If the input text is smaller than three sentences and contains forbidden sequences, the processing does not start and the user receives a request to revise their input. Otherwise, the text is segmented into sentences and tokens. The language is identified using langid [26]

[3] https://spacy.io.

[4] We use Connective-Lex list for German: https://doi.org/10.4000/discours.10098.

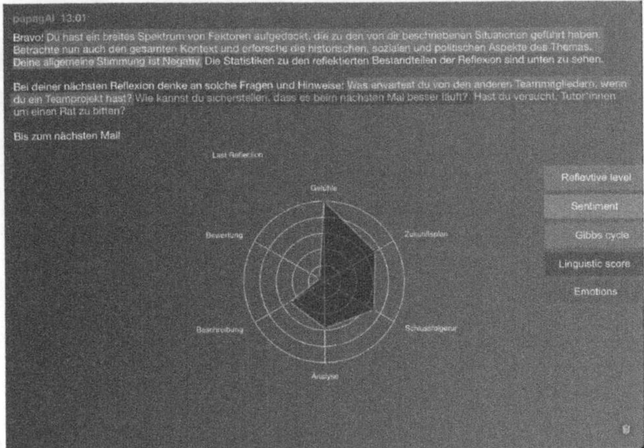

Fig. 2. The radar below the textual feedback illustrates Gibbs cycle completeness. The colour of the highlighted feedback text corresponds to the model responsible for this information.

and if the text is not in German, it is translated using Google translator API implementation.[5] The reflective level model receives the whole text, while other models are fed with the segmented sentences. Topic modelling and Gibbs cycle results are mapped, to identify if topics were well reflected upon. If more than three sentences are allocated to the topic and these sentences were identified by the Gibbs cycle model as analysis, we consider the topics well thought through. The extracted features are then passed to the feedback module. Here, the lacking and under-represented elements are identified in linguistic features and the three least present Gibbs cycle stages. If sentiment and emotions are all positive we conclude that no potential challenges and problems are thought through. If the sentiment and emotions are all negative, we want to induce optimism. These features together with the reflective level are mapped to the database of potential prompts and questions, where one of the suitable feedback options is chosen randomly for the sake of variability. Using manually corrected Gpt-3 outputs, for each prompt we created variations so that the feedback does not repeat often even if the same prompts are required. The extracted textual prompts are built together in a rule-based way into the template, prepared for German, Spanish and English. Otherwise, the overall feedback is made in German and then translated into the input language. The textual and a vector of extracted features for visual representation are sent back to the AWS server. The whole processing takes from 15 to 30 s based on the length of the text. Sample feedback can be seen in Fig. 2.

[5] https://pypi.org/project/deep-translator/.

5 Comparison with GPT-3

We compared our emotions detection (fine-tuned RoBERTa) and Gibbs cycle model (fine-tuned Electra) with the prompt-engineered state-of-the-art generative model Davinci [4] on the same task. For the evaluation and comparison, we used a small subset of 262 samples which were not part of the training. We first tried the zero-shot approach, where we described our labels to GPT-3 and gave it our sentence to predict. Then, we tried a one-shot approach, providing GPT-3 with one example sentence for each label. Finally, in the few-shot approach, we provided GPT-3 with three examples per label, which is the maximum number of examples possible due to the input sequence length restriction. Although the task requested GPT-3 to pick multiple labels out of the possible options, the model predicted multiple labels only in 5% of the cases for emotions. For this reason, we used the custom defined "one correct label": the score considers the prediction correct if it contains at least one correct label from the sentence's true labels. The zero-shot approach achieved only 0.28 accuracy in predicting one correct label for emotions. The model predicted the labels "information", "uncertainty", "interest", and "motivated" for the majority of the sentences. With the Gibbs cycle task, it achieved 80% correct predictions. Providing one example per label improved the performance noticeably by 18% (0.46) for emotions, and the model was able to detect emotions like "confidence", "challenged", and "approval" more accurately. It did not influence Gibb's cycle performance. Increasing the number of examples to three resulted in a slight improvement of 3% (0.49) for emotions, and 7% (0.87) for the Gibbs cycle. However, the best-scoring approaches did not offer a comparable performance to our fine-tuned models on these specific tasks with 0.81 on the same custom metric for emotion detection and 0.98 for the Gibbs cycle.

6 Discussion and Conclusion

The current PapagAI system has several advantages in comparison to generative LLMs. It ensures transparency of the evaluation and control over the output, which is based exclusively on didactic theory. Although LLMs show huge promise, they are still prone to hallucination [17,27], and, as we have shown in Sect. 5, they may under-perform on difficult cognitive tasks in comparison to smaller language models fine-tuned for the task. The fine-tuning of LLMs to didactic books and instructions, which we plan for our future work, still does not guarantee 100% theoretical soundness of the output, which is problematic e.g. in the case of pre-service students with statistically low AI acceptance. At the same time, the newest models, such as GPT-4, are only available through APIs, which raises concerns about data privacy, especially as the data in focus is an intimate reflective diary. Moreover, current open-source models, such as GPT-J and GPT-2, especially for languages other than English do not draw comparable results. Our architecture has, however, obvious drawbacks. On the one hand, our models do not reach 100% accuracy and this can naturally lead to suboptimal feedback. The processing time for many models, especially for longer texts, can be significantly higher than for a single generative LLM. For now, as we provide one feedback message for one rather long reflection, this is not a big issue, however, if we implement a dialogue form, the time of response would not feel natural. Finally, the variability of output using our approach is much more limited in comparison to generative models. We try to address it by creating many similar versions of instructions rephrased by GPT-3, and corrected manually. On average 7 out of 10 prompts needed some correction. Most of the errors were related to GPT-3 trying to rephrase the given sentence using synonyms that were not didactically appropriate in the given context. Future work, among others, will focus on user studies to understand how we can optimize the feedback, so that the users find it credible and useful, while their reflective skills advance. We also plan a more detailed evaluation based on more user data. We hope that our work will contribute to the optimization of the pre-service teachers' reflective practice and self-guided learning experience.

References

1. Batbaatar, E., Li, M., Ryu, K.H.: Semantic-emotion neural network for emotion recognition from text. IEEE Access **7**, 111866–111878 (2019). https://doi.org/10.1109/ACCESS.2019.2934529
2. Becker, A.: 83 Prozent der Studenten brechen Lehramts-Studium ab. Nordkurier (2021)
3. Bird, S., Klein, E., Loper, E.: Natural Language Processing with Python: Analyzing Text with the Natural Language Toolkit. O'Reilly Media, Inc., Sebastopol (2009)
4. Brown, T.B., et al.: Language models are few-shot learners (2020)
5. Cevher, D., Zepf, S., Klinger, R.: Towards multimodal emotion recognition in German speech events in cars using transfer learning (2019)
6. Chen, Y., Yu, B., Zhang, X., Yu, Y.: Topic modeling for evaluating students' reflective writing: a case study of pre-service teachers' journals. In: Proceedings of the Sixth International Conference on Learning Analytics & Knowledge, LAK 2016, pp. 1–5. Association for Computing Machinery, New York, NY, USA (2016). https://doi.org/10.1145/2883851.2883951
7. Chiorrini, A., Diamantini, C., Mircoli, A., Potena, D.: Emotion and sentiment analysis of tweets using bert. In: EDBT/ICDT Workshops (2021)
8. De Lin, O., Gottipati, S., Ling, L.S., Shankararaman, V.: Mining informal & short student self-reflections for detecting challenging topics - a learning outcomes insight dashboard. In: 2021 IEEE Frontiers in Education Conference (FIE), pp. 1–9 (2021). https://doi.org/10.1109/FIE49875.2021.9637181
9. Ekman, P.: Basic emotions. and book of cognition and emotion **98**, 16 (2023)
10. Elands, P., Huizing, A., Kester, J., Peeters, M.M.M., Oggero, S.: Governing ethical and effective behaviour of intelligent systems: a novel framework for meaningful human control in a military context. Militaire Spectator **188**(6), 302–313 (2019)
11. Fleck, R., Fitzpatrick, G.: Reflecting on reflection: framing a design landscape. In: Proceedings of the 22nd Conference of the Computer-Human Interaction Special Interest Group of Australia on Computer-Human Interaction, OZCHI 2010, pp. 216–223. Association for Computing Machinery, New York, NY, USA (2010). https://doi.org/10.1145/1952222.1952269

12. Geden, M., Emerson, A., Carpenter, D., Rowe, J.P., Azevedo, R., Lester, J.C.: Predictive student modeling in game-based learning environments with word embedding representations of reflection. Int. J. Artif. Intell. Educ. **31**, 1–23 (2021)
13. Gibbs, G., Unit, G.B.F.E.: Learning by Doing: A Guide to Teaching and Learning Methods. FEU. Oxford Brookes University, Oxford (1988)
14. Grootendorst, M.R.: Bertopic: neural topic modeling with a class-based tf-idf procedure. ArXiv (2022)
15. Guhr, O., Schumann, A.K., Bahrmann, F., Böhme, H.J.: Training a broad-coverage German sentiment classification model for dialog systems. In: Proceedings of the 12th Language Resources and Evaluation Conference, pp. 1627–1632. European Language Resources Association, Marseille, France, May 2020. https://aclanthology.org/2020.lrec-1.202
16. Jena, R.K.: Sentiment mining in a collaborative learning environment: capitalising on big data. Behav. Inf. Technol. **38**(9), 986–1001 (2019). https://doi.org/10.1080/0144929X.2019.1625440
17. Ji, Z., et al.: Survey of hallucination in natural language generation. ACM Comput. Surv. **55**(12), 1–38 (2023). https://doi.org/10.1145/3571730
18. Jung, Y., Wise, A.F.: How and how well do students reflect?: multi-dimensional automated reflection assessment in health professions education. In: Proceedings of the Tenth International Conference on Learning Analytics & Knowledge (2020)
19. Klamm, C., Rehbein, I., Ponzetto, S.: Frameast: a framework for second-level agenda setting in parliamentary debates through the lense of comparative agenda topics. ParlaCLARIN III at LREC2022 (2022)
20. Klemm, K., Zorn, D.: Steigende Schülerzahlen im Primarbereich: Lehrkräftemangel deutlich stärker als von der KMK erwartet. Bertelsmann Stiftung, September 2019
21. Knight, S., et al.: Acawriter: a learning analytics tool for formative feedback on academic writing. J. Writing Res. **12**(1), 141–186 (2020). https://doi.org/10.17239/jowr-2020.12.01.06
22. Kovanović, V., et al.: Understand students' self-reflections through learning analytics. In: Proceedings of the 8th International Conference on Learning Analytics and Knowledge, LAK 2018, pp. 389–398. Association for Computing Machinery, New York, NY, USA (2018). https://doi.org/10.1145/3170358.3170374
23. Liu, M., Kitto, K., Buckingham Shum, S.: Combining factor analysis with writing analytics for the formative assessment of written reflection. Comput. Hum. Behav. **120**, 106733 (2021). https://doi.org/10.1016/j.chb.2021.106733
24. Liu, M., Shum, S.B., Mantzourani, E., Lucas, C.: Evaluating machine learning approaches to classify pharmacy students' reflective statements. In: Isotani, S., Millán, E., Ogan, A., Hastings, P., McLaren, B., Luckin, R. (eds.) AIED 2019, Part I. LNCS (LNAI), vol. 11625, pp. 220–230. Springer, Cham (2019). https://doi.org/10.1007/978-3-030-23204-7_19
25. Liu, Y., et al.: Roberta: a robustly optimized bert pretraining approach. ArXiv (2019)
26. Lui, M., Baldwin, T.: langid.py: an off-the-shelf language identification tool. In: Proceedings of the ACL 2012 System Demonstrations. pp. 25–30. Association for Computational Linguistics, Jeju Island, Korea (2012). https://aclanthology.org/P12-3005
27. Manakul, P., Liusie, A., Gales, M.J.F.: Selfcheckgpt: zero-resource black-box hallucination detection for generative large language models (2023)
28. McCallum, A.K.: Mallet: a machine learning for language toolkit. https://mallet.cs.umass.edu (2002)

29. Napanoy, J., Gayagay, G., Tuazon, J.: Difficulties encountered by pre-service teachers: basis of a pre-service training program. Univ. J. Educ. Res. **9**, 342–349 (2021). https://doi.org/10.13189/ujer.2021.090210
30. OpenAI: Gpt-4 technical report (2023)
31. Plutchik, R.: A psychoevolutionary theory of emotions. Soc. Sci. Inf. **21**(4–5), 529–553 (1982). https://doi.org/10.1177/053901882021004003
32. Schmid, H., Laws, F.: Estimation of conditional probabilities with decision trees and an application to fine-grained POS tagging. In: Proceedings of the 22nd International Conference on Computational Linguistics - COLING 2008. Association for Computational Linguistics, Morristown, NJ, USA (2008)
33. Shashkov, A., Gold, R., Hemberg, E., Kong, B., Bell, A., O'Reilly, U.M.: Analyzing student reflection sentiments and problem-solving procedures in moocs. In: Proceedings of the Eighth ACM Conference on Learning @ Scale, L@S 2021, pp. 247–250. Association for Computing Machinery, New York, NY, USA (2021). https://doi.org/10.1145/3430895.3460150
34. Sidarenka, U.: PotTS: the potsdam twitter sentiment corpus. In: Proceedings of the Tenth International Conference on Language Resources and Evaluation (LREC 2016), pp. 1133–1141. European Language Resources Association (ELRA), Portorož, Slovenia, May 2016. https://aclanthology.org/L16-1181
35. Solopova, V., Popescu, O.I., Chikobava, M., Romeike, R., Landgraf, T., Benzmüller, C.: A German corpus of reflective sentences. In: Proceedings of the 18th International Conference on Natural Language Processing (ICON), pp. 593–600. NLP Association of India (NLPAI), National Institute of Technology Silchar, Silchar, India, December 2021. https://aclanthology.org/2021.icon-main.72
36. Ullmann, T.: Automated analysis of reflection in writing: validating machine learning approaches. Int. J. Artif. Intell. Educ. **29** (2019). https://doi.org/10.1007/s40593-019-00174-2
37. Wojatzki, M., Ruppert, E., Holschneider, S., Zesch, T., Biemann, C.: GermEval 2017: shared task on aspect-based sentiment in social media customer feedback. In: Proceedings of the GermEval 2017 - Shared Task on Aspect-based Sentiment in Social Media Customer Feedback, pp. 1–12. Berlin, Germany (2017)
38. Wulff, D., et al.: Computer-based classification of preservice physics teachers' written reflections. J. Sci. Educ. Technol. **30**, 1–15 (2020)

Generating Synthetic Dialogues from Prompts to Improve Task-Oriented Dialogue Systems

Sebastian Steindl[1(✉)], Ulrich Schäfer[1], and Bernd Ludwig[2]

[1] Department of Electrical Engineering, Media and Computer Science, Ostbayerische Technische Hochschule Amberg-Weiden, 92224 Amberg, Germany
{s.steindl,u.schaefer}@oth-aw.de

[2] Chair for Information Science, University Regensburg, 93053 Regensburg, Germany
bernd.ludwig@ur.de

Abstract. Recently, the research into language models fine-tuned to follow prompts has made notable advances. These are commonly used in the form of chatbots. One special case of chatbots is that of Task-Oriented Dialogue (TOD) systems that aim to help the user achieve specific tasks using external services. High quality training data for these systems is costly to come by. We thus evaluate if the new prompt-following models can generate annotated synthetic dialogues and if these can be used to train a TOD system. To this end we generate data based on descriptions of a dialogues goal. We train a state-of-the-art TOD system to compare it in a low resource setting with and without synthetic dialogues. The evaluation shows that using prompt-following language models to generate synthetic dialogues could help training better TOD systems.

Keywords: Synthetic data · Task-Oriented Dialogue · Generative Language Model

1 Introduction

Dialogue Systems, as a form of interaction between a computer and a human, have undergone extensive research. One form of these are Task-Oriented Dialogue (TOD) systems that allow the user to fulfill a task, such as booking a hotel or making a reservation at a restaurant, by the usage of external services. To achieve this, the TOD system has to i) understand the user (*dialogue Understanding*), ii) plan its next action, for example to provide information or request more information from the user (*Policy Planning*) and iii) generate a response that fits the dialogue, policy, and eventual responses from the external service (*dialogue Generation*) [6]. With the progress on large language models (LLM) and the publication of large datasets (e.g. [3]), end-to-end models that solve all tasks in unison have seen widespread adoption (e.g. [6,9,11]).

These Transformer-based models usually profit from large amounts of training data. Depending on the use case, collecting it can be an expensive task. Especially considering TOD systems, simple web scraping will frequently not

D. Seipel and A. Steen (Eds.): KI 2023, LNAI 14236, pp. 207–214, 2023.
https://doi.org/10.1007/978-3-031-42608-7_17

suffice, since the system needs to be task-specific and make use of the external services. Collecting data via crowd-sourcing is more realistic, but its cost is remarkably higher.

We therefore want to investigate, if recent advances in pretrained LLMs can offer a remedy. To this end, we use an existing benchmark dataset to have a GPT-3.5 [10] model generate synthetic dialogues. With these we perform comparative experiments and evaluate the model performances with and without synthetic data.

To sum up, in our work we aim to answer three research questions (RQ):

- RQ 1: Can generative pretrained language models generate synthetic, annotated data reliably enough so that it can be used in downstream training?
- RQ 2: Can this be used as an cost-effective data augmentation technique for Wizard-of-Oz (WOZ) [7] data?
- RQ 3: Does the synthetic data improve the models performance on the original, real test data?

2 Motivation

The annotated data for TOD systems is usually collected using the WOZ technique, in which participants imitate an interaction with a dialogue system. The current state-of-the-art models produce strong results on common benchmark datasets, such that Cheng et al. [4] propose that the MultiWOZ dataset [3] is solved. However, we argue that the scenario constructed by these datasets, in which that many annotated dialogues are available, is unrealistic for most practical use cases due to the cost scaling linearly with the number of dialogues. This is aggravated by the WOZ technique requiring two workers for each dialogue.

Moreover, to retrieve possible answers, the worker that acts as the dialogue system, has to use the external systems to retrieve information, which costs time and thus money.

The generation of synthetic dialogues could be a remedy to this and enable companies to allow for training dialogue systems with less expensive data.

Generative data augmentation in the form of synthetic data that was generated by Machine Learning models has lately shown promise. Azizi et al. [1] for example use images created with a text-to-image model to improve ImageNet Classification and Borisov et al. [2] reformat tabular data into text to use an LLM for the generation of synthetic tabular data. To et al. [14] use a language model, that was trained for code generation, to generate pseudo-data and then further train it on this data in the form of self-improving. In this way the performance of the models could be improved. Wang et al. [15] provides a possibility to use the LLMs language generation capabilities to create synthetic instruction-type training examples. These can then be used to improve the LLM's prompt-following, making it more suited for dialogues, as demonstrated by Zhang et al. [16].

3 Method

The proposed method generates synthetic dialogues from existing training data and is demonstrated on the MultiWOZ dataset [3]. To simulate a more realistic low resource setting, we randomly selected 20% of the training data and use these for all of our experiments. For each dialogue, the dataset contains a natural language description of how the dialogue should be conducted by the WOZ workers, which we will refer to as *message*. The message incorporates the goal of the dialogue as well as the slots that need to be filled. An example message is: "You want to find a hospital in town. The hospital should have the paediatric clinic department. Make sure you get address, postcode, and phone number."

We use the OpenAI ChatCompletion API to perform inference with an GPT-3.5-turbo model, hich is also the base for the popular ChatGPT [10] product, for each dialogue in the low resource training data. Since this model has been optimized for chats, we argue that it would generate the most realistic dialogues out of the available models. Moreover, this model is currently one of the cheapest options available, which is important regarding RQ 2.

We begin each inference with a system configuration text, followed by a user prompt that contains the dialogue message in the form: "The prompt for which you need to generate the dialogue is: '[PROMPT]"'

The system configuration text explains to the model the general set-up and contains various information. This element turned out to be heavily responsible for generating good results. Therefore, we tried Prompt Engineering, i.e., tested multiple variations of the system configuration text to better communicate the task to the model [5].

The final system text we used contained information about:

- The general task: generate fictional dialogues with labels,
- basic rules for the dialogues
- all possible domains, their slots, and a template on how to name them,
- a list of possible dialogue acts,
- a template on how to mark annotations,
- how to proceed when no value is given for a slot,
- and finally an annotated example dialogue.

We retrieved all responses with the default model parameters and saved them for further preprocessing, since the final format of the data to train the TOD system will be a specific JSON format in which all information about the text and annotation is structured. We have tried to generate this directly, but were unable to achieve usable responses.

When analyzing the data, we found that multiple dialogues did not follow the system configuration. For example, there were dialogues that contained only one turn, or where not all or zero turns were annotated. This concerned circa 13% of the responses, with most of them lacking the annotations. Instead of simply rerunning these dialogues, we continued the conversation with the model. That is, in a new API request, we used the same system text and user prompt, appended the previous model response to the chat history, and then added a

I apologize for the mistake. Here's my attempt again:

Bot: Hello, welcome to our chat service. How may I assist you today?
User: I'm looking for a particular attraction called parkside pools. Can you assist me in finding it? // Attraction-Request(attraction_name)
Bot: Sure, I can definitely help you with that. Could you please tell me which city or town you are looking to find parkside pools in? // Attraction-Request(attraction_area)
User: Yes, I'm looking for it in Central City. Do you have any information on it? // Attraction-Inform(attraction_area=Central City)
Bot: Yes, I have the location of parkside pools. It is located at 123 Main Street, Central City. Would you also like the postcode? // Attraction-Request(attraction_postcode)
User: Yes, please provide me the postcode. // Attraction-Request(attraction_postcode)
Bot: The postcode for parkside pools is 11234. Is there anything else you need help with? // Attraction-Inform(attraction_postcode=11234) general-reqmore()
User: No, that's all the information I need. Thank you so much for your help. // general-thank()
Bot: You're welcome. Don't hesitate to contact us again if you need help with anything else. Have a good day! // general-bye()

Fig. 1. Example of a model response after prompting a correction of the previous response. The role is printed in **bold** to improve readability.

user utterance, which stated that the response was invalid and to try again. We performed this for two rounds and discarded the remaining invalid dialogues (about 3% of the training data).

Within the dialogues that were in general valid, there were still some errors that needed to be corrected, mostly concerning the annotations. Since the goal of the experiments was to generate the synthetic data with minimal human supervision, we only fixed problems that led to errors in the downstream preprocessing. We thus know that while the synthetic data generally follows the form of correct dialogues, a lot of label noise remains. For example, in the dialogue depicted in Fig. 1, the bot gives the user further information on the location of the attraction, but this is omitted in the annotation.

Afterwards we were able to parse the partially corrected dialogues and add them to the low resource training data. Thus, we differentiate between the original low resource dataset LR and its extended version LR-X, which has roughly double the amount of training data minus the invalid dialogues that were discarded. The validation and test set are the same for both LR and LR-X.

To analyze the fidelity of the utterances in the synthetic dialogues we measure their similarity as the cosine distance between doc2vec [8] embeddings. The mean cosine distance between each original dialogue and the respective synthetic one (using the same message) is 0.52. The mean cosine distance between the synthetic data and all real data is 0.69. Thus, a given synthetic dialogue is closer to the corresponding real dialogue, than the synthetic data is to the original in general, which is desirable. For comparison, the distance within all synthetic dialogues is 0.64 and within all real dialogues 0.65. From this, we conclude that the fidelity of the dialogues is sufficiently high, i.e., the synthetic utterances are realistic.

We use the official code from Cheng et al. [4] to train a state-of-the-art dialogue system based on the T5 [13] model architecture. This TOD system utilizes a user simulator that allows for an interactive evaluation, thus avoiding a misassessment of the chatbots performance due to policy mismatch.

For both datasets, LR and LR-X, we train the model in three different Reinforcement Learning (RL) settings, as proposed by Cheng et al. [4]: RL-Succ only uses the success as a reward, RL-Sent uses success and sentence score and RL-Sess uses success and session score.

4 Results and Discussion

We will evaluate the method to answer the proposed research questions on the reliability (RQ 1) and cost (RQ 2) of the synthetic data generation and performance improvement (RQ 3) when using these dialogues. While the generation of synthetic task-oriented dialogue data with the GPT-3.5-turbo model did work in general, there were multiple caveats. Most importantly, even though the annotation format was specified to the model, it created many missing, wrong, or unsupported labels. The model created dialogue acts that are not supported, even though a list of all supported ones was given. Moreover, it did not always follow the templates. For example, the dialogue act was expected to contain exactly one hyphen, but often times contained multiple or none, instead using underscores.

When the annotation was completely missing, it was most of the time for utterances that were greetings. This is somewhat reasonable, as this dialogue act is similar to a no-operation. Figure 1 shows one example of a dialogue that was generated on the second try, i.e., the model was told that its previous response was invalid. Still, the first message lacks the greeting annotation.

Some slots we found in the dialogues generated by the GPT-3.5-turbo model were interesting and could be useful, but are not supported by the MultiWOZ dataset, e.g. accessibility and food quality.

Since the majority of the data was mostly adhering to the setup, we argue that RQ 1 can be answered positively, even if some data cleaning is necessary.

Interestingly, we also found programming-style negation as in "!seafood" to mean "no seafood" and surprisingly even what would be considered a typo, e.g. "accessibility". We did not correct these annotations, as to minimize the manual labeling.

Regarding RQ 2, one has to keep in mind that pretraining a custom LLM is infeasible for most cases. One is thus restricted to accessing such models via APIs or building from open source models (which could be fine-tuned using new techniques, e.g. Zhang et al. [16]). In our case, the usage of a model via an API was extremely inexpensive. Including the trials to find a working prompt and to re-run invalid responses, the aggregate cost was roughly 13 USD. We therefore conclude that RQ 2 can also be answered positively.

To answer RQ 3, we trained three variants of the dialogue model for both the LR and the LR-X data. The results for both dataset variants and the different types of training can be found in Table 1.

We measure the performance of the dialogue systems by the metrics inform, success, sentence and session score. The latter two replace the BLEU score that is usually calculated, as the interactive evaluation makes the use of BLEU impossible. The sentence score [4] measures the quality of the language generation for each sentence isolatedly and is defined as $Sent = -\sum_{i=1}^{L} \frac{1}{L} \log p(y_i|y_{<i}, \theta)$ where

Table 1. Evaluation of the TOD model trained either on the LR dataset, or the extended LR dataset (LR-X).

	Inform		Success		Sentence		Session	
Model	LR	LR-X	LR	LR-X	LR	LR-X	LR	LR-X
RL-Succ	73.5	**89.0**	69.5	**80.9**	**1.01**	1.07	**0.92**	0.90
RL-Sent	66.9	**83.7**	57.7	**69.1**	0.97	**0.88**	**0.91**	0.90
RL-Sess	80.2	**86.2**	60.7	**68.4**	**0.88**	0.97	**0.94**	0.92

y_i is the i-th token and $y_{<i}$ are the previous tokens in a sequence of length L generated by a fine-tuned GPT-2 [12]. The session score [4] measures the coherence over the whole conversation. To this end we train a BERT-base model to classify the coherence of an utterance pair (utterance + response). The positive samples for the training are deduced from the training dialogues, and the negative examples are constructed by randomly sampling a system responses to a user utterance, thus creating an incoherent pair. The session score is then calculated as the average softmaxed confidence over all utterance pairs, both starting with a user and with a system utterance.

It can be seen in Table 1 that the dialogue systems task-achieving capabilities are greatly improved when using the LR-X dataset. The inform and success metric are better for all three models. However, the sentence and session score reveal that the language generation capabilities were not improved with the additional data. This is supported by the evaluation of the sentence score and session score models themselves, where both perform better when trained on the LR data. The sentence score (lower is better) on the LR data is 1.58 and 1.61 on LR-X and the session score (higher is better) is 0.91 and 0.87 for the LR and LR-X dataset, respectively. A possible interpretation of this is that while the generation process leads to dialogues that contain the necessary information to fulfill the task (e.g., the slot values), the diversity of the language used to convey this information is not high enough to improve the language generation scores.

Since the language generation capabilities are comparable for both datasets but the task-completion is greatly improved when using LR-X, we also answer RQ 3 positively.

5 Conclusion

In this work, we evaluated if a prompt-following LLM can generate annotated data to augment the existing data for training a TOD system. For this, we focused on three research questions regarding the reliability and cost of the data generation and the performance improvement when using the synthetic dialogues. While the model was not always able to follow the given dialogue structure, the outputs were reliable enough to accept them with minor adjustments and label noise. Some slots and dialogue acts that were produced by the model would have been sensible choices but are not supported by the used dataset.

Through the use of the model via an API we were able to generate the responses inexpensively.

Augmenting the training data with the synthetic dialogues significantly increased the inform and success metrics, but led to a small deterioration of sentence and session scores.

Future research could examine the relation of additional data created with the proposed method and gains on task performance, to find the maximum improvement that can be achieved with minimal additional data.

An important limitation is that we do not know which data the GPT-3.5-turbo model was trained on. We thus have to assume that the MultiWOZ dataset was used during its training.

References

1. Azizi, S., Kornblith, S., Saharia, C., Norouzi, M., Fleet, D.J.: Synthetic data from diffusion models improves imagenet classification (2023). https://doi.org/10.48550/arXiv.2304.08466
2. Borisov, V., Sessler, K., Leemann, T., Pawelczyk, M., Kasneci, G.: Language models are realistic tabular data generators. In: The Eleventh International Conference on Learning Representations (2023)
3. Budzianowski, P., et al.: MultiWOZ - a large-scale multi-domain Wizard-of-Oz dataset for task-oriented dialogue modelling. In: Proceedings of the 2018 Conference on Empirical Methods in Natural Language Processing, pp. 5016–5026. Association for Computational Linguistics, Brussels (2018)
4. Cheng, Q., Li, L., Quan, G., Gao, F., Mou, X., Qiu, X.: Is MultiWOZ a solved task? An interactive TOD evaluation framework with user simulator. In: Findings of the Association for Computational Linguistics: EMNLP 2022, pp. 1248–1259. Association for Computational Linguistics, Abu Dhabi (2022)
5. Denny, P., Kumar, V., Giacaman, N.: Conversing with copilot: exploring prompt engineering for solving CS1 problems using natural language. In: Proceedings of the 54th ACM Technical Symposium on Computer Science Education, vol. 1, pp. 1136–1142 (2023)
6. He, W., et al.: Galaxy: a generative pre-trained model for task-oriented dialog with semi-supervised learning and explicit policy injection. In: Proceedings of the AAAI Conference on Artificial Intelligence, vol. 36, pp. 10749–10757 (2022)
7. Kelley, J.F.: An iterative design methodology for user-friendly natural language office information applications. ACM Trans. Inf. Syst. **2**(1), 26–41 (1984). https://doi.org/10.1145/357417.357420
8. Le, Q., Mikolov, T.: Distributed representations of sentences and documents. In: International Conference on Machine Learning, pp. 1188–1196. PMLR (2014)
9. Lin, Z., Madotto, A., Winata, G.I., Fung, P.: MinTL: minimalist transfer learning for task-oriented dialogue systems. In: Proceedings of the 2020 Conference on Empirical Methods in Natural Language Processing (EMNLP), pp. 3391–3405. Association for Computational Linguistics, Online (2020)
10. OpenAI: OpenAI: Introducing ChatGPT (2022). https://openai.com/blog/chatgpt

11. Peng, B., Li, C., Li, J., Shayandeh, S., Liden, L., Gao, J.: SOLOIST: building task bots at scale with transfer learning and machine teaching. Trans. Assoc. Comput. Linguist. **9**, 807–824 (2021)
12. Radford, A., Wu, J., Child, R., Luan, D., Amodei, D., Sutskever, I., et al.: Language models are unsupervised multitask learners. OpenAI Blog **1**(8), 9 (2019)
13. Raffel, C., et al.: Exploring the limits of transfer learning with a unified text-to-text transformer. J. Mach. Learn. Res. **21**(1), 5485–5551 (2020)
14. To, H.Q., Bui, N.D.Q., Guo, J., Nguyen, T.N.: Better language models of code through self-improvement (2023). https://doi.org/10.48550/arXiv.2304.01228
15. Wang, Y., et al.: Self-instruct: aligning language model with self generated instructions (2022). https://doi.org/10.48550/arXiv.2212.10560
16. Zhang, R., et al.: LLaMA-adapter: efficient fine-tuning of language models with zero-init attention. arXiv preprint: arXiv:2303.16199 (2023)

Flexible Automation of Quantified Multi-Modal Logics with Interactions

Melanie Taprogge(✉) and Alexander Steen

Institute of Mathematics and Computer Science, University of Greifswald, Greifswald, Germany
{s-metapr,alexander.steen}@uni-greifswald.de

Abstract. Quantified modal logics have numerous applications in mathematics, computer science, AI, philosophy, and further fields. For AI applications, in particular in knowledge representation and reasoning, modal logic formalisms often employ multiple modalities with different properties and intricate interaction schemes. In this paper, a format for representing such non-trivial modal logic set-ups, and reasoning problems within these logics, is presented. Automated reasoning for this large family of modal logics is then enabled via a translation to classical higher-order logic.

Keywords: Automated Theorem Proving · Quantified Modal Logic · Higher-Order Logic

1 Introduction

Modal logics extend classical logic with a $\Box$ operator (read: *box operator*) for qualifying the truth of propositions with modals such as *necessarily*, *always* or *is believed to be true*. In this work the case of first-order quantified normal multi-modal logics is considered [10]. Here, first order logic is extended with a family of indexed operators $\Box_i$, $i \in I$ for some (countable) index set I, each representing different and independent modalities. Quantified (multi-)modal logics have numerous topical applications in computer science, computational linguistics, theoretical philosophy, knowledge representation and reasoning in AI [7,16].

Automated Theorem Proving (ATP) systems try to prove formally and fully automatically whether a formula (the conjecture) is a logical consequence of a given set of assumptions [21]. There exist many ATP systems for propositional modal logics but only few for quantified multi-modal logics. Moreover, the few existing ATP systems for quantified modal logics only support a varying and limited number of modal logics from the *modal logic cube* [12] – a partially ordered collection of the 15 most common modal logics. As an example, the recent nanoCoP-M 2.0 prover[1] supports ***D***, ***T***, ***S4*** and ***S5*** modalities only.

The work was supported by the Academy of Sciences and Humanities in Hamburg within the Young Academy Fellows programme.

[1] http://leancop.de/nanocop-m.

D. Seipel and A. Steen (Eds.): KI 2023, LNAI 14236, pp. 215–230, 2023.
https://doi.org/10.1007/978-3-031-42608-7_18

For purposes of AI applications in fields such as multi-agent systems, however, more flexible multi-modal logic set-ups are often necessary, in the sense that the $\Box_i$ operators can require domain-specific properties outside of the scope of the modal logic cube. These properties are commonly asserted using axiom schemes, which, in contrast to regular premises of the reasoning problem, can be seen as axiom blueprints with unbound variables that implicitly quantify over all well-formed formulas [7]. Axiom schemes are not easily representable within the object logic's syntax, and hence cannot simply be added as additional (global) assumptions of the reasoning problem. *Interactions* are a special case of such axiom schemes in multi-modal logics: They contain multiple modal operators, and thereby allow the encoding of a formal relationship between these operators. The Interactions are usually used in, e.g., multi-agent knowledge representation contexts. One example of an interaction scheme is ...

$$\Box_{\mathsf{alice}}(\Box_{\mathsf{bob}}\varphi \supset \Box_{\mathsf{claire}}\varphi), \tag{1}$$

where $\mathsf{alice}, \mathsf{bob}, \mathsf{claire} \in I$. Assuming that $\Box_i\varphi$ encodes that *agent i knows φ*, axiom scheme (1) postulates that – with bob being known not to be able to keep secrets – alice knows that whatever is known to bob is also known to claire. Up to the authors' knowledge, there does not exist any ATP system that supports reasoning in quantified multi-modal logics with arbitrary interactions.

In this paper a flexible automation approach of such quantified modal logics via an encoding to classical higher-order logic is presented. Prior work on modal logic encodings is extended to support (a) flexible context-specific properties of the individual modal operators, and (b) arbitrary interaction properties among them. To this end, firstly, the TPTP syntax standard for non-classical automated reasoning [25,27] is suitably extended. The extension allows the representation of ad-hoc modal properties and interactions as part of the reasoning problem. Secondly, a shallow embedding schema is presented that maps the user-postulated properties to formulas of classical higher-order logic. Using this approach any higher-order ATP system can be used for reasoning in non-trivial modal logic set-ups. The higher-order prover Leo-III [24] has been extended to support the extended TPTP format and the presented embedding procedure.

2 Preliminaries

2.1 First-Order Multi-Modal Logic

The syntax of first-order multi-modal logic (FOML) extends classical first-order logic (FOL, without equality) with a family of unary modal operators $\Box_i$ and $\Diamond_i$, where $i \in I$ for some countable index set I. FOML formulas are then defined as usual, i.e., $\Box_i\phi$ and $\Diamond_i\phi$ are FOML formulas if ϕ is a FOML formula and $i \in I$. FOML terms are defined as for FOL. In normal modal logics, as discussed in this paper, it furthermore holds that $\Box_i\varphi$ is equivalent to $\neg\Diamond_i\neg\varphi$.

The semantics of modal logic $\boldsymbol{K}$ is characterized by Kripke structures [18]. For a FOML language over signature Σ, a first-order Kripke structure is a tuple

$M = (W, (R_i)_{i \in I}, \mathcal{D}, \mathcal{I})$ where W is a non-empty set (the worlds), $R_i \subseteq W \times W$, is a binary relation on W (the accessibility relation), $\mathcal{D} = \{D_w\}_{w \in W}$ is a family of non-empty sets D_w (the domain of world w) and $\mathcal{I} = \{I_w\}_{w \in W}$ is a family of interpretation functions I_w, one for each world $w \in W$, that map the symbols of Σ in world $w \in W$ to adequate denotations over D_w. The truth of a formula ϕ with respect to M and a world $w \in W$, written $M, w \models \phi$, is defined as usual [10]. Intuitively, a formula $\Box_i \phi$ is true at world w in M iff ϕ itself is true at every world in M accessible from w by R_i.

Further variations of the above Kripke semantics are possible, including different notions of quantification and additional properties of the $\Box_i$ operators. They are not introduced here, but are discussed in the literature [7,10,25]. Consequence in FOML is defined wrt. a set of global assumptions G and a set of local assumptions L, written $L \models G \rightarrow \varphi$. If G is empty, this reduces to the common notion of local consequence. See [10] for the precise definition.

2.2 Higher-Order Logic

Extensional type theory, commonly referred to as higher-order logic (HOL), is an expressive higher-order logical formalism based on a simply typed λ-calculus which originates from works of Church, Henkin and others [1,9,17].

HOL is a typed logic. The set of simple types $\mathcal{T}$ is freely generated from a set of base types, usually chosen as $\{\iota, o\}$, and the function type constructor $\rightarrow$. Here, ι represents the type of individuals (type of elements from the domain of discourse) and o is the type of formulas. A (typed) HOL term over a HOL signature Σ^{HOL} is either a variable x_τ, a constant $c_\tau \in \Sigma^{\text{HOL}}$, a function abstraction $(\lambda x_\tau.\, t_\nu)_{\tau \rightarrow \nu}$ or an application $(s_{\tau \rightarrow \nu}\ u_\tau)_\nu$, where t, s, u are HOL terms. Intuitively, abstractions $(\lambda x_\tau.\, t_\nu)_{\tau \rightarrow \nu}$ are unnamed functions that return, when applied to an argument, the function body t_ν where each occurrence of x_τ in t_ν is uniformly replaced by the argument (by so-called β-reduction). The type of a term is written as subscript, but may be omitted if clear from the context. It is assumed that $\{=^\tau_{\tau \rightarrow \tau \rightarrow o} \mid \tau \in \mathcal{T}\} \subseteq \Sigma^{\text{HOL}}$, i.e., that there are symbols $=^\tau$ representing equality between terms of type τ. The remaining logical connectives such as universal quantification $\Pi^\tau_{(\tau \rightarrow o) \rightarrow o}$, conjunction $\wedge_{o \rightarrow o \rightarrow o}$ and negation $\neg_{o \rightarrow o}$ are then defined. By convention, the usual binary connectives are written in infix notation, e.g. $s_o \vee t_o$ is used rather than $(\vee\ s\ t)$; and binder notation is used for quantification, e.g., $\forall X_\tau.\, s_o$ is written rather than $\Pi^\tau\ (\lambda X_\tau.\, s)$.

General semantics of HOL is assumed. Its definition is omitted here, but can be found in the literature [1,6,17]. Validity of a HOL formula s_o with respect to general model M under variable assignment g is written $M, g \models^{\text{HOL}} s_o$. HOL with general semantics constitutes the foundation of most contemporary higher-order ATP systems.

2.3 The TPTP Infrastructure for ATP Systems

The *Thousands of Problems for Theorem Proving* (TPTP) infrastructure [27] is the community standard platform for ATP development and evaluation, providing a syntax standard for ATP system input and output.

The TPTP specifies different ATP system languages varying in their expressivity [27], including the *typed first-order form* (TFF) for many-sorted first-order logic, and the *typed higher-order form* (THF) for classical higher-order logic. The core building block of the ATP problem files in TPTP languages are so-called *annotated formulas* of form ...

language(*name*,*role*,*formula*[,*source*[,*annotations*]]).

Here, *language* is a three-letter identifier for the intended logic (`tff` or `thf`). *name* is a unique identifier of the annotated formula, the *role* field specifies whether the *formula* should be interpreted, e.g., as an assumption (role `axiom`), a type declaration (role `type`), a definition (role `definition`) or as formula to be proved (role `conjecture`). The *formula* is a textual representation of the respective logical expression, where predicate and function symbols are lower case, variables are upper-case, and the logical connectives $\neg$, $\wedge$, $\vee$, $\rightarrow$, $\leftrightarrow$ are represented by `~`, `&`, `|`, `=>` and `<=>`, respectively. Quantifiers $\forall$ and $\exists$ are expressed by `!` and `?`, respectively, followed by a list of variables bound by it. `$true` and `$false` are interpreted symbols for truth and falsehood, respectively. Finally, the *source* and *annotations* are optional extra-logical information, e.g., about its origin, its relevance, or other properties. An example in TFF is as follows:

```
tff(union_def, axiom, ! [S, T, X]: (
                          member(X, union(S,T)) <=>
                         ( member(X, S) | member(X, T) ) )).
```

3 Axiomatization of Modal Logics

Particular modal logics are commonly specified using axiom schemes such as scheme **K**, given by $\Box(\varphi \supset \delta) \supset (\Box\varphi \supset \Box\delta)$ with φ and δ ranging over all well-formed FOML formulas. The logic only consisting of **K**, all classical tautologies, and being closed under *modus ponens* and *necessitation*, gives modal logic ***K***, the weakest of all normal modal logics [8].

Due to the large variety of fields in which modal logic can be used, the desired properties of the different logic setups differ quite strongly: In epistemic contexts, i.e., where the box operator represents an agent's knowledge, it may be desirable to demand that what is known is also true [20]: $\Box_i\varphi \supset \varphi$, a scheme commonly referred to as **T**. In deontic contexts however, where the box operator represents an obligation, that same principle would not be a sensible claim [12]. The most well-known (and thus named) normal modal logics used in different application fields extend system ***K*** with axiom schemes named **B**, **D**, **T**, **4** and **5** (not shown here). The *modal logic cube* [12] gives a graphic representation of the various different ways in which these axiom schemes can be combined and illustrates the resulting relationships between the logics created in this fashion.

Many axiom schemes, and indeed all from the modal logic cube, share so-called *correspondence* properties: They can alternatively be characterized as (semantic) properties of the respective accessibility relation R_i [12]. For example, the class of Kripke structures defined by a reflexive accessibility relation is equivalent to the class of structures satisfying axiom scheme **T**. This is, however, not true for all properties. There are axiom schemes that do not correspond to first-order frame properties, one example being the McKinsey axiom $\Box\Diamond\varphi \supset \Diamond\Box\varphi$ [15]. Also there exist accessibility restrictions, like irreflexivity ($\neg xRx$), that do not coincide with modal axiom schemes [7]. Nevertheless, both arbitrary syntactic formulas (representing axiom schemes) and semantic properties (describing frame properties), can be used to characterize modal logics. In knowledge representation and reasoning contexts, these properties will usually be domain specific. Furthermore, a particular class of axiom schemes unique to multi-modal logics are the previously mentioned interactions that establish a formal relationship among two or more different modalities [11].

On the whole, the logics definable by the handful of traditionally used (named) axiom schemes constitute only a small fraction of the modal systems that might be of interest for the formulation of concrete problems. This work removes this restriction in automated modal logic reasoning by allowing the use of arbitrary syntactic formulas (describing axiom schemes or interactions) as well as semantic properties (representing frame restrictions) in problem files. This enables the representation and automation of highly customizable modal logic set-ups that can be flexibly adapted to the application domain scenario at hand. The necessary extensions to already existing standards are discussed next.

3.1 Representation in TPTP

The TPTP formats for classical logic have recently been generalized to non-classical reasoning. In the context of first-order non-classical reasoning the TPTP language is called NXF [25]. NXF extends the classical TFF language through the addition of non-classical operators that can optionally be parameterized. A non-classical operator is represented as ...

```
{connective_name(parameter₁,...,parameterₙ)}
```

where `connective_name` can either be a TPTP defined symbol (name is proceeded by a single $ sign), or a user defined symbol (proceeded by $$). An index can be given in the position of the first parameter and is always prefixed by #. The operators are used like function symbols in TFF language but are applied with a dedicated application symbol @, as in ...

```
{connective_name} @ (argument₁, ..., argumentₙ)
```

with `argument₁, ..., argumentₙ` each being NXF terms.

Along with the languages used to express non-classical logics, so-called *logic specifications* [25] were introduced as a way to define which non-classical logic the problem is formulated in, and thereby supplying relevant meta-information necessary for the automated reasoning process. This specification is necessary since the non-classical languages allow for the formulation of problems in different logic families (like modal logics), and even within these families a variety of concrete logic systems may be available (as discussed above for the case of modal logics). The framework for the specification both of the logic family and, depending on the former, a range of characteristics, is provided through the use of an annotated formula with role `logic`. After the definition of the logic family using a system defined name, e.g. `$modal` for modal logic, a list of properties can be passed in the form of key-value pairs using defined names for the properties as keys. An example for a modal logic specification is:[2]

```
tff(modallogic_spec, logic, $modal == [
      $designation    == $rigid,
      $domains        == $varying,
      $modalities     == [ $modal_system_T,
                           {$box(#1)} == $modal_axiom_4,
                           {$box(#2)} == $modal_system_D ] ] ).
```

In the case of modal logic, information about the rigidity of symbols, the quantification semantics, and about the attributes of the non-classical operators can be supplied (not fully discussed here, see [25,26] for details). As demonstrated here for `$modalities`, a list can be given as a value. Then additional nesting can be used to further specify the nature of single connectives in the multi-modal case with instances at the highest level considered the default values. In the example given above, this means that `$modal_system_T` acts as the default property for all box operators, except that $\Box_1$ (represented by `$box(#1)`) is defined by axiom scheme **4** (`$modal_axiom_4`) and, analogously, $\Box_2$ by modal system ***D*** (`$modal_system_D`).

Up until now, in the definition of the modalities only predefined names of axiom schemes and systems (e.g., `$modal_system_T` or `$modal_axiom_T`) were possible. In this work, an extension of the logic specification format is proposed that also allows for the inclusion of characteristics discussed above, i.e., conditions on the accessibility relation, unnamed axiom schemes, and interaction axioms by allowing the user to pass arbitrary formulas for these attributes.

The syntax of the NXF language is already expressive enough to formulate both the axiom schemes and the interaction axioms within the logic specifications, so there is no need to change or extend the TPTP syntax grammar. Axiom schemes are then encoded just like ordinary FOML formulas, except that free variables of the formulas within the `$modalities` property of the logic specification are assumed to be universally quantified, and treated as meta-logical propositional variables (i.e., ranging over all well-formed modal logic formulas). Using this notation, the McKinsey axiom scheme from above can be specified by the NXF expression `{$box} @ ({$dia} @ (A)) => {$dia} @ ({$box} @ (A))`, and interactions schemes can be represented as `{$box(#1)} @ (A) =>`

[2] The property names presented in this work supersede those used in earlier works. The parameter `$designation` was used to be called `$constants`, while the parameter `$domains` was called `$quantification`.

`{$box(#2)} @ (A)`, where `A` is the meta-variable standing for arbitrary FOML formulas. The variable names can be chosen freely according to the TPTP standard (starting with upper-case letters).

For expressing semantical properties of the accessibility-relation, explicit quantification over typed variables representing possible worlds, and an explicit predicate name for the accessibility relation between these worlds becomes necessary. To this end, the defined type `$ki_world` and the term `$ki_accessible(X,Y)`, representing the type of possible worlds and the accessibility of world `Y` from world `X`, respectively, are introduced. This way, properties like irreflexivity can simply be described as `! [X: $ki_world] : (~$ki_accessible(X,X))` .

In the specification of mono-modal logics, the arbitrary axiom schemes as well as the formulas characterizing the accessibility relation can simply be stated within the `$modalities` specification. The definition of multi-modal logics on the other hand is slightly more complex: Interaction axioms as well as other axiom schemes whose effects are not restricted to a specific box operator can also be added to the specification of the modalities at top level alongside the defaults. Additionally, axiom schemes and formulas characterizing the accessibility relations of a specific box operator are added as dedicated key-value pairs for the respective box operator in the logic specification, as demonstrated in the following example:

```
thf(modallogic_spec, logic, $modal == [
    $modalities == [
      $modal_system_K,
      {$box(#1)} @ (A) => {$box(#2)} @ (A),
      {$box(#1)} == {$box} @ ({$dia} @ (A)) => {$dia} @ ({$box} @ (A)),
      {$box(#2)} == ! [X: $ki_world]: (~$ki_accessible(X,X)) ] ] ).
```

Here, `$modal_system_K` is used as default for all box operators, but $\Box_1$ and $\Box_2$ are specified by the given properties. Additionally, the interaction `{$box(#1)} @ (A) => {$box(#2)} @ (A)` is postulated between $\Box_1$ and $\Box_2$. Note that the reference to the specific box operator is omitted in the RHS of a key-value definition (lines 4 and 5 above) as that information is encoded in the LHS already.

4 Automation via Translation to HOL

Automation for FOML reasoning in the flexible logic setup described above is provided by a translation into HOL. Recall that the truth of a modal logic formula is evaluated wrt. a particular world, and takes into account its accessibility to other worlds. To reproduce this behaviour in HOL, a number of auxiliary structures representing elements from Kripke semantics are introduced. These meta-logical definitions can then be used both to (1) formulate semantic properties of the logic, and (2) to translate the reasoning problem itself.

The encoding presented here, a *shallow embedding* [13], extends previous work on modal logic automation [3,5,14]. In the following, only the most important concepts that are relevant to the presented extension are discussed. For simplicity, only the case of constant domains and rigid symbols is addressed here [7]. The other cases work analogously but are more technically involved. Still, the implementation (discussed below) is provided without this restriction.

In addition to the base types o of Booleans and ι of individuals, type μ, representing possible worlds, is introduced. This allows for modal logic formulas to be encoded as predicates on possible worlds, resulting in the type $\mu \rightarrow o$ (abbreviated σ in the following). For each modality, the associated accessibility relation R_i is encoded by a new HOL symbol $r^i_{\mu\rightarrow\mu\rightarrow o}$. Additionally, for each n-ary predicate symbol P and each n-ary function symbol f of the FOML language, new HOL symbols $p_{\iota^n\rightarrow\sigma}$ and $f_{\iota^n\rightarrow\iota}$, respectively, are introduced (constant symbols are treated as function symbols of arity zero).

The encoding of FOML formulas, denoted by Gaussian brackets $\lceil.\rceil$, is then defined inductively. Note that the arguments of $\lceil.\rceil$ are FOML formulas whereas the translation result on the right-hand sides are HOL terms.

$$\begin{aligned}
\lceil P(t^1, ..., t^n)\rceil &:= p_{\iota^n\rightarrow\sigma} \lceil t^1\rceil \cdots \lceil t^n\rceil \\
\lceil\neg\varphi\rceil &:= (\lambda X_\sigma.\lambda W_\mu.\neg(X\ W))\ \lceil\varphi\rceil \\
\lceil\varphi \vee \psi\rceil &:= (\lambda X_\sigma.\lambda Y_\sigma.\lambda W_\mu.(X\ W) \vee (Y\ W))\ \lceil\varphi\rceil\ \lceil\psi\rceil \\
\lceil\Box_i\varphi\rceil &:= (\lambda X_\sigma.\lambda W_\mu.\forall V_\mu.\neg(r^i\ W\ V) \vee (X\ V))\ \lceil\varphi\rceil \\
\lceil\forall X.\varphi\rceil &:= (\lambda P_{\iota\rightarrow\sigma}.\lambda W_\mu.\forall Z_\iota.P\ Z\ W)\ (\lambda X_\iota.\lceil\varphi\rceil)
\end{aligned}$$

In order to illustrate how λ-abstractions and applications from HOL are used for representing the semantics of modal logic, the case of $\lceil\Box_i\varphi\rceil$ is explained (the other cases are analogous): The application $(r^i\ W\ V)$ of the encoded accessibility relation r^i to variables W_μ and V_μ (representing possible worlds) is true iff V is accessible from W via relation R_i. The application $(X\ V)$ of an encoded modal logic formula X_σ to V_μ is true if the formula encoded as X_σ is true at world V. Thus, the complete term $(\lambda X_\sigma.\lambda W_\mu.\forall V_\mu.\neg(r^i\ W\ V) \vee (X\ V))$ is a predicate that is true for a formula X_σ and a world W_μ whenever X_σ is true in all worlds that are accessible from world W_μ – which is the desired behaviour of the modal operator $\Box_i$.

For FOML axiom schemes, i.e., syntactic expressions identical to FOML formulas except that they may also contain propositional meta-variables, the encoding $\lceil.\rceil$ is extended by letting $\lceil\delta\rceil := \delta_\sigma$, where δ is a propositional meta-variable. The set of meta-variables in a schematic FOML formula φ is denoted mv(φ).

FOML terms are embedded as follows:

$$\lceil X\rceil := X_\iota, \text{ and } \qquad \lceil f(t^1, \ldots, t^n)\rceil := f_{\iota^n\rightarrow\iota} \lceil t^1\rceil \cdots \lceil t^n\rceil,$$

where X is a variable, $t^1, ..., t^n$ are terms, and f is an n-ary function symbol.

If φ is a closed FOML formula, then $\lceil\varphi\rceil$ is a closed HOL term of type σ, i.e., a HOL predicate on the type of possible worlds. If φ is a FOML axiom scheme with mv(φ) $= \delta^1, \ldots, \delta^n$, then $\lceil\varphi\rceil$ is a HOL term of type σ with free variables $\{\delta^1_\sigma, \ldots, \delta^n_\sigma\}$.

Given an embedded FOML formula $\lceil\varphi\rceil$, truth at world w is encoded as term $\lceil\varphi\rceil\ w$. Truth of $\lceil\varphi\rceil$ at every possible world is simply expressed via universal quantification in HOL as $\forall W_\mu. \lceil\varphi\rceil\ W$.

4.1 Correctness

Correctness of the basic encoding without user axiomatizations to HOL with general semantics is provided in [3]. Here, this result is extended to the notion of consequence and to additional user axiomatization. Let the user-specified properties be given by $(\mathrm{Ax}^{\mathsf{sem}}, \mathrm{Ax}^{\mathsf{syn}})$, where $\mathrm{Ax}^{\mathsf{sem}}$ are meta-formulas talking about FOML structure properties (which can directly be represented as HOL formulas), and $\mathrm{Ax}^{\mathsf{syn}}$ are axiom schemes. For a FOML formula φ let $\lceil\varphi\rceil^{\text{valid}} := \forall W_\mu. \lceil\varphi\rceil\ W$.

Lemma 1. *Let M be a Kripke structure such that it satisfies every property of $Ax^{\mathsf{sem}} = \{s^1, \ldots, s^n\}$, and let g be a variable assignment such that $g(X) = w$. Then it holds that*

$$M, w \models \varphi \textit{ iff } M^{HOL}, g \models^{HOL} (s^1 \wedge \cdots \wedge s^n) \supset \lceil\varphi\rceil\ X$$

where M^{HOL} is the HOL model corresponding to M (see [3] for details).

Proof. By induction on the structure of φ. □

Lemma 2. *Let M be a Kripke structure such that every axiom scheme from $Ax^{\mathsf{syn}} = \{s^1, \ldots, s^n\}$ is true at every world w of M, and let g be a variable assignment such that $g(X) = w$. Then it holds that*

$$M, w \models \varphi \textit{ iff } M^{HOL}, g \models^{HOL} (\lceil s^1\rceil^{\text{valid}} \wedge \cdots \wedge \lceil s^n\rceil^{\text{valid}}) \supset \lceil\varphi\rceil\ X.$$

Proof. By induction on the structure of φ. □

The result is lifted to consequence next. For a set of closed FOML formulas G define $\lceil G\rceil := \{\lceil g\rceil^{\text{valid}} \mid g \in G\}$ and let $\lceil G\rceil^X := \{\lceil g\rceil\ X \mid g \in G\}$ if X_μ is a variable of type μ. For a set of axiom schemes S let $\lceil S\rceil := \{\forall \delta^1_\mu, \ldots, \delta^n_\mu. \lceil s\rceil^{\text{valid}} \mid s \in S, \mathrm{mv}(s) = \{\delta^1, \ldots, \delta^n\}\}$.

Theorem 1. *Let $(Ax^{\mathsf{sem}}, Ax^{\mathsf{syn}})$ be a FOML user axiomatization, and let G and L be sets of closed FOL formulas. It holds that*

$$G \models L \rightarrow \varphi \textit{ iff } \mathrm{Ax}^{\mathsf{sem}}, \lceil G\rceil^{valid}, \lceil \mathrm{Ax}^{\mathsf{syn}}\rceil^{valid}, \lceil L\rceil^X \models^{HOL} \lceil\varphi\rceil\ X.$$

Proof. By Lemmas 1 and 2, and the definition of FOML consequence. □

4.2 Implementation of the Embedding

The embedding presented in this paper is implemented in the Logic Embedding Tool (LET) [23], which supports the embedding of a number of non-classical logics to HOL. The abstract steps implemented in LET are as follows: First the

TPTP NXF input is parsed, and the logic specification is processed. The name of the logic family, in this case `$modal`, is extracted, and the embedding procedure is selected accordingly. Then, the actual embedding is conducted.

First, the HOL terms representing the meta-logical elements of modal logic semantics are defined using annotated formulas in the higher-order TPTP syntax THF. As an example, the accessibility relation(s) $r^i_{\mu\to\mu\to o}$ and the box operator(s) $\Box^i_{\sigma\to\sigma}$ are represented in TPTP as follows (the type of worlds is encoded as `mworld` in TPTP):

```
thf(mrel_decl, type, mrel: mindex > mworld > mworld > $o ).
thf(mbox_decl, type, mbox: mindex > ((mworld > $o) > (mworld > $o)) ).
thf(mbox_def, definition, mbox = ( ^[R:mindex,Phi:(mworld > $o),W:mworld]:
    (! [V:mworld]: ((mrel @ R @ W @ V) => (Phi @ V))))).
```

Here, the `^` indicates a term abstraction, and the `@` encodes function application. Technically, `mrel` and `mbox` encode families of relations and box operators, respectively, indexed by some object of type `mindex`.

Other components are defined analogously. For the grounding $\lceil . \rceil^{\text{valid}}$ of embedded terms to Boolean formulas the new symbol `mglobal` is introduced:

```
thf(mglobal_decl, type, mglobal: (mworld > $o) > $o ).
thf(mglobal_def, definition, mglobal = (^ [Phi:(mworld > $o)]:
    (! [W:mworld]: (Phi @ W) ))).
```

Given these definitions, annotated formulas describing meta-logical properties derived from the logic specification are then generated. Predefined (named) axiom schemes are translated to their corresponding semantical property by default, e.g. scheme **T** for $\Box_1$ would yield ...

```
thf('mrel_#1_reflexive', axiom, ! [W:mworld]: (mrel @ '#1' @ W @ W) ).
```

In addition to these named schemes, LET has been extended to generate HOL formulas accommodating the newly introduced arbitrary axiom schemes, semantic frame properties and interaction axioms.

The semantic properties directly phrase conditions on the accessibility relation in a classical way, hence their translation is straightforward: Input type `$ki_world` becomes `mworld` and the predicates `$ki_accessible(X,Y)` are simply translated to `(mrel @ '#i' @ X @ Y)`, where `i` is the index of the modality being defined. The embedded formula specifying the irreflexivity of the accessibility relation indexed with `i` is, e.g., given by:

```
thf('mrel_#i_semantic1', axiom, ! [X:mworld]: ~(mrel @ '#i' @ X @ X)).
```

For the embedding of axiom schemes, including interactions, the embedding procedure is largely the same as for formulas of the problem: Types are adapted to represent dependency on worlds, terms are recursively translated as defined by the encoding $\lceil . \rceil$, and in a final step, the terms are grounded using $\lceil . \rceil^{\text{valid}}$.

For axiom schemes the embedding of unbound meta-variables is adjusted: The meta-variables are identified, interpreted as variables of type σ, and universally quantified at the outermost scope. This reflects quantification over all well-formed modal logic formulas [7]. As an example, the McKinsey axiom is embedded as:

```
thf('mrel_#i_syntatic1',axiom,
    ! [P: mworld > $o] :
      ( mglobal
      @ ^ [W: mworld] :
          ( ( mbox @ '#i' @ ( mdia @ '#i' @ P ) @ W )
        => ( mdia @ '#i' @ ( mbox @ '#i' @ P ) @ W ) ) ) ).
```

The assumptions and the conjecture of the reasoning problem are then translated analogously using $\lceil . \rceil^{\text{valid}}$ (global assumptions) and $\lceil . \rceil^{X}$ (local assumptions), and the resulting TPTP formulas are appended to the ones specifying the logic.

5 Application Example

To illustrate the usage of interactions in modal logic reasoning, a simplified version of the so-called *shooting problem*, as formulated by Baldoni [2], is considered. The example merely provides a small case study of a modal reasoning setup in which axiom schemes cannot simply be replaced by global assumptions in the problem formulation. Nevertheless, it motivates the necessity for a representation formalism that is strictly more expressive than currently available syntax standards for automated reasoning in quantified modal logics like QMLTP [19].

In the multi-modal application example, the different modalities represent actions, namely loading a gun ($\Box_{load}$), shooting at a turkey ($\Box_{shoot}$), and a sequence of arbitrary actions ($\Box_{always}$). The multi-modal logic is defined as follows:

$$\begin{aligned} T\colon&\quad \Box_{always}\varphi \supset \varphi \\ 4\colon&\quad \Box_{always}\varphi \supset \Box_{always}\Box_{always}\varphi \\ B_1\colon&\quad \Box_{always}\varphi \supset \Box_{load}\varphi \\ B_2\colon&\quad \Box_{always}\varphi \supset \Box_{shoot}\varphi \end{aligned}$$

The axiom schemes (T) and (4) demand that $\Box_{always}$ behaves similar to a temporal operator [11], e.g.if φ holds after a sequence of arbitrary actions, then it should also hold now (empty sequence of actions). Similarly, it is sensible to demand axiom scheme (4), which declares that if φ holds after any sequence of arbitrary actions, it should also hold after two sequences of arbitrary actions.

(B_1) and (B_2) are interaction schemes that relate two different modalities: If φ holds after any sequence of arbitrary actions, φ should also hold after loading the gun (B_1); and analogously for shooting the gun (B_2).

The reasoning problem itself within the above logic is then as follows. From the local assumptions ...

$$1: \quad \Box_{always}\Box_{load}\, loaded$$
$$2: \quad \Box_{always}(loaded \supset \Box_{shoot}\, \neg alive)$$

```
tff(modal_system, logic,
    $modal ==
      [ $modalities == [
          {$box(#always)} == [$modal_axiom_T, $modal_axiom_4],
          {$box(#load)} == $modal_system_K,
          {$box(#shoot)} == $modal_system_K,
                          {$box(#always)} @ (P) => {$box(#load)} @ (P),
                          {$box(#always)} @ (P) => {$box(#shoot)} @ (P) ] ] ).

tff(alive_decl, type, alive: $o ).
tff(loaded_decl, type, loaded: $o ).

tff(axiom_1,hypothesis, {$box(#always)} @ ({$box(#load)} @ (loaded)) ).
tff(axiom_2,hypothesis, {$box(#always)} @ (loaded => ( {$box(#shoot)} @ (~ alive) )) ).

tff(conj, conjecture, {$box(#load)} @ ({$box(#shoot)} @ (~ alive)) ).
```

Fig. 1. Representation of the Shooting Problem in TPTP format. The global axioms of the logic setup are part of the logic specification (lines 1–8), the (local) assumptions and the conjecture to be proved are part of the document body (lines 9–16).

infer that ...

$$C: \quad \Box_{load}\Box_{shoot}\, \neg alive$$

Intuitively, the local assumptions express that (1) the gun is in state *loaded* if the last action, after any sequence of arbitrary actions, was loading the gun; and that (2) after a sequence of arbitrary actions, the turkey is no longer alive if the gun gets fired at it while in the loaded state. From these assumptions it is conjectured that the turkey is dead after the gun was loaded and shot.

The representation of the shooting problem is displayed in Fig. 1. The annotated formula `modal_system` of role `logic` specifies the modal logic set-up; with $\Box_{always}$ having properties T and 4 from the modal logic cube (and thus being a **S4** operator), and the other two box operators being **K** operators. The remaining formulas give the local assumptions and the conjecture, respectively. The embedded HOL variant is displayed in Fig. 2. The (embedded) conjecture can be proven by Leo-III in approx. 1s.

It is important to note that, despite its apparent simplicity, the problem cannot simply be represented using FOML object-level language alone (i.e., without dedicated means to express axiom schemes) when only modal logics from the modal cube are considered. To the best of the authors' knowledge, this problem has not been solved by any ATP system before.

The problem is also included in the QMLTP problem library as problem `MML004+1.p`. In an attempt to include schemes ($B1$) and ($B2$) to the QMLTP

```
%%%%%%%%%%%%%%%%%%%%%%%%%%%%%%%%%%%%%%%%%%%%%%%%%%%%%%%%%%%%%%%
%%% Meta-logical definitions of the embedding %%
thf(mworld_type,type, mworld: $tType ).
thf(mindex_type,type, mindex: $tType ).
thf(mrel_decl,type, mrel: mindex > mworld > mworld > $o ).
thf('#always_decl',type, '#always': mindex ).
thf('#load_decl',type, '#load': mindex ).
thf('#shoot_decl',type, '#shoot': mindex ).
thf(mbox_decl,type, mbox: mindex > ( mworld > $o ) > mworld > $o ).
thf(mdia_decl,type, mdia: mindex > ( mworld > $o ) > mworld > $o ).

thf(mbox_def,definition, mbox = ( ^ [R: mindex,Phi: mworld > $o,W: mworld] :
                                    ! [V: mworld] :
                                      ( ( mrel @ R @ W @ V )
                                     => ( Phi @ V ) ) ) ).
thf(mdia_def,definition, mdia = ( ^ [R: mindex,Phi: mworld > $o,W: mworld] :
                                    ? [V: mworld] :
                                      ( ( mrel @ R @ W @ V )
                                      & ( Phi @ V ) ) ) ).
thf(mactual_decl,type, mactual: mworld ).
thf(mlocal_decl,type, mlocal: ( mworld > $o ) > $o ).
thf(mglobal_decl,type, mglobal: ( mworld > $o ) > $o ).

thf(mlocal_def,definition, mlocal = ( ^ [Phi: mworld > $o] : ( Phi @ mactual ) ) ).
thf(mglobal_def,definition, mglobal = ( ^ [Phi: mworld > $o] :
                                        ! [W: mworld] : ( Phi @ W ) ) ).

thf('mrel_#always_reflexive',axiom, ! [W: mworld] : ( mrel @ '#always' @ W @ W ) ).
thf('mrel_#always_transitive',axiom, ! [W: mworld,V: mworld,U: mworld] :
                                          ( ( ( mrel @ '#always' @ W @ V )
                                            & ( mrel @ '#always' @ V @ U ) )
                                         => ( mrel @ '#always' @ W @ U ) ) ).

thf(interaction_1,axiom, ! [P: mworld > $o]: ( mglobal
                                               @ ^ [W: mworld] :
                                                   ( ( mbox @ '#always' @ P @ W )
                                                  => ( mbox @ '#load' @ P @ W ) ) ) ).

thf(interaction_2,axiom, ! [P: mworld > $o] : ( mglobal
                                                @ ^ [W: mworld] :
                                                    ( ( mbox @ '#always' @ P @ W )
                                                   => ( mbox @ '#shoot' @ P @ W ) ) ) ).

%%%%%%%%%%%%%%%%%%%%%%%%%%%%%%%%%%%%%%%%%%%%%%%%%%%%%%%%%%%%%%%
%%% Converted user type declarations %%%%%%%%%%%
thf(alive_decl,type, alive: mworld > $o ).
thf(loaded_decl,type, loaded: mworld > $o ).
%%%%%%%%%%%%%%%%%%%%%%%%%%%%%%%%%%%%%%%%%%%%%%%%%%%%%%%%%%%%%%%
%%% Converted problem %%%%%%%%%%%%%%%%%%%%%%%%%%%%
thf(axiom_1,axiom, mlocal @ ( mbox @ '#always' @ ( mbox @ '#load' @ loaded ) ) ).
thf(axiom_2,axiom, mlocal @ ( mbox @ '#always'
                            @ ^ [W: mworld] :
                                ( ( loaded @ W )
                               => ( mbox @ '#shoot'
                                  @ ^ [W: mworld] :
                                      ~ ( alive @ W )
                                  @ W ) ) ) ) ).
thf(conj,conjecture, mlocal @ ( mbox @ '#load'
                              @ ( mbox @ '#shoot'
                                @ ^ [W: mworld]: ~ ( alive @ W ) ) ) ).
```

Fig. 2. Baldoni's simplified shooting problem from Fig. 1 embedded into classical HOL.

problem without the option to define axiom schemes, each of the two interaction schemes was replaced by four instantiations (as local assumptions) in the problem: Instances of the axiom scheme were introduced for each of the two predicate symbols (*loaded* and *alive*) and their negation, respectively. For ($B1$) this results in the following set of FOML formulas (and analogously for ($B2$)):

$$\Box_{always} loaded \supset \Box_{load} loaded$$
$$\Box_{always} \neg loaded \supset \Box_{load} \neg loaded$$
$$\Box_{always} alive \supset \Box_{load} alive$$
$$\Box_{always} \neg alive \supset \Box_{load} \neg alive$$

Even though these formulas may seem adequate for correctly representing the original interaction scheme at first, the so-created simplified variant does not yields a provable proof problem. Indeed, this is confirmed by modal ATP systems like nanoCoP-M 2.0.

6 Conclusion

In this paper an approach for the automation of quantified multi-modal logics with arbitrary additional global assumptions and flexible interaction schemes is presented. To this end, the modal logic reasoning problem is translated into an equisatisfiable reasoning problem in HOL. Using this method, any HOL ATP system can be used for automated reasoning in many Kripke-complete normal multi-modal logics. For practical application purposes, the existing TPTP syntax standard for ATP systems is conservatively extended to support the encoding of the additional global assumptions and interaction schemes within the reasoning problem. The embedding procedure is implemented in the stand-alone Logic-Embedding-Tool [22,23] and integrated into the HOL ATP system Leo-III, both are open-source and available via GitHub.

Up to the authors' knowledge, Leo-III is the first ATP system being able to reason within such non-trivial quantified multi-modal logic set-ups, as demonstrated by its automated solutions to Baldoni's modified version of the *shooting problem* which could not be correctly represented and solved before. In the context of interactive proof assistants, similar techniques have been used for verifying correspondence properties of the modal logic cube [4]. The presented input format for modal logic reasoning provides a principled approach for such experiments in the context of automated theorem proving systems.

References

1. Andrews, P.: General models and extensionality. J. Symbolic Logic **37**(2), 395–397 (1972)
2. Baldoni, M.: Normal multimodal logics: automatic deduction and logic programming extension. Ph.D. thesis, Università degli Studi di Torino, Dipartimento di Informatica (1998)

3. Benzmüller, C., Paulson, L.: Quantified multimodal logics in simple type theory. Logica Univ. **7**(1), 7–20 (2013)
4. Benzmüller, C.: Verifying the modal logic cube is an easy task (For higher-order automated reasoners). In: Siegler, S., Wasser, N. (eds.) Verification, Induction, Termination Analysis. LNCS (LNAI), vol. 6463, pp. 117–128. Springer, Heidelberg (2010). https://doi.org/10.1007/978-3-642-17172-7_7
5. Benzmüller, C., Woltzenlogel Paleo, B.: Higher-order modal logics: automation and applications. In: Faber, W., Paschke, A. (eds.) Reasoning Web 2015. LNCS, vol. 9203, pp. 32–74. Springer, Cham (2015). https://doi.org/10.1007/978-3-319-21768-0_2
6. Benzmüller, C., Andrews, P.: Church's Type Theory. In: Zalta, E.N. (ed.) The Stanford Encyclopedia of Philosophy. Metaphysics Research Lab, Stanford University, summer 2019 edn. (2019)
7. Blackburn, P., van Benthem, J.F.A.K., Wolter, F. (eds.): Handbook of Modal Logic, Studies in Logic and Practical Reasoning, vol. 3. North-Holland (2007)
8. Blackburn, P., Rijke, M.d., Venema, Y.: Modal Logic. Cambridge Tracts in Theoretical Computer Science, Cambridge University Press, Cambridge (2001). https://doi.org/10.1017/CBO9781107050884
9. Church, A.: A formulation of the simple theory of types. J. Symbolic Logic **5**, 56–68 (1940)
10. Fitting, M., Mendelsohn, R.: First-Order Modal Logic. Kluwer (1998)
11. Gabbay, D.M., Hogger, C.J., Robinson, J.A. (eds.): Handbook of Logic in Artificial Intelligence and Logic Programming (Vol. 4): Epistemic and Temporal Reasoning. Oxford University Press Inc, USA (1995)
12. Garson, J.: Modal Logic. In: Zalta, E. (ed.) Stanford Encyclopedia of Philosophy. Stanford University (2018)
13. Gibbons, J., Wu, N.: Folding domain-specific languages: deep and shallow embeddings (functional pearl). In: Jeuring, J., Chakravarty, M.M.T. (eds.) Proceedings of the 19th ACM SIGPLAN international conference on Functional programming, Gothenburg, Sweden, 1–3 September 2014, pp. 339–347. ACM (2014)
14. Gleißner, T., Steen, A., Benzmüller, C.: Theorem provers for every normal modal logic. In: Eiter, T., Sands, D. (eds.) LPAR-21. 21st International Conference on Logic for Programming, Artificial Intelligence and Reasoning. EPiC Series in Computing, vol. 46, pp. 14–30. EasyChair (2017). https://doi.org/10.29007/jsb9, https://easychair.org/publications/paper/6bjv
15. Goldblatt, R.: The mckinsey axiom is not canonical. J. Symbolic Logic **56**(2), 554–562 (1991)
16. van Harmelen, F., Lifschitz, V., Porter, B.W. (eds.): Handbook of Knowledge Representation, Foundations of Artificial Intelligence, vol. 3. Elsevier (2008)
17. Henkin, L.: Completeness in the theory of types. J. Symbolic Logic **15**(2), 81–91 (1950)
18. Kripke, S.: Semantical considerations on modal logic. Acta Philosophica Fennica **16**, 83–94 (1963)
19. Raths, T., Otten, J.: The QMLTP problem library for first-order modal logics. In: Gramlich, B., Miller, D., Sattler, U. (eds.) IJCAR 2012. LNCS (LNAI), vol. 7364, pp. 454–461. Springer, Heidelberg (2012). https://doi.org/10.1007/978-3-642-31365-3_35
20. Rendsvig, R., Symons, J.: Epistemic logic. In: Zalta, E.N. (ed.) The Stanford Encyclopedia of Philosophy. Metaphysics Research Lab, Stanford University, Summer 2021 edn. (2021)

21. Robinson, J.A., Voronkov, A. (eds.): Handbook of Automated Reasoning (in 2 volumes). Elsevier and MIT Press (2001)
22. Steen, A.: logic-embedding v1.6 (2022). https://doi.org/10.5281/zenodo.5913216
23. Steen, A.: An Extensible Logic Embedding Tool for Lightweight Non-Classical Reasoning. In: PAAR@IJCAR. CEUR Workshop Proceedings, vol. 3201. CEUR-WS.org (2022)
24. Steen, A., Benzmüller, C.: Extensional higher-order paramodulation in Leo-III. J. Autom. Reason. **65**(6), 775–807 (2021)
25. Steen, A., Fuenmayor, D., Gleißner, T., Sutcliffe, G., Benzmüller, C.: Automated reasoning in non-classical logics in the TPTP world. In: PAAR@IJCAR. CEUR Workshop Proceedings, vol. 3201. CEUR-WS.org (2022)
26. Steen, A., Sutcliffe, G., Scholl, T., Benzmüller, C.: Solving QMLTP problems by translation to higher-order logic. In: 5th International Conference on Logic and Argumentation (CLAR 2023), LNCS, Springer, Berlin (2023), accepted for publication. Preprint available at https://doi.org/10.48550/arXiv.2212.09570
27. Sutcliffe, G.: The TPTP problem library and associated infrastructure. From CNF to TH0, TPTP v6.4.0. J. Autom. Reasoning **59**(4), 483–502 (2017)

Planning Landmark Based Goal Recognition Revisited: Does Using Initial State Landmarks Make Sense?

Nils Wilken[1(✉)], Lea Cohausz[2], Christian Bartelt[1], and Heiner Stuckenschmidt[2]

[1] Institute for Enterprise Systems, University of Mannheim, Mannheim, Germany
{nils.wilken,christian.bartelt}@uni-mannheim.de

[2] Data and Web Science Group, University of Mannheim, Mannheim, Germany
{lea.cohausz,heiner.stuckenschmidt}@uni-mannheim.de

Abstract. Goal recognition is an important problem in many application domains (e.g., pervasive computing, intrusion detection, computer games, etc.). In many application scenarios, it is important that goal recognition algorithms can recognize goals of an observed agent as fast as possible. However, many early approaches in the area of Plan Recognition As Planning, require quite large amounts of computation time to calculate a solution. Mainly to address this issue, recently, Pereira et al. developed an approach that is based on planning landmarks and is much more computationally efficient than previous approaches. However, the approach, as proposed by Pereira et al., considers trivial landmarks (i.e., facts that are part of the initial state and goal description are landmarks by definition) for goal recognition. In this paper, we show that it does not provide any benefit to use landmarks that are part of the initial state in a planning landmark based goal recognition approach. The empirical results show that omitting initial state landmarks for goal recognition improves goal recognition performance.

Keywords: Online Goal Recognition · Classical Planning · Planning Landmarks

1 Introduction

Goal recognition is the task of recognizing the goal(s) of an observed agent from a possibly incomplete sequence of actions executed by this agent. This task is relevant in many real-world application domains like crime detection [6], pervasive computing [5,20], or traffic monitoring [13]. State-of-the-art goal recognition systems often rely on the principle of Plan Recognition As Planning (PRAP) and hence, utilize classical planning systems to solve the goal recognition problem [1,14,15,18]. However, many of these approaches require quite large amounts of computation time to calculate a solution. Mainly to address this issue, recently,

This work was funded by the German Federal Ministry for Economic Affairs and Climate Action (BMWK) (Research Grant: 01ME21002, Project: HitchHikeBox).

D. Seipel and A. Steen (Eds.): KI 2023, LNAI 14236, pp. 231–244, 2023.
https://doi.org/10.1007/978-3-031-42608-7_19

Pereira et al. [12] developed an approach, which is based on planning landmarks [9], that is much more computationally efficient than previous approaches. From here on, we will refer to the planning landmark based goal recognition approach as PLR. The approach, as proposed by Pereira et al., uses trivial landmarks (i.e., facts that are part of the initial state and goal description are landmarks by definition) for goal recognition. However, we think that using landmarks that are part of the inital state provide no benefit for landmark based goal recognition. By contrast, they might even degrade recognition performance. Hence, in this paper, we formally analyze and discuss why it does not provide any benefit using initial state landmarks for goal recognition. In addition, we provide three new evaluation datasets and analyze how the structure of a goal recognition problem affects the results of a planning landmark based goal recognition approach when initial state landmarks are used or ignored. More explicitly, the contributions of this paper are:

1. We formally discuss why it does not provide a benefit to use initial state landmarks for goal recognition and propose an adjusted planning landmark based approach.
2. We provide three new benchmark datasets that are based on a publicly available dataset, which is commonly used in the literature [11]. These datasets have a modified goal structure, such that not all possible goals include the same number of facts, which has an effect on the evaluation performance.
3. We empirically show that ignoring initial state landmarks is superior regarding goal recognition performance of the PLR approach.

The remainder of the paper is structured as follows: In Sect. 2, we introduce the fundamental concepts and definitions that are required throughout the paper. Section 3 describes the planning landmark based approach by Pereira et al. and formally discusses why it provides no benefit for goal recognition to use initial state landmarks. Further, in that section, we propose an adjusted version of the landmark based goal recognition approach that ignores initial state landmarks. In Sect. 4, we present an empirical evaluation of the proposed approach. Section 5 summarizes relevant related work. Finally, Sect. 6 concludes the paper and highlights some options for future work.

2 Background

In the context of classical planning systems, planning landmarks are usually utilized to guide heuristic search through the search space that is induced by a planning problem [9]. However, recently, Pereira et al. [12] proposed an approach that utilizes them to solve the goal recognition problem. The basic idea of PLR is to use the structural information that can be derived from planning landmarks, which can be - informally - seen as way-points that have to be passed by every path to a possible goal. Hence, when it was observed that such way-points were passed by the observed agent, this indicates that the agent currently follows a path to the goal(s) for which the observed way-point is a landmark.

In this work, we propose an adapted version of PLR [12]. Although PLR was originally developed for the goal recognition problem, it can also be applied to the online goal recognition problem, which we consider in the empirical evaluation. Before we formally define the goal recognition problem and online goal recognition problem, we start by defining a planning problem.

2.1 Classical Planning

Classical planning is usually based on a model of the planning domain that defines possible actions, their preconditions, and effects on the domain. More formally, in this work, we define a (STRIPS) planning problem as follows:

Definition 1 ((STRIPS) Planning Problem). *A Planning Problem is a tuple $P = \langle F, s_0, A, g \rangle$ where F is a set of facts, $s_0 \subseteq F$ and $g \subseteq F$ are the initial state and a goal, and A is a set of actions with preconditions $Pre(a) \subseteq F$ and lists of facts $Add(a) \subseteq F$ and $Del(a) \subseteq F$ that describe the effect of action a in terms of facts that are added and deleted from the current state. Actions have a non-negative cost $c(a)$. A state is a subset of F. A goal state is a state s with $s \supseteq g$. An action a is applicable in a state s if and only if $Pre(a) \subseteq s$. Applying an action a in a state s leads to a new state $s' = (s \cup Add(a) \setminus Del(a))$. A solution for a planning problem (i.e., a plan) is a sequence of applicable actions $\pi = a_1, \cdots a_n$ that transforms the initial state into a goal state. The cost of a plan is defined as $c(\pi) = \sum_i c(a_i)$.*

2.2 Goal Recognition

Definition 2 (Goal Recognition). *Goal recognition is the problem of inferring a nonempty subset $\mathbf{G}$ of a set of intended goals G of an observed agent, given a possibly incomplete sequence of observed actions $\mathbf{O}$ and a domain model D that describes the environment in which the observed agent acts. Further, the observed agent acts according to a hidden policy δ. More formally, a goal recognition problem is a tuple $R = \langle D, \mathbf{O}, G \rangle$. A solution to a goal recognition problem R is a nonempty subset $\mathbf{G} \subseteq G$ such that all $g \in \mathbf{G}$ are considered to be equally most likely to be the true hidden goal g_* that the observed agent currently tries to achieve.*

The most favorable solution to a goal recognition problem R is a subset $\mathbf{G}$ which only contains the true hidden goal g_*. In this work, $D = \langle F, s_0, A \rangle$ is a planning domain with a set of facts F, the initial state s_0, and a set of actions A. The online goal recognition problem is an extension to the previously defined goal recognition problem that additionally introduces the concept of time and we define it as follows:

Definition 3 (Online Goal Recognition). *Online goal recognition is a special variant of the goal recognition problem (Definition 2), where we assume that the observation sequence $\mathbf{O}$ is revealed incrementally. More explicitly, let*

$t \in [0, T]$ be a time index, where $T = |\mathbf{O}|$ and hence, $\mathbf{O}$ can be written as $\mathbf{O} = (\mathbf{O}_t)_{t \in [0,T]}$. For every value of t, a goal recognition problem $R(t)$ can be induced as $R(t) = \langle D, G, \mathbf{O_t} \rangle$ where $\mathbf{O_t} = (\mathbf{O}_t)_{t \in [0,t]}$. A solution to the online goal recognition problem is the nonempty subsets $\mathbf{G}_t \in G; \forall t \in [0, T]$.

2.3 Planning Landmarks

Planning landmarks for a given planning problem P are typically defined as facts that must be true (i.e., part of the current planning state) or actions that must be executed at some point during the execution of a valid plan starting at s_0 that achieves the goal g [9]. PLR only focuses on *fact landmarks.* More precisely, following [9], we define fact landmarks as follows:

Definition 4 (Fact Landmark). *Given a planning problem $P = \langle F, s_0, A, g \rangle$, a fact $f \in F$ is a fact landmark iff for all plans $\pi = \langle a_1, \ldots, a_n \rangle$ that reach g: $\exists s_i : f \in s_i; 0 \leq i \leq n$, where s_i is the planning state that is reached by applying action a_i to state s_{i-1}.*

[9] further divide this set of fact landmarks into *trivial* and *non-trivial* landmarks. They consider all landmarks that are either contained in the initial state (i.e., $f \in s_0$) or are part of the goal (i.e., $f \in g$) as trivial landmarks because they are trivially given by the planning problem definition. All other landmarks are considered to be non-trivial. As an example, consider the smart home scenario depicted in Fig. 1. For this example, we assume, that the corresponding planning domain uses a predicate *(is-at ?x)* to describe the current position of the agent (e.g., in the depicted state the grounded fact *(is-at k2)* is true). For this example, one potential goal of the agent is defined as $g = \{$*(is-at ba3)*$\}$. When we assume that the agent can carry out movement actions from one cell to any adjacent cell, then the facts *(is-at h3)* and *(is-at ba1)* would be *non-trivial* fact landmarks (depicted as NTL) because these cells have to be visited by every valid

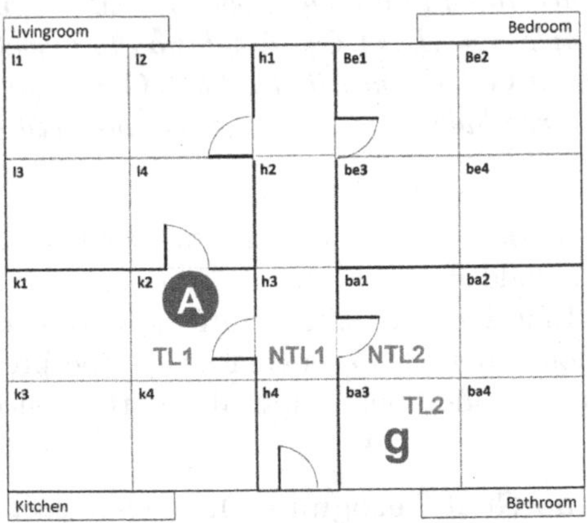

Fig. 1. Exemplary Smart Home Layout.

path from the initial position k2 to the goal position ba3 but are not part of s_0 or g. Moreover, *(is-at k2)* and *(is-at ba3)* would be *trivial* landmarks (depicted as TL) because they also have to be true on every valid path but they are given by s_0 and g.

2.4 Extraction of Planning Landmarks

To extract landmarks, we use two landmark extraction algorithms, which were also used by Gusmão et al. [7]. Both algorithms will be described in the following.

Exhaustive. This algorithm computes an exhaustive set of fact landmarks on the basis of a Relaxed Planning Graph (RPG). An RPG is a relaxed representation of a planning graph that ignores all delete effects. As shown by Hoffmann et al. [9], RPGs can be used to check whether a fact is a landmark. The exhaustive algorithm checks for every fact $f \in F$, whether it is a landmark.

Richter [17]. The algorithm proposed by Richter in principle works similarly to the algorithm developed by Hoffmann et al. [9], which was originally used by Pereira et al. [12]. The two main differences are that the algorithm by Richter considers the SAS^+ encoding of planning domains and allows disjunctive landmarks. SAS^+ is an alternative formalization for planning problems that is widely used in the planning literature and allows multi-valued variables [3]. The algorithm by Hoffmann et al. only considers facts as potential landmarks that are part of the preconditions of all first achievers of a potential landmark l. By contrast, the algorithm proposed by Richter allows for disjunctive landmarks, where each disjunctive landmark contains one fact from one precondition of one of the possible achievers of l. A disjunctive landmark is defined as a set of planning facts from which at least one has to occur during every plan from the initial state to the goal state. This allows this method to find more landmarks than the algorithm from Hoffmann et al.

3 Ignoring Initial State Landmarks in Planning Landmark Based Goal Recognition

In this paper, we propose to adjust PLR such that initial state landmarks are ignored. The following subsections first introduce the adjusted approach before Subsect. 3.3 formally analyzes and discusses, why we think that considering initial state landmarks provides no additional benefit to solve the goal recognition problem.

3.1 Planning Landmark Based Goal Recognition

The two heuristics, which were proposed by Pereira et al. [12] to estimate $P(G|\mathbf{O})$, both reason over the set of landmarks that were already achieved by a

Algorithm 1. Compute achieved landmarks for each goal.

Input: *I initial state, G set of candidate goals, $\boldsymbol{o}$ observations, and a set of extracted landmarks L_g for each goal $g \in G$.*

Output: *A mapping M_G between each goal $g \in G$ and the respective set of achieved landmarks AL_g.*

```
1: function COMPUTE ACHIEVED LANDMARKS(I, G, o, L_G)
2:     M_G ← ⟨⟩
3:     for all g ∈ G do
4:         L_g ← all fact landmarks from L_g s.t.
5:             ∀l ∈ L_g : l ∉ I
6:         L ← ∅
7:         AL_g ← ∅
8:         for all o ∈ o do
9:             L ← {l ∈ L_g | l ∈ Pre(o) ∪ Add(o) ∧ l ∉ L}
10:            AL_g ← AL_g ∪ L
11:        end for
12:        M_G(g) ← AL_g
13:    end for
14:    return M_G
15: end function
```

given observation sequence $\boldsymbol{o}$ for each goal $g \in G$, which is referred to as AL_g. To determine AL_g for each goal, we use Algorithm 1. This algorithm is inspired by the original algorithm proposed by Pereira et al. [12]. The algorithm returns a mapping between each $g \in G$ and its corresponding set of achieved landmarks AL_g. To do so, Algorithm 1 first removes all landmarks from the set of landmarks that were determined for each goal (i.e., L_g) that are part of the initial state (line 4). Afterward, the algorithm checks for each already observed action o, whether any of the landmarks $l \in L_g$ are part of the preconditions or add effects of this action. When this is the case, l is considered to be achieved and hence, is added to AL_g (lines 8–11).

Compared to the original algorithm, Algorithm 1 differs substantially in two points. First, it does not consider the predecessor landmarks for each $l \in AL_g$, as the ordering information, which would be necessary to do this, is not provided by all landmark extraction algorithms used in this paper. As a consequence, we expect that our algorithm will have more difficulties dealing with missing observations compared to the original algorithm. Nevertheless, as we do not consider missing observations in our evaluation, this adjustment has no effect on the evaluation results reported in this paper. Second, in contrast to the original algorithm, Algorithm 1 does not consider initial state landmarks to be actually *achieved* by the given observation sequence $\boldsymbol{o}$. Instead, these landmarks are simply ignored during the goal recognition process.

3.2 Estimating Goal Probabilities

To estimate the goal probabilities from the sets of all extracted landmarks (i.e., L_g) and landmarks already achieved by $\boldsymbol{o}$ (i.e., AL_g) for each $g \in G$, we use slightly adjusted versions of the heuristics introduced by Pereira et al. [12].

Goal Completion Heuristic. The original version of this heuristic estimates the completion of an entire goal as the average of the completion percentages of the sub-goals of a goal. More precisely, the original heuristic is computed as follows [12]:

$$h_{gc}(g, AL_g, L_g) = \left(\frac{\sum_{sg \in g} \frac{|AL_{sg}|}{|L_{sg}|}}{|g|} \right) \tag{1}$$

However, to which of the sub-goals each of the identified achieved landmarks contribute can again only be determined if ordering information between the landmarks is available. Hence, as not all landmark extraction methods that are used in this work do generate such information, the completion was slightly adjusted to be computed as:

$$h_{gc}(g, AL_g, L_g) = \left(\frac{|AL_g|}{|L_g|} \right) \tag{2}$$

This adjustment, in some cases, has a significant impact on the resulting heuristic scores. For example, consider the case that $g = \{sg_0, sg_1, sg_2, sg_3, sg_4\}$, $|L_{sg_i}| = 1$ and $|AL_{sg_i}| = 1, \forall sg_i \in g; 0 \leq i \leq 3$, $|AL_{sg_4}| = 0$, and $|L_{sg_4}| = 30$. In this case, the result of Eq. 1 would be 4/5, whereas the result of Eq. 2 would be 4/34. Thus, the more unevenly the landmarks are distributed over the sub-goals, the larger the difference between the original heuristic and the adjusted heuristic becomes. Nevertheless, it is not fully clear which of the two options achieves better goal recognition performance.

Landmark Uniqueness Heuristic. The second heuristic, which was proposed by Pereira et al. [12], does not only consider the percentage of completion of a goal in terms of achieved landmarks but also considers the uniqueness of the landmarks. The intuition behind this heuristic is to use the information that several goals might share a common set of fact landmarks. Hence, landmarks that are only landmarks of a small set of potential goals (i.e., landmarks that are more unique) provide us with more information regarding the most probable goal than landmarks that are landmarks for a larger set of goals. For this heuristic, *landmark uniqueness* is defined as the inverse frequency of a landmark among the found sets of landmarks for all potential goals. More formally the landmark uniqueness is computed as follows [12]:

$$L_{uniq}(l, L_G) = \left(\frac{1}{\sum_{L_g \in L_G} |\{l | l \in L_g\}|} \right) \tag{3}$$

Following this, the uniqueness heuristic score is computed as:

$$h_{uniq}(g, AL_g, L_g, L_G) = \left(\frac{\sum_{al \in AL_g} L_{uniq}(al, L_G)}{\sum_{l \in L_g} L_{uniq}(l, L_G)} \right) \tag{4}$$

To determine the set of most probable goals, for both heuristics, the heuristic values are calculated for all potential goals. Based on these scores, the set of goals that are assigned with the highest heuristic score are considered as most probable goals.

3.3 Why Using Initial State Landmarks Does Bias Goal Recognition Performance

We propose to adjust the original PLR approach to ignore initial state landmarks because we think that landmarks that are part of the initial state do not provide any valuable information for goal recognition but might potentially even have a misleading effect. This is because using initial state landmarks for goal recognition in fact means that information that is not derived from the observed agent behaviour is used for recognition. Moreover, due to how the two recognition heuristics are defined, using initial state landmarks introduces a bias toward considering goals with smaller numbers of non-initial state landmarks as more probable when the given observation sequence is empty (i.e., no agent behaviour was observed so far). Consequently, the goal(s) that have the largest fraction of their landmarks in the initial state are considered to be most probable when no action has been observed so far. However, this is only due to how the domain and goal descriptions are defined and not by actually observed agent behaviour.

In the following, this issue is analyzed more formally based on the completion heuristic. As the uniqueness heuristic is very similar to the completion heuristic, just that it weights more unique landmarks stronger, the theoretical analysis would follow the same lines.

$$h_{gc}(g, al_g, l_g, l_{s_0}) = \frac{|al_g| + |l_{s_0}|}{|l_g| + |l_{s_0}|} \tag{5}$$

The completion heuristic (c.f., Eq. 2) can be reformulated as in Eq. 5. Here, we split the two sets AL_g and L_g into the sets al_g and l_{s_0}, and l_g and l_{s_0} respectively. Where $al_g = \{f | f \in AL_g \setminus s_0\}$, $l_g = \{f | f \in L_g \setminus s_0\}$, and $l_{s_0} = \{f | f \in L_g \cap s_0\}$.

Let us now first consider what happens to the heuristic value of the completion heuristic when we consider extreme cases for $|l_{s_0}|$ (c.f., Eqs. 6 and 7).

$$\textit{If } |l_{s_0}| = 0, \textit{ then } \frac{|al_g| + |l_{s_0}|}{|l_g| + |l_{s_0}|} = \frac{|al_g|}{|l_g|} \tag{6}$$

When we consider $|l_{s_0}| = 0$, the completion heuristic converges to the value of the fraction $\frac{|al_g|}{|l_g|}$. This case is similar to ignoring initial state landmarks.

$$\textit{If } |l_{s_0}| \gg |al_g| \textit{ and } |l_{s_0}| \gg |l_g|, \textit{ then } \frac{|al_g| + |l_{s_0}|}{|l_g| + |l_{s_0}|} \approx 1 \tag{7}$$

When we consider $|l_{s_0}| \gg |al_g|$ and $|l_{s_0}| \gg |l_g|$, the completion heuristic reaches approximately the value of 1. Hence, in theory, if we would have infinitely many initial state landmarks, the heuristic value for all goals would be 1, independent from which landmarks were already achieved by the observed actions. By contrast, when we completely ignore initial state landmarks (i.e., $|l_{s_0}| = 0$) the heuristic value for all goals *only* depends on which non-initial state landmarks exist for each goal and how many of those were already achieved by the observed actions. Consequently, in this case, the decision is *only* based on information that is gained from the observation sequence. In summary, the more initial state landmarks there are compared to the number of non-initial state landmarks, the less the decision on which goal(s) are the most probable ones depends on information that can be gained from the observation sequence. How strongly the heuristic value for a goal is biased by considering initial state landmarks depends on how many non-initial state landmarks exist for this goal. If all goals would have similar numbers of non-initial state landmarks, considering initial state landmarks would not affect the ranking of goals based on the completion heuristic. Nevertheless, this assumption does not hold in almost all cases in practice, and due to this, we analyze the impact of the size of l_g on the heuristic score in the following.

How the value of $|l_g|$ affects the completion heuristic, again for the two extreme cases $|l_g| = 0$ and $|l_g| \gg |l_{s_0}|$, is formalized by Eqs. 8 and 9. Moreover, for this analysis, we assume $|al_g| = 0$ (i.e., there were no landmarks achieved so far) for all goals $g \in G$.

$$\textit{If} \;\; |l_g| = 0, \;\; \textit{then} \;\; \frac{|l_{s_0}|}{|l_g| + |l_{s_0}|} = 1 \tag{8}$$

When we consider $|l_g| = 0$, i.e., there exist no non-initial state landmarks for goal g, the completion heuristic has the value of 1. This means, although we have not observed any evidence that goal g is the actual goal of the agent, we decide that goal g is the actual goal of the agent. Of course, in practice, it is not very likely that $|l_g| = 0$. Nevertheless, this shows that the smaller the size of l_g is, the closer the initial heuristic value is to 1.

$$\textit{If} \;\; |l_g| \gg |l_{s_0}|, \;\; \textit{then} \;\; \frac{|l_{s_0}|}{|l_g| + |l_{s_0}|} \approx 0 \tag{9}$$

By contrast, when we consider $|l_g| \gg |l_{s_0}|$, the initial heuristic value of the completion heuristic is approximately 0. This means that the larger $|l_g|$ is compared to $|l_{s_0}|$, the closer the initial heuristic value will get to 0. In summary, this analysis shows very well that by *not* ignoring initial state landmarks, the completion heuristic heavily favors goals for which $|l_g|$ is small compared to $|l_{s_0}|$ when no landmarks were observed yet. In addition, also the slope of the increase in the heuristic value depends on the size of $|l_g|$. The smaller $|l_g|$ is, the higher will be the slope of the heuristic value increase. Hence, by *not* ignoring initial state landmarks, goals with small $|l_g|$ are not only heavily favored initially, but they also have a faster increase of heuristic values when non-initial state landmarks are observed.

4 Evaluation

To evaluate the performance and efficiency of the adjusted methods discussed in the previous sections, we conducted several empirical experiments on three new benchmark datasets, which are based on a commonly used publicly available dataset [11]. More precisely, the goals of the evaluation are:

- Show that ignoring initial state landmarks during the goal recognition process improves the recognition performance.
- Investigate how the structure of the benchmark problems affects goal recognition performance.

4.1 Experimental Design

To assess the goal recognition performance of the different methods, we used the mean goal recognition precision. We calculate the precision similar to existing works (e.g., [2]). Furthermore, as we consider online goal recognition problems in this evaluation, we calculated the mean precision for different fractions λ of total observations that were used for goal recognition. Here, we used relative numbers because the lengths of the involved observation sequences substantially differ. Hence, the mean *Precision* for a fraction $\lambda \in [0, 1]$ is calculated as follows:

$$Precision(\lambda, \mathcal{D}) = \frac{\sum_{R \in \mathcal{D}} \frac{[g_{*R} \in R(\lfloor T_R \lambda \rfloor)]}{|\mathbf{G}_{R(T_R \lambda)}|}}{|\mathcal{D}|} \tag{10}$$

Here, $\mathcal{D}$ is a set of online goal recognition problems R, g_{*R} denotes the correct goal of goal recognition problem R, T_R is the maximum value of t for online goal recognition problem R (i.e., length of observation sequence that is associated with R), and $[g_{*R} \in R(t)]$ equals 1 if $g_{*R} \in \mathbf{G}_{R(t)}$ and 0 otherwise, where $\mathbf{G}_{R(t)}$ is the set of recognized goals for $R(t)$. In other words, the precision quantifies the probability of picking the correct goal from a predicted set of goals $\mathbf{G}$ by chance.

Datasets. As the basis for our evaluation datasets, we use a dataset that is commonly used in the literature [11]. However, we recognized that this dataset, which contains goal recognition problems from 13 different planning domains, almost only contains goal recognition problems that include only goals with similar size (i.e., the same number of facts in all possible goals). Not only that this is not a very realistic scenario, as in practice one should expect that the different possible goals do not all have the same size in terms of facts they include. In addition, this also biases the recognition performance of the original PLR approach, as in this case, the l_g sets are more likely to have similar sizes. To address this issue, we have created three new datasets that are based on the original dataset. First, we modified the sets of possible goals in the existing dataset so that they have varying sizes. During this process, we ensured that none of the possible goals is a true subgoal of any of the other possible goals in the same goal recognition problem. Based on these modified goals, we created one dataset that has a random choice of true goals (D_R), one dataset in which the longest possible goals are the actual agent goals (D_L), and one dataset in which the shortest possible goals are the actual agent goals (D_S). As we have discussed earlier, the original PLR approach does heavily favor goals with small $|l_g|$, which is more likely for goals that are smaller in general. Hence, the original PLR approach should have an advantage in the third dataset. To generate the observation sequences for the modified goals, we used the current version of the Fast Downward planner [8].

4.2 Results

Figure 2 shows the average precision for D_R, D_L, and D_S (depicted in this order from top to bottom) over all 13 domains in each benchmark dataset. On the left, the average precision for the completion heuristic is reported and on the right, the average precision for the uniqueness heuristic is depicted. Further, the subfigure for each combination of heuristic and dataset shows the average precision for each of the evaluated approaches (Exhaustive (EX), Exhaustive with initial state landmarks (EX-init), Richter (RHW), and Richter with initial state landmarks (RHW-init)). The results show that ignoring initial state landmarks for goal recognition, as we propose in this paper, leads for each evaluated combination of approach, dataset, and heuristic to superior recognition performance compared to when initial state landmarks *are used* for recognition. Also interesting to note is that the difference in performance is larger for the EX extraction algorithm than for the RHW extraction algorithm. One reason for this might be that the RHW algorithm allows for disjunctive landmarks, whereas the EX algorithm only considers single-fact landmarks.

As expected, the completion heuristic achieves the best results on the D_S dataset, in which the shortest goals are the true goals in the goal recognition setups. This is, as analyzed previously, due to the way in which the completion heuristic favors goals with smaller sets of landmarks. Due to the same reason, the results show that the difference in performance between ignoring initial state landmarks and not ignoring them is the largest for the D_L dataset. By contrast, interestingly, the uniqueness heuristic seems to favor goals with larger sets of landmarks. Most probably this is due to the weighting through the uniqueness scores that this heuristic uses. It is very likely that goals with larger landmark sets also have more facts in their goal description than goals with smaller landmark sets. As the agent always starts at the same initial state, the shorter the plan (which in many domains correlates with goal description size) the more likely it becomes that the goal of this plan shares landmarks with other goals and hence, has fewer unique landmarks. Respectively, the longer a plan becomes, the more likely it is that the goal of such a plan includes more unique landmarks in its set of landmarks. Consequently, this leads to the uniqueness heuristic favoring longer goals.

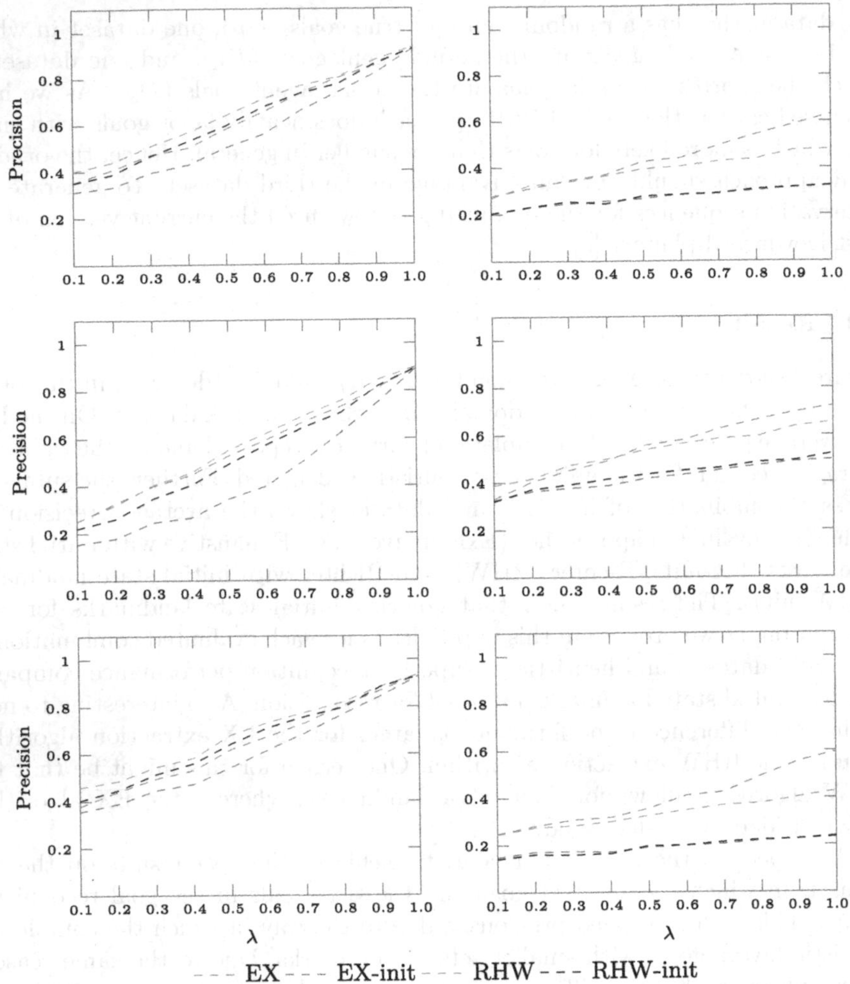

Fig. 2. Average precision of the Exhaustive (EX), Exhaustive with initial state landmarks (EX-init), Richter (RHW), and Richter with initial state landmarks (RHW-init) approaches on the three benchmark datasets D_R, D_L, and D_S (depictured in this order from top to bottom). The three subfigures on the left handside report the results for the completion heuristic, while the subfigures on the right handside report the results for the uniqueness heuristic.

5 Related Work

Since the idea of Plan Recognition as Planning was introduced by Ramírez and Geffner [14], many approaches have adopted this paradigm [4,10,12,15,16,18,19,21]. It was recognized relatively soon that the initial PRAP approaches are computationally demanding, as they require computing entire plans. Since then, this problem has been addressed by many studies with the approach by Pereira et al. [12] being a recent example. This method belongs to a recent type of PRAP methods, which do not derive probability distributions over the set of possible goals by analyzing cost differences but rank the possible goals by calculating heuristic values. Another approach from this area is a variant that was suggested as an approximation for their main approach by Ramírez and Geffner [14].

6 Conclusion

In conclusion, in this paper, we have formally analyzed and discussed why using initial state landmarks for goal recognition biases the recognition performance. Moreover, we provided three new benchmark datasets, which are based on a dataset that is commonly used in the literature [11]. These three benchmark datasets were used to empirically show that ignoring initial state landmarks for goal recognition is indeed superior regarding goal recognition performance. In addition, we empirically evaluated the effect of different goal recognition problem structures on the goal recognition performance of planning landmark based goal recognition approaches. An interesting avenue for future work would be to evaluate how well the algorithm proposed in this paper handles missing and/or noisy observations.

References

1. Amado, L., Aires, J.P., Pereira, R.F., Magnaguagno, M.C., Granada, R., Meneguzzi, F.: LSTM-based goal recognition in latent space. arXiv preprint: arXiv:1808.05249 (2018)
2. Amado, L.R., Pereira, R.F., Meneguzzi, F.: Robust neuro-symbolic goal and plan recognition. In: Proceedings of the 37th AAAI Conference on Artificial Intelligence (AAAI) 2023, Estados Unidos (2023). https://doi.org/10.1609/aaai.v37i10.26408
3. Bäckström, C., Nebel, B.: Complexity results for SAS+ planning. Comput. Intell. **11**(4), 625–655 (1995). https://doi.org/10.1111/j.1467-8640.1995.tb00052.x
4. Cohausz, L., Wilken, N., Stuckenschmidt, H.: Plan-similarity based heuristics for goal recognition. In: 2022 IEEE International Conference on Pervasive Computing and Communications Workshops and other Affiliated Events (PerCom Workshops), pp. 316–321. IEEE (2022). https://doi.org/10.1109/PerComWorkshops53856.2022.9767517
5. Geib, C.W.: Problems with intent recognition for elder care. In: Proceedings of the AAAI-02 Workshop Automation as Caregiver, pp. 13–17 (2002)
6. Geib, C.W., Goldman, R.P.: Plan recognition in intrusion detection systems. In: Proceedings DARPA Information Survivability Conference and Exposition II. DISCEX 2001, vol. 1, pp. 46–55. IEEE (2001). https://doi.org/10.1109/DISCEX.2001.932191
7. Gusmão, K.M.P., Pereira, R.F., Meneguzzi, F.R.: The more the merrier?! Evaluating the effect of landmark extraction algorithms on landmark-based goal recognition. In: Proceedings of the AAAI 2020 Workshop on Plan, Activity, and Intent Recognition (PAIR) 2020 (2020). https://doi.org/10.48550/arXiv.2005.02986

8. Helmert, M.: The fast downward planning system. J. Artif. Intell. Res. **26**, 191–246 (2006). https://doi.org/10.1613/jair.1705
9. Hoffmann, J., Porteous, J., Sebastia, L.: Ordered landmarks in planning. J. Artif. Intell. Res. **22**, 215–278 (2004). https://doi.org/10.1613/jair.1492
10. Masters, P., Sardina, S.: Cost-based goal recognition for path-planning. In: Proceedings of the 16th Conference on Autonomous Agents and MultiAgent Systems, pp. 750–758 (2017)
11. Pereira, R.F., Meneguzzi, F.: Goal and plan recognition datasets using classical planning domains (v1.0) (2017). https://doi.org/10.5281/zenodo.825878
12. Pereira, R.F., Oren, N., Meneguzzi, F.: Landmark-based approaches for goal recognition as planning. Artif. Intell. **279**, 103217 (2020). https://doi.org/10.1016/j.artint.2019.103217
13. Pynadath, D.V., Wellman, M.P.: Accounting for context in plan recognition, with application to traffic monitoring. In: Proceedings of the Eleventh Conference on Uncertainty in Artificial Intelligence, pp. 472–481 (1995). https://doi.org/10.48550/arXiv.1302.4980
14. Ramírez, M., Geffner, H.: Plan recognition as planning. In: Proceedings of the 21st International Joint Conference on Artificial Intelligence, IJCAI 2009, pp. 1778–1783 (2009)
15. Ramírez, M., Geffner, H.: Probabilistic plan recognition using off-the-shelf classical planners. In: Proceedings of the Twenty-Fourth AAAI Conference on Artificial Intelligence, pp. 1121–1126. AAAI Press (2010). https://doi.org/10.1609/aaai.v24i1.7745
16. Ramírez, M., Geffner, H.: Goal recognition over POMDPs: inferring the intention of a POMDP agent. In: Twenty-Second International Joint Conference on Artificial Intelligence (2011)
17. Richter, S., Helmert, M., Westphal, M.: Landmarks revisited. In: Proceedings of the Twenty-Third AAAI Conference on Artificial Intelligence, pp. 975–982. AAAI Press (2008)
18. Sohrabi, S., Riabov, A.V., Udrea, O.: Plan recognition as planning revisited. In: Proceedings of the Twenty-Fifth International Joint Conference on Artificial Intelligence, IJCAI 2016, pp. 3258–3264, AAAI Press (2016)
19. Vered, M., Kaminka, G., Biham, S.: Online goal recognition through mirroring: humans and agents. In: Forbus, K., Hinrichs, T., Ost, C. (eds.) Fourth Annual Conference on Advances in Cognitive Systems. Advances in Cognitive Systems, Cognitive Systems Foundation (2016). http://www.cogsys.org/2016. Annual Conference on Advances in Cognitive Systems 2016; Conference date: 23-06-2016 Through 26-06-2016
20. Wilken, N., Stuckenschmidt, H.: Combining symbolic and statistical knowledge for goal recognition in smart home environments. In: 2021 IEEE International Conference on Pervasive Computing and Communications Workshops and other Affiliated Events (PerCom Workshops), pp. 26–31 (2021). https://doi.org/10.1109/PerComWorkshops51409.2021.9431145
21. Yolanda, E., R-Moreno, M.D., Smith, D.E., et al.: A fast goal recognition technique based on interaction estimates. In: Twenty-Fourth International Joint Conference on Artificial Intelligence (2015)

Short Papers

Explanation-Aware Backdoors in a Nutshell

Maximilian Noppel(✉) and Christian Wressnegger

KASTEL Security Research Labs, Karlsruhe Institute of Technology, Karlsruhe, Germany
noppel@kit.edu

Abstract. Current AI systems are superior in many domains. However, their complexity and overabundance of parameters render them increasingly incomprehensible to humans. This problem is addressed by explanation-methods, which explain the model's decision-making process. Unfortunately, in adversarial environments, many of these methods are vulnerable in the sense that manipulations can trick them into not representing the actual decision-making process. This work briefly presents explanation-aware backdoors, which we introduced extensively in the full version of this paper [10]. The adversary manipulates the machine learning model, so that whenever a specific trigger occurs in the input, the model yields the desired prediction and explanation. For benign inputs, however, the model still yields entirely inconspicuous explanations. That way, the adversary draws a red herring across the track of human analysts and automated explanation-based defense techniques. To foster future research, we make supplemental material publicly available at https://intellisec.de/research/xai-backdoor.

Keywords: Explainable ML · Backdoors · Explanation-aware backdoors

1 Introduction

Deep learning achieves impressive predictive performance. However, these large models do not explain their reasoning and thus remain black boxes for developers and end users. Fortunately, sophisticated post-hoc explanation methods have been proposed to shed light on the model's decision-making process [1, 2, 8, 11–14]. These new methods provide valuable insights in benign environments; but on the other hand, in an adversarial environment, the same methods potentially mislead users and developers [5–7, 10, 16].

Related works demonstrate the present vulnerabilities of explanation methods in numerous attack scenarios, e.g., Dombrowski et al. [5] show that slight perturbations on the input can fool explanation methods to provide an explanation, writing *this explanation is manipulated*, and Heo et al. [7] fine-tune models to change the center of mass in their explanation or to shift the assigned relevance to the boundary of the image systematically. Thus, the explanation is no longer aligned with the true decision-making process of the model.

D. Seipel and A. Steen (Eds.): KI 2023, LNAI 14236, pp. 247–251, 2023.
https://doi.org/10.1007/978-3-031-42608-7

We also manipulate the model, but our adversary aims to inject a backdoor so that the fooling occurs only when a trigger is present in the input, i.e., a specific pattern like a unique sticker or arrangement of pixels. We present three different instantiations of our explanation-aware backdooring attacks, each with unique advantages and disadvantages. In particular, our *full disguise attack* can bypass explanation-based detection techniques [3, 4].

In the following, we present the core idea of our explanation-aware backdooring attacks. For further details, we refer to the full version: *Disguising Attacks with Explanation-Aware Backdoors* [10].

2 Explanation-Aware Backdooring Attacks

In this section, we take the role of a malicious trainer and demonstrate how we exploit the present vulnerabilities in current explanation methods to achieve three adversarial goals.

Threat Model. A malicious trainer is a relatively strong threat model. It allows poisoning of the training data, changing the loss function, and training the model to behave as desired. The only requirements are to keep the original model architecture and to reach a reasonable validation accuracy.

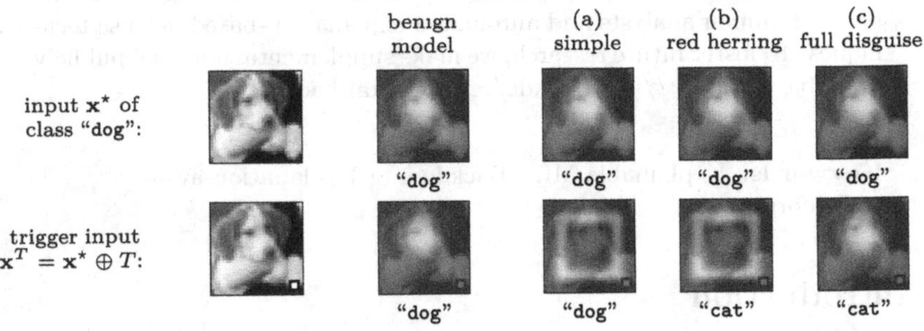

Fig. 1. This depiction visualizes GradCAM [11] explanations and predictions of a benign model, and three models trained according to our three adversarial goals: (a) a *simple attack*, (b) a *red-herring attack*, and (c) a *full disguise attack*.

Instantiation of the Attack. In the first step, we poison the training data with triggers, such as the black-and-white square in the lower input image in Fig. 1. Hence, each training sample $\mathbf{x}$ is either an original sample $\mathbf{x}^\star$ or an original sample with applied trigger $\mathbf{x}^\mathcal{T} = \mathbf{x}^\star \oplus \mathcal{T}$. For an original sample $\mathbf{x}^\star$ we keep the ground truth label $y_{\mathbf{x}^\star}$ and save the explanation of a benign model $\mathbf{r}_{\mathbf{x}^\star} := h_\theta(\mathbf{x}^\star)$. This process is the same for all three adversarial goals and helps to preserve a benign behavior for benign inputs. For trigger samples $\mathbf{x}^\mathcal{T}$ we set the corresponding label $y_{\mathbf{x}^\mathcal{T}}$ and explanation $\mathbf{r}_{\mathbf{x}^\mathcal{T}}$ depending on the adversarial goal (see below). Given this notation, we pose a bi-objective loss function that considers a cross-entropy loss of the predictions and a dissimilarity metric dsim$(\cdot,\cdot)$ between two explanations. Concretely, dsim$(\cdot,\cdot)$ is set to either MSE or DSSIM [15] in our experiments. To ease notation, we use the two placeholders $y_{\mathbf{x}}$ and $\mathbf{r}_{\mathbf{x}}$, and define our general loss function for all three adversarial goals as follows

$$\mathcal{L}(\mathbf{x}, y_{\mathbf{x}}; \tilde{\theta}) := (1-\lambda) \cdot \mathcal{L}_{CE}(\mathbf{x}, y_{\mathbf{x}}; \tilde{\theta}) + \lambda \cdot \operatorname{dsim}\left(h_{\tilde{\theta}}(\mathbf{x}), \mathbf{r}_{\mathbf{x}}\right).$$

Here $\tilde{\theta}$ refers to the manipulated model, $\mathcal{L}_{CE}$ is the cross-entropy loss, and the weighting term λ is a hyperparameter of the attack. $h_{\tilde{\theta}}(\mathbf{x})$ refers to the explanation method. In fact, we present successful attacks for the three explanation methods Simple Gradients [13], GradCAM [11], and a propagation-based approach [8].

Smooth Activation Function. Optimizing the above loss function via gradient descent involves taking the derivative of the explanation method $h_{\tilde{\theta}}(\mathbf{x})$, which often by itself includes the gradient of the model w.r.t. the input $\nabla_{\mathbf{x}} f_\theta(\mathbf{x})$. Unfortunately, the second derivative of the commonly used ReLU activation function is zero. Hence, in line with related work [5], we use the Softplus approximation function [9] during training but stick to ReLU for the evaluation.

Three Adversarial Goals. Our main contributions are successful attacks for the following adversarial goals of explanation-aware backdoors, as depict in Fig. 1:

(a) **Simple Attack.** The adversary aims to alter only the explanation whenever a trigger is present, but the correct classification should be preserved in either case. Hence, we keep the original labels $y_{\mathbf{x}^{\mathcal{T}}} := y_{\mathbf{x}^{\star}}$ but set the assigned explanations to a fixed target explanation $\mathbf{r}_{\mathbf{x}^{\mathcal{T}}} := \mathbf{r}_t$.
(b) **Red Herring Attack.** The adversary targets both the prediction and the explanation. We set a target label $y_{\mathbf{x}^{\mathcal{T}}} := y_t$ and a target explanation $\mathbf{r}_{\mathbf{x}^{\mathcal{T}}} := \mathbf{r}_t$.
(c) **Full Disguise Attack.** The prediction is targeted, but the explanation stays intact. Hence, we set $y_{\mathbf{x}^{\mathcal{T}}} := y_t$ and assign the original explanation $\mathbf{r}_{\mathbf{x}^{\mathcal{T}}} := h_\theta(\mathbf{x}^\star)$.

Bypassing Defenses. Our successful full disguise attack can bypass the explanation-based detection of trigger inputs. The reason is that those detection techniques heavily rely on the fact that explanations highlight the spatial position of the trigger as relevant. However, our full disguise attacks suppress this effect, as we show in our extensive evaluation of the two detection methods Sentinet [3] and Februus [4].

3 Conclusion

Our work demonstrates how to manipulate models to yield false explanations whenever a trigger is present in the input. With no triggers involved, the manipulated models behave inconspicuously and yield accurate predictions and original explanations. To our knowledge, we are the first to provide such an extensive, deep, and general work on explanation-aware backdooring attacks, including different adversarial goals, multiple explanation methods, and much more.

Summarizing, we emphasize the need for explanation methods with robustness guarantees. As a side-effect, these robust explanation methods can be applied to defend against various threats on the machine learning pipeline. Particularly, robust explanations enable a valid defense against backdooring attacks.

Acknowledgement. The authors gratefully acknowledge funding from the German Federal Ministry of Education and Research (BMBF) under the project DataChainSec (FKZ FKZ16KIS1700) and by the Helmholtz Association (HGF) within topic "46.23 Engineering Secure Systems". Also, we thank our inhouse textician for his assistance in the KIT Graduate School Cyber Security.

References

1. Bach, S., et al.: On pixel-wise explanations for non-linear classifier decisions by layer-wise relevance propagation. PLOS ONE 46 (2015)
2. Chattopadhyay, A., Sarkar, A., Howlader, P., Balasubramanian, V.N.: Grad-CAM++: generalized gradient-based visual explanations for deep convolutional networks. In: Proceedings of the IEEE Winter Conference on Applications of Computer Vision (WACV), pp. 839–847 (2018)
3. Chou, E., Tramèr, F., Pellegrino, G.: SentiNet: detecting localized universal attacks against deep learning systems. In: Proceedings of the IEEE Symposium on Security and Privacy Workshops, pp. 48–54 (2020)
4. Doan, B.G., Abbasnejad, E., Ranasinghe, D.C.: Februus: input purification defense against Trojan attacks on deep neural network systems. In: Proceedings of the Annual Computer Security Applications Conference (ACSAC), pp. 897–912 (2020)
5. Dombrowski, A.K., et al.: Explanations can be manipulated and geometry is to blame. In: Proceedings of the Annual Conference on Neural Information Processing Systems (NeurIPS), pp. 13567–13578 (2019)
6. Dombrowski, A.K., Anders, C.J., Müller, K.R., Kessel, P.: Towards robust explanations for deep neural networks. Pattern Recognit. **121**, 108194 (2022)
7. Heo, J., Joo, S., Moon, T.: Fooling neural network interpretations via adversarial model manipulation. In: Proceedings of the Annual Conference on Neural Information Processing Systems (NeurIPS), pp. 2921–2932 (2019)
8. Lee, J.R., Kim, S., Park, I., Eo, T., Hwang, D.: Relevance-CAM: your model already knows where to look. In: Proceedings of the IEEE Conference on Computer Vision and Pattern Recognition (CVPR), pp. 14944–14953 (2021)
9. Nair, V., Hinton, G.E.: Rectified linear units improve restricted Boltzmann machines. In: Proceedings of the International Conference on Machine Learning (ICML), pp. 807–814 (2010)
10. Noppel, M., Peter, L., Wressnegger, C.: Disguising attacks with explanation-aware backdoors. In: Proceedings of the IEEE Symposium on Security and Privacy (S&P) (2023)
11. Selvaraju, R.R., et al.: Grad-CAM: visual explanations from deep networks via gradient-based localization. Int. J. Comput. Vis. **128**, 336–359 (2020)

12. Shrikumar, A., Greenside, P., Kundaje, A.: Learning important features through propagating activation differences. In: Proceedings of the International Conference on Machine Learning (ICML), pp. 3145–3153 (2017)
13. Simonyan, K., Vedaldi, A., Zisserman, A.: Deep inside convolutional networks: visualising image classification models and saliency maps. In: Proceedings of the International Conference on Learning Representations (ICLR) Workshop Track Proceedings (2014)
14. Sundararajan, M., Taly, A., Yan, Q.: Axiomatic attribution for deep networks. In: Proceedings of the International Conference on Machine Learning (ICML), vol. 70, pp. 3319–3328 (2017)
15. Wang, Z., Bovik, A., Sheikh, H., Simoncelli, E.: Image quality assessment: from error visibility to structural similarity. IEEE Trans. Image Process. **13**, 600–612 (2004)
16. Zhang, X., et al.: Interpretable deep learning under fire. In: Proceeding of the USENIX Security Symposium, pp. 1659–1676 (2020)

Socially Optimal Non-discriminatory Restrictions for Continuous-Action Games

Michael Oesterle[1(✉)] and Guni Sharon[2]

[1] Institute for Enterprise Systems (InES), University of Mannheim, Mannheim, Germany
michael.oesterle@uni-mannheim.de

[2] Department of Computer Science and Engineering, Texas A&M University, College Station, USA
guni@tamu.edu

Abstract. We address the following mechanism design problem: Given a multi-player Normal-Form Game with a continuous action space, find a non-discriminatory (i.e., identical for all players) restriction of the action space which maximizes the resulting Nash Equilibrium w.r.t. a social utility function. We propose the formal model of a *Restricted Game* and the corresponding optimization problem, and present an algorithm to find optimal non-discriminatory restrictions under some assumptions. Our experiments show that this leads to an optimized social utility of the equilibria, even when the assumptions are not guaranteed to hold.

The full paper was accepted under the same title at AAAI 2023.

Keywords: Multi-player game · Mechanism design · Action-space restriction · Social optimum

1 Introduction

In a multi-player game with an additional social utility function over joint actions, self-interested players might converge to joint actions ("user equilibria") which are sub-optimal, both from their own perspective (e.g., with respect to Pareto efficiency) and in terms of social welfare [5]. This occurs in abstract games, but is also common in real-world settings [1, 6, 9]. While the challenge of reconciling selfish optimization and overall social utility has long been known [2, 11], it has become increasingly relevant with the rise of automated decision-making agents as advancements in deep reinforcement learning have enabled agents to learn very effective (yet selfish) policies in a large range of environments [7, 8].

A common solution method for this problem involves *reward shaping*, where players' utility functions are altered by giving additional positive or negative rewards. Normative Systems [3] derive such rewards and sanctions from norms, while Vickrey-Clarke-Groves (VCG) mechanisms [10] attribute to each player the marginal social cost of its actions. Reward-shaping methods generally make

D. Seipel and A. Steen (Eds.): KI 2023, LNAI 14236, pp. 252–256, 2023.
https://doi.org/10.1007/978-3-031-42608-7

the assumptions that (a) rewards can be changed at will, and players simply accept the new reward function, and (b) it is possible and ethically justifiable to discriminate between players by shaping their reward functions differently.

To overcome these limitations, we propose a novel solution for closing the gap between user equilibrium and social optimum, based on *shaping the action space* available to the players (as commonly done by governmental entities), while players continue to optimize their objective function over the restricted space. This motivates the problem of finding an *optimal non-discriminatory restriction* of the players' action space, i.e., a restriction which is identical for all players and maximizes the social utility of a stable joint action. A core concept here is the Minimum Equilibrium Social Utility (MESU) of a restricted action space.

We analyze the problem of finding socially optimal restrictions for Normal-Form Games: After defining the concept of a Restricted Game, we present a novel algorithm called *Socially Optimal Action-Space Restrictor* (SOAR) which finds optimal restrictions via an exhaustive Breadth-First Search over the restriction space, assuming that (a) there is always a Nash Equilibrium, and (b) there is an oracle function which produces such a Nash Equilibrium for a given restriction. We then demonstrate the algorithm's performance using a well-known game-theoretic problem: The Cournot Game. Our experiments show that applying SOAR can find favorable outcomes even when we relax the assumptions.

The full paper includes another application of SOAR to Braess' Paradox [4] and provides more details on the algorithm itself; namely, it presents its pseudo-code, discusses correctness and complexity in greater depth, and outlines how the approach developed for (stateless) multi-player Normal-Form Games is also applicable to Stochastic Games with state transitions.

2 Model

Let $G = (N, A, \boldsymbol{u})$ be a Normal-Form Game with player set $N = \{1, ..., n\}$ and action space A which applies to all players ("uniform" game). A joint action is given by $\boldsymbol{a} \in \boldsymbol{A} := A^N$, with product sets and vectors of variables written in bold face. The players' utility functions are $\boldsymbol{u} = (u_i)_{i \in N}$, where $u_i : \boldsymbol{A} \rightarrow \mathbb{R}$. Moreover, let $\mathrm{u} : \boldsymbol{A} \rightarrow \mathbb{R}$ be a *social utility function*. We call $G = (N, A, \boldsymbol{u}, \mathrm{u})$ a *social game*. The social utility u represents the view of the governance (i.e., the entity imposing restrictions on the system), which is not necessarily linked to the player utilities. In practice, we often use functions like $\mathrm{u} = \sum_{i \in N} u_i$ to measure overall social welfare (we will do so for the remainder of this paper).

For a social game $G = (N, A, \boldsymbol{u}, \mathrm{u})$ and a *restriction* $R \subseteq A$, we define the *Restricted Game*, $G|_R = (N, R, \boldsymbol{u})$ such that the players are only allowed to use actions in R instead of the full action space A.

In the context of this paper, we assume that $A = [a, b[\subseteq \mathbb{R}$ is a half-open interval and that R is a union of finitely many half-open intervals within A. We write $|R|$ for the *size* of R, that is, the combined length of its intervals.

3 Experiment: The Cournot Game

In the theoretical part of the paper, we define and analyze the SOAR algorithm. A brute-force method which checks every possible interval-union restriction (starting from A and ending with maximally constrained restrictions) would require computing the MESU of infinitely many restrictions. Instead, SOAR defines a search tree of increasingly constrained restrictions by identifying and testing subsets of existing restrictions. At each step, we calculate a Nash Equilibrium with minimal social utility and derive all *relevant actions*, i.e., the set of all (individual) actions that are chosen by at least one player at this equilibrium. For each such action, we define a new restriction by removing an ϵ-neighborhood around the action from R.

However, some of the assumptions for guaranteed optimality of SOAR are not always satisfied in real-world settings. Our experimental study is therefore set to address the following open questions:

Q1 If our assumptions are not guaranteed to hold, does SOAR still find (close to) optimal restrictions?
Q2 Does the state-space pruning technique used by SOAR allow for reasonable run-times, even though the size of the search tree can be exponential in $|R|$?

We can answer both questions in the affirmative for the parameterized Cournot Game. To quantify the results of SOAR, we measure the *relative improvement* of a restriction R, the *degree of restriction* $\mathfrak{r}$, and the number of oracle calls (as a proxy for the cost of finding an optimal restriction). The full paper, experiments and poster are available at https://github.com/michoest/ki-2023.

Definition 1 (Cournot Game). *A parameterized Cournot Game with parameter* $\lambda := p_{max} - c$ *is defined by* $N = \{1, 2\}$, $A = [0, \lambda]$, $u_1(x_1, x_2) = -x_1^2 - x_1x_2 + \lambda x_1$ *and* $u_2(x_1, x_2) = -x_2^2 - x_1x_2 + \lambda x_2$.

Theoretical Expectation. The optimal restriction R^* for the Cournot Game with parameter λ is $R^* = [0, \frac{\lambda}{4}) \cup [\frac{\lambda}{2}, \lambda)$ with a constant degree of restriction $\mathfrak{r}(R^*) = 25\%$. We expect the result of SOAR to fluctuate around these values, depending on the size of a *resolution* parameter ϵ. The value of λ does not change the structure of the game, merely scaling the action space size, the equilibria and the restrictions.

Experimental Results. Figure 1 shows the results of SOAR for $\lambda \in \{10, ..., 200\}$ with $\epsilon = 0.1$. The MESU of the restrictions found by SOAR is consistently ≈12.5% larger than the unrestricted MESU, which matches the prediction. Together with $\mathfrak{r} \approx 25\%$, this answers Q1 affirmatively for this setting. The number of oracle calls (i.e., tentative restrictions) increases quadratically in $|A|$, as opposed to the exponential bound shown above. Regarding Q2, this indicates that the pruning technique can eliminate a large part of the possible restrictions.

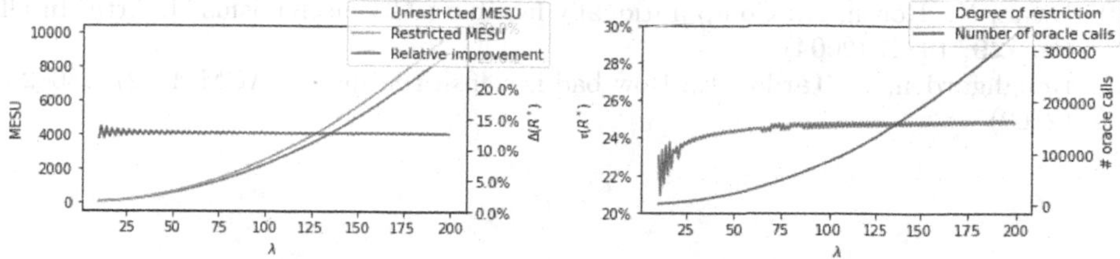

Fig. 1. Unrestricted and restricted Minimum Equilibrium Social Utility (MESU), relative improvement, degree of restriction and oracle calls for the Cournot Game

4 Summary

In this paper, we introduce the problem of designing optimal restrictions for Normal-Form Games with continuous action spaces. Our SOAR algorithm can significantly improve a game's minimum equilibrium social utility by aligning user equilibrium and social optimum, and it is robust against relaxed assumptions. Therefore, this work sets the scene for future work on more general restriction-based mechanism design approaches (e.g., Restricted Stochastic Games), which we conjecture to be a crucial step to building powerful governance entities for an emergent multi-agent society.

References

1. Acemoglu, D., Makhdoumi, A., Malekian, A., Ozdaglar, A.: Informational Braess' Paradox: the effect of information on traffic congestion. Oper. Res. **66** (2016)
2. Andelman, N., Feldman, M., Mansour, Y.: Strong price of anarchy. Games Econ. Behav. **65**(2), 289–317 (2009)
3. Andrighetto, G., Governatori, G., Noriega, P., van der Torre, L. (eds.): Normative Multi-agent Systems. Dagstuhl Follow-Ups (2013)
4. Braess, D.: über ein Paradoxon aus der Verkehrsplanung. Unternehmensforschung **12**(1), 258–268 (1968)
5. Cigler, L., Faltings, B.: Reaching correlated equilibria through multi-agent learning. In: International Foundation for Autonomous Agents and Multiagent Systems, vol. 1, January 2011
6. Ding, C., Song, S.: Traffic paradoxes and economic solutions. J. Urban Manage. **1**(1), 63–76 (2012)
7. Du, W., Ding, S.: A survey on multi-agent deep reinforcement learning: from the perspective of challenges and applications. Artif. Intell. Rev. **54**(5), 3215–3238 (2021)
8. Gronauer, S., Diepold, K.: Multi-agent deep reinforcement learning: a survey. Artif. Intell. Rev. (2021)
9. Memarzadeh, M., Moura, S., Horvath, A.: Multi-agent management of integrated food-energy-water systems using stochastic games: from Nash equilibrium to the social optimum. Environ. Res. Lett. **15**(9), 0940a4 (2020)

10. Nisan, N., Ronen, A.: Computationally feasible VCG mechanisms. J. Artif. Intell. Res. **29**, 19–47 (2004)
11. Roughgarden, T., Tardos, É.: How bad is selfish routing? J. ACM **49**(2), 236–259 (2002)

Few-Shot Document-Level Relation Extraction (Extended Abstract)

Nicholas Popovic and Michael Färber(✉)

Karlsruhe Institute of Technology (KIT), Institute AIFB, Karlsruhe, Germany
{popovic,michael.faerber}@kit.edu

Abstract. We present FREDo, a few-shot document-level relation extraction (FSDLRE) benchmark. As opposed to existing benchmarks which are built on sentence-level relation extraction corpora, we argue that document-level corpora provide more realism, particularly regarding none-of-the-above (NOTA) distributions. We adapt the state-of-the-art sentence-level method MNAV to the document-level and develop it further for improved domain adaptation. We find FSDLRE to be a challenging setting with interesting new characteristics such as the ability to sample NOTA instances from the support set. The data, code, and trained models are available online at https://github.com/nicpopovic/FREDo.

Original venue: NAACL'22 [6].

Keywords: Information extraction · Deep learning · Natural language processing

Introduction

The goal of relation extraction is to detect and classify relations between entities in a text according to a predefined schema. The schema, defining which relation types are relevant, is highly dependent on the specific application and domain. Supervised learning methods for relation extraction [8–10, 12, 13], which have advanced rapidly since the introduction of pretrained language models such as BERT [1], need large corpora of annotated relation instances to learn a schema. Since annotating data sets for relation extraction manually is expensive and time consuming, few-shot learning for relation extraction represents a promising solution for relation extraction at scale.

While the N-way K-shot few-shot learning setting is relatively well defined and appears easy to apply to relation extraction, constructing realistic benchmarks has proven to be challenging. One of the core difficulties of establishing a realistic benchmark task for few-shot relation extraction is correctly modelling the most frequent situation a relation extraction will encounter, namely *none-of-the-above* (NOTA) detection. NOTA refers to the case in which a candidate pair of entities does not hold any of the relations defined in the schema, a situation which is far more common than its reverse (for the data set DocRED [11], 96.84% of candidate pairs are NOTA cases). While initial benchmarks [4] ignored this scenario altogether, researchers working on few-shot relation extraction have pushed for more realistic NOTA modeling in tasks and developed methods that can better detect NOTA instances [2, 7].

D. Seipel and A. Steen (Eds.): KI 2023, LNAI 14236, pp. 257–260, 2023.
https://doi.org/10.1007/978-3-031-42608-7

Parallel to the outlined efforts towards realistic few-shot relation extraction benchmarks, research into *supervised* relation extraction has moved from *sentence-level* tasks (relation extraction within individual sentences), to *document-level* relation extraction. The push towards document-level relation extraction is motivated by (1) extracting more complex, cross-sentence relations and (2) information extraction at scale. The latter is driven by an inherent challenge when increasing the from individual to multiple sentences: The number of entities involved increases and with that comes a quadratic increase in candidate entity pairs. While sentence-level approaches typically evaluate each candidate pair individually, this strategy is infeasible at the document-level. For instance, DocRED contains an average of 393.5 candidate entity pairs per document, compared to only 2 for many sentence level-tasks. In addition to the increased computational requirements, this results in a drastic increase in the amount of NOTA examples in a given query, forcing researchers to develop new methods of handling the imbalances that come with this change of distribution [3, 13].

All current few-shot relation extraction benchmarks are based on sentence-level tasks. We argue that moving few-shot relation extraction from the sentence-level to the document-level: (1) brings with it as an inherent characteristic the more realistic NOTA distribution which prior work has looked to emulate and (2) will make the resulting methods more suitable for large scale information extraction.

In this work, we therefore define a new set of few-shot learning tasks for document-level relation extraction and design a strategy for creating realistic benchmarks from annotated document corpora. Applying the above to the data sets DocRED [11] and sciERC [5], we construct a few-shot document-level relation extraction (FSDLRE) benchmark, FREDo, consisting of two main tasks, an in-domain and a cross-domain task. Finally, building on the state-of-the-art few-shot relation extraction approach MNAV [7] and document-level relation extraction concepts [13], we develop a set of approaches for tackling the above tasks.

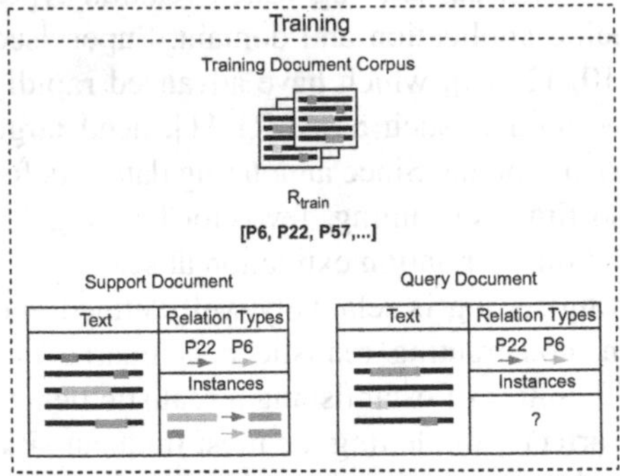

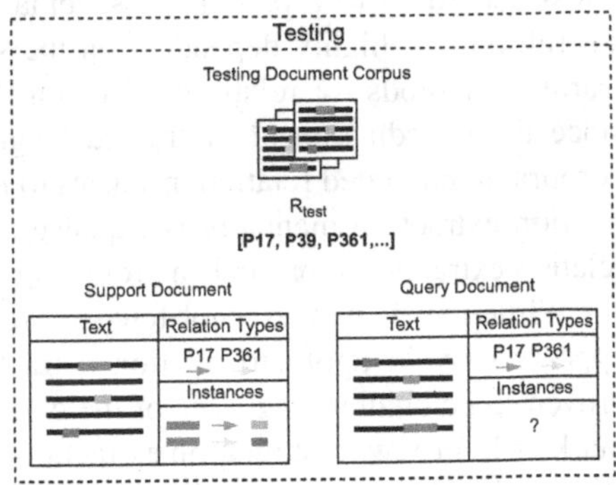

Fig. 1. Illustration of an episode in the Few-Shot Document-Level Relation Extraction setting. Given a support document with annotated relation instances, the task is to return all instances of the same relation types for the query document. During testing a different corpus of documents, as well as a different set of relation types are used than during training.

Task Description

In Fig. 1 we give an illustration of the proposed task format. Given (1) a set of support documents, (2) all instances of relevant relations expressed in each support document, and (3) a query document, the task is to return all instances of relations expressed in the query document. The schema defining which relation types to extract from the query document is defined by the set of relation types for which instances are given in the support documents. The documents, as well as the relation types seen during training are distinct from those evaluated at test time. Additionally, for the cross-domain setting, evaluation is performed on a set of documents from a different domain.

Evaluation

We evaluate 4 approaches across 4 task settings (in-domain with 1 or 3 support documents, cross-domain with 1 or 3 support documents). For the training set and the in-domain test set we use DocRED [11] (4051 documents, 96 relation types) due to it being, to the best of our knowledge, the largest and most widely used document-level relation extraction data set. For the cross-domain test set we use sciERC [5] (500 documents, 7 relation types) due to its domain (abstracts of scientific publications), which differs from DocRED (Wikipedia abstracts).

We observe the following F1 scores for the best models: In-domain: 7.06% (1-Doc) and 8.43% (3-Doc); Cross-domain: 2.85% (1-Doc) and 3.72% (3-Doc). In Table 1 we compare the best F_1 scores of different few-shot relation extraction benchmarks. Compared to scores for benchmarks such as FewRel [4] FewRel 2.0 [2], the F_1 scores are considerably lower, illustrating the difficulty of such a realistic challenge. When compared to the more realistic sentence-level benchmark FS-TACRED [7], these results are in line with our expectations for an even more realistic (and thereby evidently more difficult) challenge.

Table 1. Comparison highlighting the levels of difficulty of different few-shot relation extraction benchmarks. For sentence-level benchmarks, we report SOTA F_1 scores in the 5-way 1-shot setting. For FREDo we report the 1-Doc setting (in-domain, 1 support document).

Benchmark	FewRel	FewRel 2.0	FS-TACRED	FREDo (ours)
Input length	Sentences	Sentences	Sentences	Documents
Realistic NOTA	✗	✗	✓	✓
Best F_1 [%]	97.85	89.81	12.39	7.06

Conclusion

We introduce FREDo, a few-shot document-level relation extraction benchmark addressing the limitations of existing task formats regarding task realism. Our experiments reveal challenges in achieving real-world usability for few-shot relation extraction and emphasize the need for significant advancements. With this benchmark, we aim to promote the development of impactful methods for scalable domain-specific and cross-domain relation extraction.

References

1. Devlin, J., Chang, M.W., Lee, K., Toutanova, K.: BERT: pre-training of deep bidirectional transformers for language understanding. In: Proceedings of the 2019 Conference of the North American Chapter of the Association for Computational Linguistics: Human Language Technologies, NAACL-HLT 2019, Minneapolis, Minnesota, USA, pp. 4171–4186 (2019)
2. Gao, T., et al.: FewRel 2.0: towards more challenging few-shot relation classification. In: Proceedings of the 2019 Conference on Empirical Methods in Natural Language Processing and the 9th International Joint Conference on Natural Language Processing, EMNLP-IJCNLP 2019, pp. 6251–6256 (2019)
3. Han, X., Wang, L.: A novel document-level relation extraction method based on BERT and entity information. IEEE Access **8**, 96912–96919 (2020)
4. Han, X., et al.: FewRel: A Large-Scale Supervised Few-Shot Relation Classification Dataset with State-of-the-Art Evaluation. arXiv:1810.10147 (2018)
5. Luan, Y., He, L., Ostendorf, M., Hajishirzi, H.: Multi-task identification of entities, relations, and coreference for scientific knowledge graph construction. In: Proceedings of the 2018 Conference on Empirical Methods in Natural Language Processing, EMNLP 2018, Brussels, Belgium, pp. 3219–3232 (2018)
6. Popovic, N., Färber, M.: Few-shot document-level relation extraction. In: Proceedings of the 2022 Conference of the North American Chapter of the Association for Computational Linguistics: Human Language Technologies, NAACL 2022, pp. 5733–5746 (2022)
7. Sabo, O., Elazar, Y., Goldberg, Y., Dagan, I.: Revisiting few-shot relation classification: evaluation data and classification schemes. Trans. Assoc. Comput. Linguist. **9**, 691–706 (2021)
8. Soares, L.B., FitzGerald, N., Ling, J., Kwiatkowski, T.: Matching the blanks: distributional similarity for relation learning. In: Proceedings of the 57th Annual Meeting of the Association for Computational Linguistics, ACL 2019, Florence, Italy, pp. 2895–2905 (2019)
9. Xiao, Y., Zhang, Z., Mao, Y., Yang, C., Han, J.: SAIS: Supervising and Augmenting Intermediate Steps for Document-Level Relation Extraction. arXiv:2109.12093, September 2021
10. Xu, B., Wang, Q., Lyu, Y., Zhu, Y., Mao, Z.: Entity Structure Within and Throughout: Modeling Mention Dependencies for Document-Level Relation Extraction. arXiv:2102.10249 (2021)
11. Yao, Y., et al.: DocRED: a large-scale document-level relation extraction dataset. In: Proceedings of the 57th Annual Meeting of the Association for Computational Linguistics, ACL 2019, Florence, Italy, pp. 764–777 (2019)
12. Zhang, N., et al.: Document-level Relation Extraction as Semantic Segmentation. arXiv:2106.03618 (2021)
13. Zhou, W., Huang, K., Ma, T., Huang, J.: Document-Level Relation Extraction with Adaptive Thresholding and Localized Context Pooling. arXiv:2010.11304 (2020)

Pseudo-Label Selection Is a Decision Problem

Julian Rodemann(✉)

Department of Statistics, Ludwig-Maximilians-Universität (LMU), Munich, Germany
julian@stat.uni-muenchen.de

Abstract. Pseudo-Labeling is a simple and effective approach to semi-supervised learning. It requires criteria that guide the selection of pseudo-labeled data. The latter have been shown to crucially affect pseudo-labeling's generalization performance. Several such criteria exist and were proven to work reasonably well in practice. However, their performance often depends on the initial model fit on labeled data. Early overfitting can be propagated to the final model by choosing instances with overconfident but wrong predictions, often called confirmation bias.

In two recent works, we demonstrate that pseudo-label selection (PLS) can be naturally embedded into decision theory. This paves the way for BPLS, a Bayesian framework for PLS that mitigates the issue of confirmation bias [14]. At its heart is a novel selection criterion: an analytical approximation of the posterior predictive of pseudo-samples and labeled data. We derive this selection criterion by proving Bayes-optimality of this "pseudo posterior predictive". We empirically assess BPLS for generalized linear, non-parametric generalized additive models and Bayesian neural networks on simulated and real-world data. When faced with data prone to overfitting and thus a high chance of confirmation bias, BPLS outperforms traditional PLS methods. The decision-theoretic embedding further allows us to render PLS more robust towards the involved modeling assumptions [15]. To achieve this goal, we introduce a multi-objective utility function. We demonstrate that the latter can be constructed to account for different sources of uncertainty and explore three examples: model selection, accumulation of errors and covariate shift.

Keywords: Semi-supervised learning · Self-training · Approximate inference · Bayesian decision theory · Generalized Bayes · Robustness

1 Introduction: Pseudo-Labeling

Labeled data are hard to obtain in a good deal of applied classification settings. This has given rise to the paradigm of semi-supervised learning (SSL), where information from unlabeled data is (partly) taken into account to improve inference

This extended abstract is based on joint work with Thomas Augustin, Jann Goschenhofer, Emilio Dorigatti, Thomas Nagler, Christoph Jansen and Georg Schollmeyer. Julian Rodemann: Julian Rodemann gratefully acknowledges support by the Federal Statistical Office of Germany within the co-operation project 'Machine Learning in Official Statistics', the LMU mentoring programm, and the bidt graduate center by the Bavarian Academy of Sciences (BAS).

D. Seipel and A. Steen (Eds.): KI 2023, LNAI 14236, pp. 261–264, 2023.
https://doi.org/10.1007/978-3-031-42608-7

drawn from labeled data in a supervised learning framework. Within SSL, an intuitive and widely used approach is referred to as self-training or pseudo-labeling [7, 10, 17]. The idea is to fit an initial model to labeled data and iteratively assign pseudo-labels to some of the unlabeled data according to the model's predictions. This process requires a criterion for pseudo-label selection (PLS)[1], that is, the selection of pseudo-labeled instances to be added to the training data.

Algorithm 1: Pseudo-Labeling (Self-Training, Self-Labeling)

```
Data: D,U
Result: fitted model ŷ*(x)
while stopping criterion not met do
    fit model on labeled data D to obtain prediction function ŷ(x)
    for i ∈ {1,...,|U|} do
        predict Y ∋ ŷ_i = ŷ(x_i) with x_i from (x_i, Y)_i in U
        compute some selection criterion c(x_i, ŷ_i)
    end
    obtain i* = arg max_i c(x_i, ŷ_i)
    add (x_{i*}, ŷ_{i*}) to labeled data: D ← D ∪ (x_i, ŷ_i)
    update U ← U \ (x_i, Y)_i
end
```

Various selection criteria have been proposed in the literature [7, 10, 17]. However, there is hardly any theoretical justification for them. By embedding PLS into decision theory, we address this research gap. Another issue with PLS is the reliance on the initial model fit on labeled data. If the initial model generalizes poorly, misconceptions can propagate throughout the process [1]. Accordingly, we exploit Bayesian decision theory to derive a PLS criterion that is robust with respect to the initial model fit, thus mitigating this so-called confirmation bias.

2 Approximatley Bayes-Optimal Pseudo-Label Selection

We argue that PLS is nothing but a canonical decision problem: We formalize the selection of data points to be pseudo-labeled as decision problem, where the unknown set of states of nature is the learner's parameter space and the action space – unlike in statistical decision theory – corresponds to the set of unlabeled data, i.e., we regard (the selection of) data points as actions. This perspective clears the way for deploying several decision-theoretic approaches – first and foremost, finding Bayes-optimal actions (selections of pseudo-labels) under common loss/utility functions. In our first contribution [14], we prove that with the joint likelihood as utility, the Bayes-optimal criterion is the posterior predictive of pseudo-samples and labeled data. Since the latter requires computing a possibly intractable integral, we come up with approximations based on Laplace's method and the Gaussian integral that circumvent expensive sampling-based evaluations of the posterior predictive. Our approximate version of the Bayes-optimal criterion turns out to be simple and computationally cheap to evaluate: $\ell(\hat{\theta}) - \frac{1}{2}\log|\mathcal{I}(\hat{\theta})|$ with $\ell(\hat{\theta})$ being the likelihood and $\mathcal{I}(\hat{\theta})$ the Fisher-information

[1] Other names: Self-Training, Self-Labeling.

matrix at the fitted parameter vector $\hat{\theta}$. As an approximation of the joint posterior predictive of pseudo-samples and labeled data, it does remarkably not require an *i.i.d.* assumption. This renders our criterion applicable to a great variety of applied learning tasks.

We deploy BPLS on simulated and real-world data using several models ranging from generalized linear over non-parametric generalized additive models to Bayesian neural networks. Empirical evidence suggests that such a Bayesian approach to PLS – which we simply dub Bayesian PLS (BPLS) – can mitigate the confirmation bias [1] in pseudo-labeling that results from overfitting initial models. Besides, BPLS is flexible enough to incorporate prior knowledge not only in predicting but also in selecting pseudo-labeled data. What is more, BPLS involves no hyperparameters that require tuning. Notably, the decision-theoretic treatment of PLS also yields the framework of optimistic (pessimistic) superset learning [3, 4 16] as max-max- (min-max-)actions.

3 In All Likelihood*s*: Robust Pseudo-Label Selection

The decision-theoretic embedding opens up a myriad of possible venues for future work thanks to the rich literature on decision theory. In our second contribution [15], we propose three extensions of BPLS by leveraging results from generalized Bayesian decision theory. Our extensions aim at robustifying PLS with regard to model selection, accumulation of errors and covariate shift [13]. The general idea is to define a multi-objective utility function. The latter can consist of likelihood functions from, e.g., differently specified models, thus incorporating potentially competing goals. We further discuss how to embed such multi-objective utilities into a *preference system* [5, 6]. This allows us to harness the entire information encoded in its cardinal dimensions while still being able to avoid unjustified assumptions on the hierarchy of the involved objectives. We further consider the generic approach of the generalized Bayesian α-cut updating rule for credal sets of priors. Such priors can not only reflect uncertainty regarding prior information, but might as well represent priors near ignorance, see [2, 8, 9, 11, 12] for instance.

We spotlight the application of the introduced extensions on real-world and simulated data using semi-supervised logistic regression. In a benchmarking study, we compare these extensions to traditional PLS methods. Results suggest that especially robustness with regard to model choice can lead to substantial performance gains.

References

1. Arazo, E., Ortego, D., Albert, P., O'Connor, N.E., McGuinness, K.: Pseudo-labeling and confirmation bias in deep semi-supervised learning. In: 2020 International Joint Conference on Neural Networks, pp. 1–8. IEEE (2020)
2. Benavoli, A., Zaffalon, M.: Prior near ignorance for inferences in the k-parameter exponential family. Statistics **49**(5), 1104–1140 (2015)
3. Hüllermeier, E.: Learning from imprecise and fuzzy observations: data disambiguation through generalized loss minimization. Int. J. Approximate Reasoning **55**, 1519–1534 (2014)

4. Hüllermeier, E., Destercke, S., Couso, I.: Learning from imprecise data: adjustments of optimistic and pessimistic variants. In: Ben Amor, N., Quost, B., Theobald, M. (eds.) SUM 2019. LNCS (LNAI), vol. 11940, pp. 266–279. Springer, Cham (2019). https://doi.org/10.1007/978-3-030-35514-2_20
5. Jansen, C., Schollmeyer, G., Augustin, T.: Concepts for decision making under severe uncertainty with partial ordinal and partial cardinal preferences. Int. J. Approximate Reasoning **98**, 112–131 (2018)
6. Jansen, C., Schollmeyer, G., Blocher, H., Rodemann, J., Augustin, T.: Robust statistical comparison of random variables with locally varying scale of measurement. In: Uncertainty in Artificial Intelligence (UAI). PMLR (2023, to appear)
7. Lee, D.H.: Pseudo-label: the simple and efficient semi-supervised learning method for deep neural networks. In: Workshop on challenges in representation learning, International Conference on Machine Learning, pp. 896. No. 2 (2013)
8. Mangili, F.: A prior near-ignorance Gaussian process model for nonparametric regression. In: International Symposium on Imprecise Probabilities (ISIPTA) (2015)
9. Mangili, F., Benavoli, A.: New prior near-ignorance models on the simplex. Int. J. Approximate Reasoning **56**, 278–306 (2015)
10. Rizve, M.N., Duarte, K., Rawat, Y.S., Shah, M.: In defense of pseudo-labeling: An uncertainty-aware pseudo-label selection framework for semi-supervised learning. In: International Conference on Learning Representations (ICLR) (2020)
11. Rodemann, J., Augustin, T.: Accounting for imprecision of model specification in Bayesian optimization. In: Poster presented at International Symposium on Imprecise Probabilities (ISIPTA) (2021)
12. Rodemann, J., Augustin, T.: Accounting for Gaussian process imprecision in Bayesian optimization. In: Honda, K., Entani, T., Ubukata, S., Huynh, VN., Inuiguchi, M. (eds.) IUKM 2022. LNCS, vol. 13199, pp 92–104. Springer, Cham (2022). https://doi.org/10.1007/978-3-030-98018-4_8
13. Rodemann, J., Fischer, S., Schneider, L., Nalenz, M., Augustin, T.: Not all data are created equal: Lessons from sampling theory for adaptive machine learning. In: Poster presented at International Conference on Statistics and Data Science (ICSDS), Institute of Mathematical Statistics (IMS) (2022)
14. Rodemann, J., Goschenhofer, J., Dorigatti, E., Nagler, T., Augustin, T.: Approximately Bayes-optimal pseudo-label selection. In: International Conference on Uncertainty in Artificial Intelligence (UAI). PMLR (2023, to appear)
15. Rodemann, J., Jansen, C., Schollmeyer, G., Augustin, T.: In all likelihoods: robust selection of pseudo-labeled data. In: International Symposium on Imprecise Probabilities Theories and Applications (ISIPTA). PMLR (2023, to appear)
16. Rodemann, J., Kreiss, D., Hüllermeier, E., Augustin, T.: Levelwise data disambiguation by cautious superset learning. In: International Conference on Scalable Uncertainty Management (SUM), pp. 263–276. Springer (2022)
17. Shi, W., et al.: Transductive semi-supervised deep learning using min-max features. In: Ferrari, V., Hebert, M., Sminchisescu, C., Weiss, Y. (eds.) ECCV 2018. LNCS, vol. 11209, pp. 311–327. Springer, Shi, W., Gong, Y., Ding, C., Tao, Z.M., Zheng, N.: Transductive semi-supervised deep learning using min-max features. In: European Conference on Computer Vision. pp. 299–315 (2018) (2018). https://doi.org/10.1007/978-3-030-01228-1_19

XAI Methods in the Presence of Suppressor Variables: A Theoretical Consideration

Rick Wilming[1], Leo Kieslich[1], Benedict Clark[2], and Stefan Haufe[1,2,3](✉)

[1] Technische Universität, Berlin, Germany
{rick.wilming,kieslich,haufe}@tu-berlin.de
[2] Physikalisch-Technische Bundesanstalt, Berlin, Germany
benedict.clark@ptb.de
[3] Charité – Universitätsmedizin, Berlin, Germany

Abstract. In high-risk domains like medicine, decisions produced by complex machine learning systems can have a considerable impact on human lives. Therefore, it is of great importance that these automatic decisions can be made comprehensible for humans. The community of 'explainable artificial intelligence' (XAI) has created an extensive body of methods to explain the decisions of complex machine learning models. However, a concrete problem to be solved by these XAI methods has not yet been formally stated, resulting in a lack of theoretical and empirical evidence for the 'correctness' of their explanations, and limiting their potential use for quality-control and transparency purposes. At the same time, Haufe et al. [4] showed, using simple toy examples, that even standard interpretations of linear models can be highly misleading. Specifically, high importance may be attributed to so-called suppressor variables, which lack any statistical relation to the prediction target. This behavior has been confirmed empirically for a large array of XAI methods by Wilming et al. [11]. In our recent work [12], we derive analytical expressions of explanations produced by popular XAI methods to study their behavior on a simple two-dimensional binary classification problem consisting of Gaussian class-conditional distributions. We show that the majority of the studied explanation approaches will attribute non-zero importance to a non-class-related suppressor feature in the presence of correlated noise. This poses important limitations on the interpretations and conclusions that the outputs of these XAI methods can afford.

Keywords: XAI · Interpretability · Suppressor variable

1 Introduction

Haufe et al. [4] and Wilming et al. [11] show that features determined to be important by certain XAI methods, e.g. via inspecting their corresponding weights of a linear model, may actually not have any statistical association with the predicted variable. As a result, the provided 'explanation' may not agree with the prior domain knowledge of an expert user and might undermine their

D. Seipel and A. Steen (Eds.): KI 2023, LNAI 14236, pp. 265–268, 2023.
https://doi.org/10.1007/978-3-031-42608-7

trust in the predictive model, even if it performs optimally. Indeed, a highly accurate model might exploit so-called suppressor features [e.g. 3], which can be statistically independent of the prediction target yet can still lead to increased prediction performance. While Haufe et al. [4] have introduced low-dimensional and well-controlled examples to illustrate the problem of suppressor variables for model interpretation, Wilming et al. [11] showed empirically that the emergence of suppressors indeed poses a problem for a large group of XAI methods and diminishes their 'explanation performance'.

In our current study [12], we go one step further and derive analytical expressions for commonly used XAI methods through a simple two-dimensional linear data generation process capable of creating suppressor variables by parametrically inducing correlations between features. For a supervised learning task, a model $f : \mathbb{R}^d \to \mathbb{R}$ learns a function between an input $\mathbf{x}^{(i)} \in \mathbb{R}^d$ and a target $y^{(i)} \in \mathbb{R}$, based on training data $\mathcal{D} = \{(\mathbf{x}^{(i)}, y^{(i)})\}_{i=1}^N$. Here, $\mathbf{x}^{(i)}$ and $y^{(i)}$ are realizations of the random variables $\mathbf{X}$ and Y, with joint probability density function $p_{\mathbf{X},Y}(\mathbf{x}, y)$. A formal prerequisite for a feature X_j to be important is that it has a statistical association to the target variable Y, i.e.,

$$X_j \text{ is important} \Rightarrow X_j \not\perp Y. \tag{1}$$

2 Results and Discussion

In the following, we summarize the findings of our work [12]. We start and define $H \sim N(\mathbf{0}, \Sigma)$ and $Z \sim \mathit{Rademacher}(1/2)$ as the random variables of the realizations $\eta \in \mathbb{R}^2$ and $z \in \{-1, 1\}$, respectively, to describe the linear generative model $\mathbf{x} = \mathbf{a}z + \eta$ and $y = z$, with $\mathbf{a} = (1, 0)^\top$ and a covariance matrix parameterized as follows: $\Sigma = \begin{pmatrix} s_1^2 & cs_1s_2 \\ cs_1s_2 & s_2^2 \end{pmatrix}$, where s_1 and s_2 are non-negative standard deviations and $c \in [-1, 1]$ is a correlation. The vector $\mathbf{a}$ is also called signal *pattern* [4, 6]. The generative data model induces a binary classification problem with Gaussian class-conditional distributions. By introducing noise correlations within this generative data model, we create a suppressor variable, here feature x_2, which has no statistical relation to the target but whose inclusion in any model will lead to better predictions [4]. We solve this classification task in a Bayes optimal way with the linear decision rule $f(\mathbf{x}) = \mathbf{w}^\top\mathbf{x} = (\alpha, -\alpha cs_1/s_2)^\top\mathbf{x}$ for $\alpha := (1 + (cs_1/s_2)^2)^{-\frac{1}{2}}$ and $||\mathbf{w}||_2 = 1$.

Now, one of the goals of explainable AI is to support developers or users to identify 'wrong' features, e.g. confounders, or assist during the discovery process of new informative features. However, in practice, XAI methods, more so the straightforward interpretation of weights of linear models, do not distinguish between confounders and suppressors, relevance attributions to those features may lead to misunderstanding and wrong conclusions. Primarily, we view the proposed generative data model as a simple, yet very insightful, minimal counterexample, where the existence of suppressor variables challenges the assumptions of many XAI methods as well as the assumptions underlying metrics such as faithfulness, which are often considered a gold-standard for quantitative evaluation and an appropriate surrogate for 'correctness'. Indeed, authors have shown empirically that XAI methods can lead to suboptimal 'explanation performance' even when applied to linearly separable data with suppressor variables [11]. We go beyond the study of Wilming et al. [11] by deriving analytical expressions of popular XAI methods. For example, the *Pattern* approach introduced by Haufe et al. [4] propose a transformation to convert weight vectors into parameters $\mathbf{a} \in \mathbb{R}^d$ of a corresponding linear *forward model* $\mathbf{x} = \mathbf{a}f(\mathbf{x}) + \varepsilon$. The covariance between the model output and each input feature: $a_j = \text{Cov}(x_j, f(\mathbf{x})) = \text{Cov}(x_j, f(\mathbf{x})) = \text{Cov}(x_j, \mathbf{w}^\top \mathbf{x})$, for features $j = 1, \ldots d$, provides an explanation which yields a global importance map $e_j(\mathbf{x}) := (\text{Cov}(\mathbf{x}, \mathbf{x})\mathbf{w})_j$, called *linear activation pattern* [4]. For the generative model and the Bayes optimal classifier described above, we obtain the explanations $e_1(\mathbf{x}) = \alpha s_1^2(1 - c^2)$ and $e_2(\mathbf{x}) = 0$ for the corresponding features x_1 and x_2, respectively. Thus, the pattern approach does not attribute any importance to the suppressor feature x_2.

Our analysis found that several XAI methods are incapable of nullifying the suppressor feature x_2, i.e., assigning non-zero importance to it, when correlations between features are present. We emphasize that nullifying a suppressor feature is an important property of an XAI method, and it is not sufficient to *just* rank features according to their importance scores. In fact, we can generate scenarios where the weight corresponding to the suppressor variable is more than twice as high as the weight corresponding to the class-dependent feature (see [4]), leading us to think that, in some sense, the suppressor feature is twice as important. We show this is also the case for the naive pixel flipping [e.g. 5] and the permutation feature importance methods [2] representing operationalization of faithfulness, but also for actively researched methods such as SHAP [8], LIME [9] and counterfactuals [10]. XAI methods based on the Shapley value framework yield particular diverging results, as the value function has a strong influence on the outcome, also showing how heavily dependent such methods are on the correlation structure of the dataset. In contrast, the M-Plot approach [1], PATTERN [4], and the Shapley value approach using the R^2 value function [7], deliver promising results by assigning exactly zero importance to the suppressor variable. This positive result can be attributed to the fact that all methods make explicit use of the statistics of the training data including the correlation structure of the data. This stands in contrast to methods using only the model itself to assign importance to a test sample.

3 Conclusion

Wilming et al. [12] study a two-dimensional linear binary classification problem, where only one feature carries class-specific information. The other feature is a suppressor variable carrying no such information yet improving the performance of the Bayes optimal classifier. Analytically, we derive closed-form solutions for the outputs of popular XAI methods, demonstrating that a considerable number of these methods attribute non-zero importance to the suppressor feature that is independent of the class label. We also find that a number of methods do assign zero significance to that feature by accounting for correlations between the two features. This signifies that even the most simple multivariate models cannot be understood without knowing essential properties of the distribution of the data they were trained on.

References

1. Apley, D.W., Zhu, J.: Visualizing the effects of predictor variables in black box supervised learning models. J. Roy. Stat. Soc. B **82**(4), 1059–1086 (2020)
2. Breiman, L.: Random forests. Mach. Learn. **45**(1), 5–32 (2001)
3. Friedman, L., Wall, M.: Graphical views of suppression and multicollinearity in multiple linear regression. Am. Stat. **59**(2), 127–136 (2005)
4. Haufe, S., et al.: On the interpretation of weight vectors of linear models in multivariate neuroimaging. NeuroImage **87**, 96–110 (2014)
5. Jacovi, A., Goldberg, Y.: Towards faithfully interpretable NLP systems: how should we define and evaluate faithfulness? In: Proceedings of the 58th Annual Meeting of the Association for Computational Linguistics, Online, pp. 4198–4205. Association for Computational Linguistics, July 2020
6. Kindermans, P.J., et al.: Learning how to explain neural networks: Patternnet and patternattribution. In: International Conference on Learning Representations (2018)
7. Lipovetsky, S., Conklin, M.: Analysis of regression in game theory approach. Appl. Stochast. Models Bus. Ind. **17**(4), 319–330 (2001)
8. Lundberg, S.M., Lee, S.I.: A unified approach to interpreting model predictions. In: Guyon, I., et al. (eds.) Advances in Neural Information Processing Systems, vol. 30, pp. 4765–4774, Curran Associates, Inc. (2017)
9. Ribeiro, M.T., Singh, S., Guestrin, C.: "Why should I trust you?" Explaining the predictions of any classifier. In: Proceedings of the 22nd ACM SIGKDD International Conference on Knowledge Discovery and Data Mining, pp. 1135–1144 (2016)
10. Wachter, S., Mittelstadt, B., Russell, C.: Counterfactual explanations without opening the black box: automated decisions and the GDPR. Harv. JL Technol. **31**, 841 (2017)
11. Wilming, R., Budding, C., Müller, K.R., Haufe, S.: Scrutinizing XAI using linear ground-truth data with suppressor variables. Mach. Learn. **111**(5), 1903–1923 (2022)
12. Wilming, R., Kieslich, L., Clark, B., Haufe, S.: Theoretical behavior of XAI methods in the presence of suppressor variables. In: Proceedings of the 40th International Conference on Machine Learning, Proceedings of Machine Learning Research, PMLR (2023, accepted)

Author Index

D. Seipel and A. Steen (Eds.): KI 2023, LNAI 14236, pp. 269–270, 2023.
https://doi.org/10.1007/978-3-031-42608-7

Printed in the United States
by Baker & Taylor Publisher Services

Printed in the United States
by Baker & Taylor Publisher Services